Graph Theory and Additive Combinatorics

Using the dichotomy of structure and pseudorandomness as a central theme, this accessible text provides a modern introduction to extremal graph theory and additive combinatorics. Readers will explore central results in additive combinatorics – notably the cornerstone theorems of Roth, Szemerédi, Freiman, and Green–Tao – and will gain additional insights into these ideas through graph theoretic perspectives. Topics discussed include the Turán problem, Szemerédi's graph regularity method, pseudorandom graphs, graph limits, graph homomorphism inequalities, Fourier analysis in additive combinatorics, the structure of set addition, and the sum-product problem. Important combinatorial, graph theoretic, analytic, Fourier, algebraic, and geometric methods are highlighted. Students will appreciate the chapter summaries, many figures and exercises, and freely available lecture videos on MIT OpenCourseWare. Meant as an introduction for students and researchers studying combinatorics, theoretical computer science, analysis, probability, and number theory, the text assumes only basic familiarity with abstract algebra, analysis, and linear algebra.

YUFEI ZHAO is Associate Professor of Mathematics at the Massachusetts Institute of Technology. His research tackles a broad range of problems in discrete mathematics, including extremal, probabilistic, and additive combinatorics, graph theory, and discrete geometry, as well as applications to computer science. His honors include the SIAM Dénes Kõnig prize (2018), the Sloan Research Fellowship (2019), and the NSF CAREER Award (2021). This book is based on an MIT graduate course, which he has taught and developed over the last five years.

Graph Theory and Additive Combinatorics

Exploring Structure and Randomness

Yufei Zhao
Massachusetts Institute of Technology

Shaftesbury Road, Cambridge CB2 8EA, United Kingdom

One Liberty Plaza, 20th Floor, New York, NY 10006, USA

477 Williamstown Road, Port Melbourne, VIC 3207, Australia

314–321, 3rd Floor, Plot 3, Splendor Forum, Jasola District Centre, New Delhi – 110025, India

103 Penang Road, #05–06/07, Visioncrest Commercial, Singapore 238467

Cambridge University Press is part of Cambridge University Press & Assessment,
a department of the University of Cambridge.

We share the University's mission to contribute to society through the pursuit of
education, learning and research at the highest international levels of excellence.

www.cambridge.org
Information on this title: www.cambridge.org/9781009310949

DOI: 10.1017/9781009310956

Image Credit (page iii): GTAC bridge by Anne Ma, 2022:

First published 2023

A catalogue record for this publication is available from the British Library

A Cataloging-in-Publication data record for this book is available from the Library of Congress.

ISBN 978-1-009-31094-9 Hardback

To Lu
for your constant love and support
and Andi
who arrived in time to get on this page

Contents

Preface		xi
Notation and Conventions		xv
0	**Appetizer: Triangles and Equations**	1
0.1	Schur's Theorem	1
0.2	Progressions	5
0.3	What's Next in the Book?	9
1	**Forbidding a Subgraph**	11
1.1	Forbidding a Triangle: Mantel's Theorem	12
1.2	Forbidding a Clique: Turán's Theorem	14
1.3	Turán Density and Supersaturation	19
1.4	Forbidding a Complete Bipartite Graph: Kővári–Sós–Turán Theorem	22
1.5	Forbidding a General Subgraph: Erdős–Stone–Simonovits Theorem	27
1.6	Forbidding a Cycle	31
1.7	Forbidding a Sparse Bipartite Graph: Dependent Random Choice	33
1.8	Lower Bound Constructions: Overview	37
1.9	Randomized Constructions	38
1.10	Algebraic Constructions	39
1.11	Randomized Algebraic Constructions	46
2	**Graph Regularity Method**	52
2.1	Szemerédi's Graph Regularity Lemma	53
2.2	Triangle Counting Lemma	61
2.3	Triangle Removal Lemma	63
2.4	Graph Theoretic Proof of Roth's Theorem	66
2.5	Large 3-AP-Free Sets: Behrend's Construction	69
2.6	Graph Counting and Removal Lemmas	70
2.7	Exercises on Applying Graph Regularity	75
2.8	Induced Graph Removal and Strong Regularity	76
2.9	Graph Property Testing	83
2.10	Hypergraph Removal and Szemerédi's Theorem	85
2.11	Hypergraph Regularity	86
3	**Pseudorandom Graphs**	89
3.1	Quasirandom Graphs	90
3.2	Expander Mixing Lemma	101

3.3	Abelian Cayley Graphs and Eigenvalues	104
3.4	Quasirandom Groups	109
3.5	Quasirandom Cayley Graphs and Grothendieck's Inequality	116
3.6	Second Eigenvalue: Alon–Boppana Bound	119
4	**Graph Limits**	127
4.1	Graphons	128
4.2	Cut Distance	131
4.3	Homomorphism Density	135
4.4	W-Random Graphs	137
4.5	Counting Lemma	140
4.6	Weak Regularity Lemma	142
4.7	Martingale Convergence Theorem	146
4.8	Compactness of the Graphon Space	148
4.9	Equivalence of Convergence	152
5	**Graph Homomorphism Inequalities**	158
5.1	Edge versus Triangle Densities	161
5.2	Cauchy–Schwarz	166
5.3	Hölder	174
5.4	Lagrangian	182
5.5	Entropy	186
6	**Forbidding 3-Term Arithmetic Progressions**	197
6.1	Fourier Analysis in Finite Field Vector Spaces	197
6.2	Roth's Theorem in the Finite Field Model	202
6.3	Fourier Analysis in the Integers	209
6.4	Roth's Theorem in the Integers	210
6.5	Polynomial Method	216
6.6	Arithmetic Regularity	220
6.7	Popular Common Difference	226
7	**Structure of Set Addition**	230
7.1	Sets of Small Doubling: Freiman's Theorem	231
7.2	Sumset Calculus I: Ruzsa Triangle Inequality	233
7.3	Sumset Calculus II: Plünnecke's Inequality	234
7.4	Covering Lemma	237
7.5	Freiman's Theorem in Groups with Bounded Exponent	239
7.6	Freiman Homomorphisms	240
7.7	Modeling Lemma	242
7.8	Iterated Sumsets: Bogolyubov's Lemma	244
7.9	Geometry of Numbers	248
7.10	Finding a GAP in a Bohr Set	251
7.11	Proof of Freiman's Theorem	253
7.12	Polynomial Freiman–Ruzsa Conjecture	254
7.13	Additive Energy and the Balog–Szemerédi–Gowers Theorem	257

8 Sum-Product Problem 264
8.1 Multiplication Table Problem 265
8.2 Crossing Number Inequality and Point-Line Incidences 266
8.3 Sum-Product via Multiplicative Energy 270

9 Progressions in Sparse Pseudorandom Sets 273
9.1 Green–Tao Theorem 273
9.2 Relative Szemerédi Theorem 275
9.3 Transference Principle 279
9.4 Dense Model Theorem 280
9.5 Sparse Counting Lemma 286
9.6 Proof of the Relative Roth Theorem 293

References 297
Index 313

Preface

Who Is This Book For?

This textbook is intended for graduate and advanced undergraduate students as well as researchers in mathematics, computer science, and related areas. The material should appeal to anyone with an interest in combinatorics, theoretical computer science, analysis, probability, and number theory. It can be used as a textbook for a class or self-study, or as a research reference.

Why This Book?

There have been many exciting developments in graph theory and additive combinatorics in recent decades. This is the first introductory graduate-level textbook to focus on a unifying set of topics connecting graph theory and additive combinatorics.

This textbook arose from a one-semester graduate-level course that I developed at Massachusetts Institute of Technology (and still teach regularly) to introduce students to a spectrum of beautiful mathematics in the field.

Lecture Videos

A complete set of video lectures from my Fall 2019 class is available for free through MIT OpenCourseWare and YouTube (search for *Graph Theory and Additive Combinatorics* and *MIT OCW*). The lecture videos are a useful resource and complement this book.

What Is This Book About?

This book introduces the readers to classical and modern developments in graph theory and additive combinatorics, with a focus on topics and themes that connect the two subjects.

A foundational result in additive combinatorics is **Roth's theorem**, which says that every subset of $\{1, 2, \ldots, \}$ without a 3-term arithmetic progression has at most $o(N)$ elements. We will see different proofs of Roth's theorem, using tools from graph theory and Fourier analysis. A key idea in both approaches is the *dichotomy of structure versus pseudorandomness*.

Roth's theorem laid the groundwork for many important later developments, such as

- **Szemerédi's theorem:** Every set of integers of positive density contains arbitrarily long arithmetic progressions; and
- **Green–Tao theorem:** The primes contain arbitrarily long arithmetic progressions.

A core thread throughout the book is the connection bridging graph theory and additive combinatorics. The book opens with Schur's theorem, which is an early example whose proof illustrates this connection. Graph theoretic perspectives are presented throughout the book.

Here are some of the topics and questions considered in this book:

CHAPTER 1: **Forbidding a subgraph**
What is the maximum number of edges in a triangle-free graph on n vertices? What if instead we forbid some other subgraph? This is known as the Turán problem.

CHAPTER 2: **Graph regularity method**
Szemerédi introduced this powerful tool that provides an approximate structural description for every large graph.

CHAPTER 3: **Pseudorandom graphs**
What does it mean for some graph to resemble a random graph?

CHAPTER 4: **Graph limits**
In what sense can a sequence of graphs, increasing in size, converge to some limit object?

CHAPTER 5: **Graph homomorphism inequalities**
What are possible relationships between subgraph densities?

CHAPTER 6: **Forbidding a 3-term arithemtic progression**
Roth's theorem and Fourier analysis in additive combinatorics.

CHAPTER 7: **Structure of set addition**
What can one say about a set of integer A with small sumset $A + A = \{a + b \colon a, b \in A\}$? Freiman's theorem is a foundational result that gives an answer.

CHAPTER 8: **Sum-product problem**
Can a set A simultaneously have both small sumset $A + A$ and product set $A \cdot A$?

CHAPTER 9: **Progressions in sparse pseudorandom sets**
Key ideas in the proof of the Green–Tao theorem. How can we apply a dense setting result, namely Szemerédi's theorem, to a sparse set?

For a more detailed list of topics, see the highlights and summary boxes at the beginning and the end of each chapter.

The book is roughly divided into two parts, with graph theory the focus of Chapters 1 to 5 and additive combinatorics the focus of Chapters 6 to 9. These are not disjoint and separate subjects. Rather, graph theory and additive combinatorics are interleaved throughout the book. We emphasize their interactions. Each chapter can be enjoyed independently, as there are very few dependencies between chapters, though one gets the most out of the book by appreciating the connections.

Using the Textbook for a Class

The contents may be taught as a fast-paced one-semester class, or as a two-semester sequence, with each term focusing on one half of the book: the first on extremal graph theory, and the second on additive combinatorics.

For a one-semester class (which is how I teach it at MIT; see my website or MIT OCW for syllabus, lecture videos, homework, and further information), I suggest skipping some more technical or advanced topics and proofs, such as the following: (Chapter 1) the proofs of the Erdős–Stone–Simonovits theorem, the $K_{s,t}$ construction, randomized algebraic construction; (Chapter 2) the proof of the graph counting lemma, induced graph removal and strong regularity, hypergraph regularity and removal; (Chapter 3) quasirandom groups, quasirandom Cayley graphs; (Chapter 4) most technical proofs on graph limits; (Chapter 5) Hölder, entropy; (Chapter 6) arithmetic regularity and popular common difference; (Chapter 7) proofs later in the chapter if short on time; (Chapter 9) proof details.

For a class focused on one part of the book, one may wish to explore further topics as suggested in *Further Reading* at the end of each chapter.

Prerequisites

The prerequisites are minimal – primarily mathematical maturity and an interest in combinatorics. Some basic concepts from abstract algebra, analysis, and linear algebra are assumed.

Exercises

The book contains around 150 carefully selected exercises. They are scattered throughout each chapter. Some exercises are embedded in the middle of a section – these exercises are meant as routine tests of understanding of the concepts just discussed. For example, they sometimes ask you to fill in missing proof details or think about easy generalizations and extensions. The exercises at the end of each section are carefully selected problems that reinforce the techniques discussed in the chapter. Hopefully they are all interesting. Most of them are intended to test your mastery of the techniques taught in the chapter. Many of these end-of-chapter exercises are quite challenging, with starred problems intended to be more difficult but still doable by a strong student given the techniques taught. Many of these exercises are adapted from lemmas and results from research papers. (I apologize for omitting references for the exercises so that they can be used as homework assignments.)

Spending time with the exercises is essential for mastering the techniques. I used many of these exercises in my classes. My students often told me that they thought that they had understood the material after a lecture, only to discover their incomplete mastery when confronted with the exercises. Struggling with these exercises led them to newfound insight.

Further Reading

This is a massive and rapidly expanding subject. The book is intended to be introductory and enticing rather than comprehensive. Each chapter concludes with recommendations for further reading for anyone who wishes to learn more. Additionally, references are given generously throughout the text for anyone who wishes to dive deeper and read the original sources.

Acknowledgments

I thank all my teachers and mentors who have taught me the subject starting from when I was a graduate student, with a special shout-out to my Ph.D. advisor Jacob Fox for his dedicated mentorship. I first encountered this subject at the University of Cambridge, when I took a Part III class on extremal graph theory taught by David Conlon. Over the years, I learned a lot from various researchers thanks to their carefully and insightfully written lecture notes scattered on the web, in particular those by David Conlon, Tim Gowers, Andrew Granville, Ben Green, Choongbum Lee, László Lovász, Imre Ruzsa, Asaf Shapira, Adam Sheffer, K. Soundararajan, Terry Tao, and Jacques Verstraete.

This book arose from a one-semester course that I taught at MIT in Fall 2017, 2019, and 2021. I thank all my amazing and dedicated students who kept up their interest in my teaching – they were instrumental in motivating me to complete this book project. Students from the 2017 and 2019 classes took notes based on my lectures, which I subsequently rewrote and revised into this book. My 2021 class used an early draft of this book and gave valuable comments and feedback. There are many students whom I wish to thank, and here is my attempt at listing them (my apologies to anyone whose name I have inadvertently omitted): Dhroova Aiylam, Ganesh Ajjanagadde, Shyan Akmal, Ryan Alweiss, Morris Ang Jie Jun, Adam Ardeishar, Matt Babbitt, Yonah Borns-Weil, Matthew Brennan, Brynmor Chapman, Evan Chen, Byron Chin, Ahmed Chowdhury Zawad, Anlong Chua, Travis Dillon, Jonathan Figueroa Rodríguez, Christian Gaetz, Shengwen Gan, Jiyang Gao, Yibo Gao, Swapnil Garg, Benjamin Gunby, Meghal Gupta, Kaarel Haenni, Milan Haiman, Linus Hamilton, Carina Hong Letong, Vishesh Jain, Pakawut Jiradilok, Sujay Kazi, Dain Kim, Elena Kim, Younhun Kim, Yael Kirkpatrick, Daishi Kiyohara, Frederic Koehler, Keiran Lewellen, Anqi Li, Jerry Li, Allen Liu, Michael Ma, Nitya Mani, Olga Medrano, Holden Mui, Eshaan Nichani, Yuchong Pan, Minjae Park, Alan Peng, Saranesh Prembabu, Michael Ren, Dhruv Rohatgi, Diego Roque, Ashwin Sah, Maya Sankar, Mehtaab Sawhney, Carl Schildkraut, Tristan Shin, Mihir Singhal, Tomasz Slusarczyk, Albert Soh, Kevin Sun, Sarah Tammen, Jonathan Tidor, Paxton Turner, Danielle Wang, Hong Wang, Nicole Wein, Jake Wellens, Chris Xu, Max Wenqiang Xu, Yinzhan Xu, Zixuan Xu, Lisa Yang, Yuan Yao, Richard Yi, Hung-Hsun Yu, Lingxian Zhang, Kai Zheng, Yunkun Zhou. Additionally, I would like to thank Thomas Bloom and Zilin Jiang for carefully reading the book draft and sending in many suggestions for corrections and improvements.

The title page illustration (with the bridge) was drawn by my friend Anne Ma.

I also wish to acknowledge research funding support during the writing of this book, including from the National Science Foundation, the Sloan Research Foundation, as well as support from Massachusetts Institute of Technology including the Solomon Buchsbaum Research Fund, the Class of 1956 Career Development Professorship, and the Edmund F. Kelly Research Award.

Finally, I am grateful to all my students, colleagues, friends, and family for their encouragement throughout the writing of the book, and most importantly to Lu for her unwavering support through the whole process, especially in the late stages of the book writing, which coincided with the arrival of our baby daughter Andi.

Notation and Conventions

We use standard notation in this book. The comments here are mostly for clarification. You might want to skip this section and return to it only as needed.

Sets

We write $\boldsymbol{[N]} := \{1, 2, \ldots, N\}$. Also $\mathbb{N} := \{1, 2, \ldots\}$.

Given a finite set S and a positive integer r, we write $\binom{S}{r}$ for the set of r-element subsets of S.

If S is a finite set and f is a function on S, we use the expectation notation $\mathbb{E}_{x \in S} f(x)$, or more simply $\mathbb{E}_x f(x)$ (or even $\mathbb{E} f$ if there is no confusion) to mean the average $|S|^{-1} \sum_{x \in S} f(x)$. We also use the symbol $\mathbb{E}$ for its usual meaning as the expectation for some random variable.

A ***k-term arithmetic progression*** (abbreviated ***k-AP***) in an abelian group is a sequence of the form

$$a, a + d, a + 2d, \ldots, a + (k - 1)d.$$

Here d is called the common difference. The progression is called ***nontrivial*** if $d \neq 0$, and ***trivial*** if $d = 0$. When we say that a set A contains a k-AP, we mean that it contains a nontrivial k-AP. Likewise, when we say that A is ***k-AP-free***, we mean that it contains no nontrivial k-APs.

Graphs

We write a graph as $\boldsymbol{G = (V, E)}$, where V is a finite set of vertices, and E is the set of edges. Each edge is an unordered pair of distinct vertices. Formally, $E \subseteq \binom{V}{2}$.

Given a graph G, we write $\boldsymbol{V(G)}$ for the set of vertices, and $\boldsymbol{E(G)}$ for the set of edges, and denote their cardinalities by $\boldsymbol{v(G)} := |V(G)|$ and $\boldsymbol{e(G)} := |E(G)|$.

In a graph G, the ***neighborhood*** of a vertex x, denoted $\boldsymbol{N_G(x)}$ (or simply $N(x)$ if there is no confusion), is the set of vertices y such that xy is an edge. The ***degree*** of x is the number of neighbors of x, denoted $\boldsymbol{\deg_G(x)} := |N_G(x)|$ (or simply written as $\boldsymbol{\deg(x)}$).

Given a graph G, for each $A \subseteq V(G)$, we write $\boldsymbol{e(A)}$ to denote the number of edges with both endpoints in A. Given $A, B \subseteq V(G)$ (not necessarily disjoint), we write

$$\boldsymbol{e(A, B)} := |\{(a, b) \in A \times B : ab \in E(G)\}|.$$

Note that when A and B are disjoint, $e(A, B)$ is the number of the edges between A and B. On the other hand, $e(A, A) = 2e(A)$ as each edge within A is counted twice.

Here are some standard graphs:

- $\boldsymbol{K_r}$ is the complete graph on r vertices, also known as an ***r*-clique**;
- $\boldsymbol{K_{s,t}}$ is the complete bipartite graph with s vertices in one vertex part and t vertices in the other vertex part;
- $\boldsymbol{K_{r,s,t}}$ is a complete tripartite graph with vertex parts having sizes r, s, t respectively (e.g., $K_{1,1,1} = K_3$); and so on analogously for complete multipartite graphs with more parts;
- $\boldsymbol{C_\ell}$ ($\ell \geq 3$) is a cycle with ℓ vertices and ℓ edges.

Some examples are shown below.

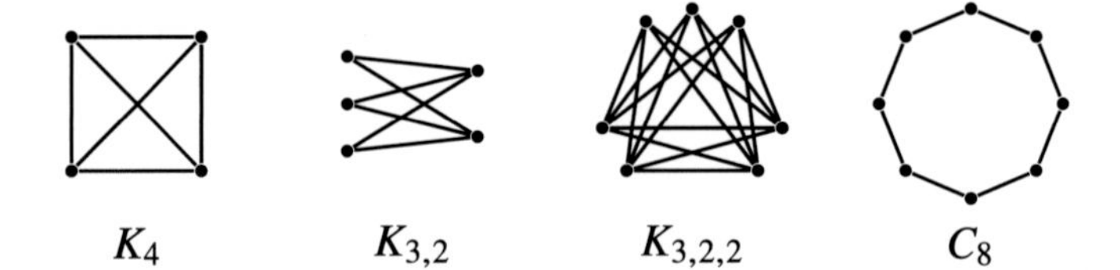

Given two graphs H and G, we say that H is a ***subgraph*** of G if one can delete some vertices and edges from G to obtain a graph isomorphic to H. A ***copy*** of H in G is a subgraph G that is isomorphic to H. A ***labeled copy*** of H in G is a subgraph of G isomorphic to H where we also specify the isomorphism from H. Equivalently, a labeled copy of H in G is an injective graph homomorphism from H to G. For example, if G has q copies of K_3, then G has $6q$ labeled copies of K_3.

We say that H is an ***induced subgraph*** of G if one can delete some vertices of G (when we delete a vertex, we also remove all edges incident to the vertex) to obtain H – note that in particular we are not allowed to remove additional edges other than those incident to a deleted vertex. If $S \subseteq V(G)$, we write $\boldsymbol{G[S]}$ to denote the subgraph of G induced by the vertex set S, that is, $G[S]$ is the subgraph with vertex set S and keeping all the edges from G among S.

As an example, the following graph contains the 4-cycle as an induced subgraph. It contains the 5-cycle as a subgraph but not as an induced subgraph.

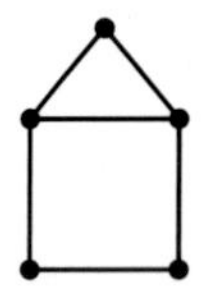

In this book, when we say ***H*-free**, we always mean not containing H as a subgraph. On the other hand, we say ***induced H*-free** to mean not containing H as an induced subgraph.

Given two graphs F and G, a ***graph homomorphism*** is a map $\phi\colon V(F) \to V(G)$ (not necessarily injective) such that $\phi(u)\phi(v) \in E(G)$ whenever $uv \in E(F)$. In other words, ϕ is a map of vertices that sends edges to edges. A key difference between a copy of F in G and a graph homomorphism from F to G is that the latter does not have to be an injective map of vertices.

The ***chromatic number*** $\chi(G)$ of a graph G is the smallest number of colors needed to color the vertices of G so that no two adjacent vertices receive the same color (such a coloring is called a ***proper coloring***).

The ***adjacency matrix*** of a graph $G = (V, E)$ is a $v(G) \times v(G)$ matrix whose rows and columns both are indexed by V, and such that the entry indexed by $(u, v) \in V \times V$ is 1 if $uv \in E$ and 0 if $uv \notin E$.

An ***r-uniform hypergraph*** (also called ***r-graph*** for short) consists of a finite vertex set V along with an edge set $E \subseteq \binom{V}{r}$. Each edge of the r-graph is an r-element subset of vertices.

Asymptotics

We use the following standard asymptotic notation. Given nonnegative quantities f and g, in each of the following items, the various notations have the same meaning (as some parameter, usually n, tends to infinity):

- $\boldsymbol{f \lesssim g}$, $\boldsymbol{f = O(g)}$, $\boldsymbol{g = \Omega(f)}$, $f \le Cg$ for some constant $C > 0$
- $\boldsymbol{f = o(g)}$, $f/g \to 0$
- $\boldsymbol{f = \Theta(g)}$, $\boldsymbol{f \asymp g}$, $g \lesssim f \lesssim g$
- $\boldsymbol{f \sim g}$, $f = (1 + o(1))g$

Subscripts (e.g., $\boldsymbol{O_s(\ )}$, $\boldsymbol{\lesssim_s}$), are used to emphasize that the hidden constants may depend on the subscripted parameters. For example, $f(s,x) \lesssim_s g(s,x)$ means that for every s there is some constant C_s so that $f(s,x) \le C_s g(s,x)$ for all x.

We avoid using $\ll$ since this notation carries different meanings in different communities and by different authors. In analytic number theory, $f \ll g$ is standard for $f = O(g)$ (this is called Vinogradov notation). In combinatorics and probability, $f \ll g$ sometimes means $f = o(g)$, and sometimes means that f is sufficiently small depending on g.

When asymptotic notation is used in the hypothesis of a statement, it should be interpreted as being applied to a sequence rather than a single object. For example, given functions f and g, we write

if $f(G) = o(1)$, then $g(G) = o(1)$

to mean

if a sequence G_n satisfies $f(G_n) = o(1)$, then $g(G_n) = o(1)$,

which is also equivalent to

for every $\varepsilon > 0$ there is some $\delta > 0$ such that, if $|f(G)| \le \delta$, then $|g(G)| \le \varepsilon$.

0

Appetizer: Triangles and Equations

Chapter Highlights

- Schur's theorem on monochromatic solutions to $x + y = z$ and its graph theoretic proof
- Problems and results on progressions (e.g., Szemerédi's theorem, the Green–Tao theorem)
- Introduction to the connection between graph theory and additive combinatorics

0.1 Schur's Theorem

Can we prove Fermat's Last Theorem by reducing the equation $X^n + Y^n = Z^n$ modulo a prime p?

It turns out this approach can never work. Dickson (1909) showed that the equation mod p can always be solved for sufficiently large primes p, no matter what n is. Schur (1916) gave a simpler proof of this result by proving the following theorem, showing that Dickson's result is much more about combinatorics than about number theory.

> **Theorem 0.1.1** (Schur's theorem)
> If the positive integers are colored using finitely many colors, then there is always a monochromatic solution to $x + y = z$ (i.e., x, y, z all have the same color).

We will prove Schur's theorem shortly.

Finitary Versus Infinitary

Many theorems in this book can be stated in multiple equivalent ways. For instance, Schur's theorem was stated in Theorem 0.1.1 in an **infinitary** form. Following is an equivalent **finitary** version. We write $[N] := \{1, 2, \ldots, N\}$.

> **Theorem 0.1.2** (Schur's theorem, finitary version)
> For every positive integer r, there exists a positive integer $N = N(r)$ such that if each element of $[N]$ is colored using one of r colors, then there is a monochromatic solution to $x + y = z$.

The finitary formulation leads to quantitative questions. For example, how large does $N(r)$ have to be as a function of r? Questions of this type are often quite difficult to resolve, even approximately. There are lots of open questions concerning quantitative bounds.

Proof that the preceding two formulations of Schur's theorem are equivalent. First, the finitary version (Theorem 0.1.2) of Schur's theorem easily implies the infinitary version (Theorem 0.1.1). Indeed, in the infinitary version, given a coloring of the positive integers, we can consider the colorings of the first $N(r)$ integers and use the finitary statement to find a monochromatic solution.

To prove that the infinitary version implies the finitary version, we use a **diagonalization argument**. Fix r, and suppose that for every N there is some coloring $\phi_N : [N] \to [r]$ that avoids monochromatic solutions to $x + y = z$. We can take an infinite subsequence of (ϕ_N) such that, for every $k \in \mathbb{N}$, the value of $\phi_N(k)$ stabilizes to a constant as N increases along this subsequence. (We can do this by repeatedly restricting to convergent infinite subsequences.) Then the ϕ_N's, along this subsequence, converge pointwise to some coloring $\phi: \mathbb{N} \to [r]$, avoiding monochromatic solutions to $x+y = z$, but ϕ contradicts the infinitary statement. □

Fermat's Equation Modulo a Prime

Let us show how to deduce the existence of solutions to $X^n + Y^n \equiv Z^n \pmod{p}$ using Schur's theorem.

Theorem 0.1.3 (Fermat's Last Theorem mod p)
Let n be a positive integer. For all sufficiently large prime p, there exist $X, Y, Z \in \{1, \ldots, p-1\}$ such that $X^n + Y^n \equiv Z^n \pmod{p}$.

Proof assuming Schur's theorem (Theorem 0.1.2). Let $(\mathbb{Z}/p\mathbb{Z})^\times$ denote the group of nonzero residues mod p under multiplication. Let $H = \{x^n : x \in (\mathbb{Z}/p\mathbb{Z})^\times\}$ be the subgroup of nth powers in $(\mathbb{Z}/p\mathbb{Z})^\times$. Since $(\mathbb{Z}/p\mathbb{Z})^\times$ is a cyclic group of order $p - 1$ (due to the existence of primitive roots mod p, a fact from elementary number theory), the index of H in $(\mathbb{Z}/p\mathbb{Z})^\times$ is equal to $\gcd(n, p-1) \le n$. So, the cosets of H partition $\{1, 2, \ldots, p-1\}$ into $\le n$ sets. Viewing each of the $\le n$ cosets of H as a "color," by the finitary statement of Schur's theorem (Theorem 0.1.2), for p large enough as a function of n, there exists a solution to

$$x + y = z \quad \text{in } \mathbb{Z}$$

in some coset of H, say $x, y, z \in aH$ for some $a \in (\mathbb{Z}/p\mathbb{Z})^\times$. Since H consists of nth powers, we have $x = aX^n$, $y = aY^n$, and $z = aZ^n$ for some $X, Y, Z \in (\mathbb{Z}/p\mathbb{Z})^\times$. Thus

$$aX^n + aY^n \equiv aZ^n \pmod{p}.$$

Since $a \in (\mathbb{Z}/p\mathbb{Z})^\times$ is invertible mod p, we have $X^n + Y^n \equiv Z^n \pmod{p}$ as desired. □

Ramsey's Theorem

Now let us prove Schur's theorem (Theorem 0.1.2) by deducing it from an analogous result about edge-coloring of a complete graph. We write K_N for the complete graph on N vertices.

Theorem 0.1.4 (Multicolor triangle Ramsey theorem)
For every positive integer r, there is some integer $N = N(r)$ such that if each edge of K_N is colored using one of r colors, then there is a monochromatic triangle.

Proof. Define

$$N_1 = 3, \quad \text{and} \quad N_r = r(N_{r-1} - 1) + 2 \text{ for all } r \geq 2. \tag{0.1}$$

We show by induction on r that every coloring of the edges of K_{N_r} by r colors has a monochromatic triangle. The case $r = 1$ holds trivially.

Suppose the claim is true for $r - 1$ colors. Consider any edge-coloring of K_{N_r} using r colors. Pick an arbitrary vertex v. Of the $N_r - 1 = r(N_{r-1} - 1) + 1$ edges incident to v, by the pigeonhole principle, at least N_{r-1} edges incident to v have the same color, say red. Let V_0 be the vertices joined to v by a red edge.

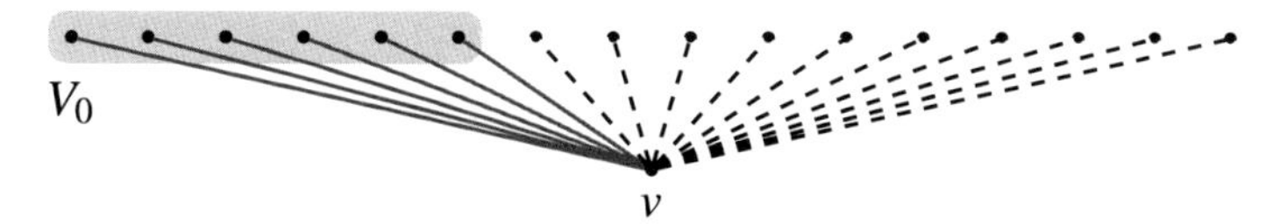

If there is a red edge inside V_0, we obtain a red triangle. Otherwise, there are at most $r - 1$ colors appearing among $|V_0| \geq N_{r-1}$ vertices, and we have a monochromatic triangle inside V_0 by the induction hypothesis. □

Exercise 0.1.5. Show that N_r from (0.1) satisfies $N_r = 1 + r! \sum_{i=0}^{r} 1/i! = \lceil r!e \rceil$.

***Remark* 0.1.6** (Ramsey's theorem). The preceding recursive/inductive pigeonhole argument can be easily adapted to prove Ramsey's theorem in general.

Theorem 0.1.7 (Graph Ramsey theorem)
For every k and r there exists some $N = N(k, r)$ such that if each edge of K_N is colored using one of r colors, then there is a monochromatic K_k.

Exercise 0.1.8. Prove the graph Ramsey theorem (Theorem 0.1.7).

Ramsey's theorem extends even more generally to hypergraphs.

Theorem 0.1.9 (Hypergraph Ramsey theorem)
For every k, r, s there exists some $N = N(k, r, s)$ such that if each edge of a complete s-uniform hypergraph on N vertices is colored using one of r colors, then there is a monochromatic clique on k vertices.

Exercise 0.1.10. Prove the hypergraph Ramsey theorem (Theorem 0.1.9).

Remark* 0.1.11** (Bounds for multicolor triangle Ramsey numbers). The smallest $N(r)$ in Theorem 0.1.4 is also known as the ***multicolor triangle Ramsey number, denoted $\boldsymbol{R(3, 3, \ldots, 3)}$ with 3 repeated r times. It is a major open problem in Ramsey theory to determine the rate of growth of this Ramsey number. Here is an easy argument showing an exponential lower bound. (Compare it to the upper bound from Exercise 0.1.5.)

Proposition 0.1.12 (Multicolor triangle Ramsey numbers: exponential lower bound)
For each positive integer r, there exists an edge-coloring of K_{2^r} using r colors with no monochromatic triangle.

Proof. Label the vertices by elements of $\{0,1\}^r$. Assign an edge color i if i is the smallest index such that the two endpoint vertices differ on coordinate i. This coloring does not have monochromatic triangles. Indeed, suppose x, y, z form a monochromatic triangle with color i, then $x_i, y_i, z_i \in \{0,1\}$ must be all distinct, which is impossible. □

Schur (1916) had actually given an even better lower bound: see Exercise 0.1.14. One of Erdős' favorite problems asks whether there is an exponential upper bound. This is a major open problem in Ramsey theory, and it is related to to other important topics in combinatorics such as the Shannon capacity of graphs (see, e.g., the survey by Nešetřil and Rosenfeld 2001).

Open Problem 0.1.13 (Multicolor triangle Ramsey numbers: exponential upper bound) Is there a constant $C > 0$ so that if $N \geq C^r$, then every edge-coloring of K_N using r colors contains a monochromatic triangle?

Graph Theoretic Proof of Schur's Theorem

We set up a graph whose triangles correspond to solutions to $x + y = z$, and then apply the multicolor triangle Ramsey theorem.

Proof of Schur's theorem (Theorem 0.1.2). Let $\phi\colon [N] \to [r]$ be a coloring. Color the edges of a complete graph with vertices $\{1, \ldots, N+1\}$ by giving the edge $\{i, j\}$ with $i < j$ the color $\phi(j - i)$.

$\phi(k-i)$

i $\phi(j-i)$ j $\phi(k-j)$ k

By Theorem 0.1.4, if N is large enough, then there is a monochromatic triangle, say on vertices $i < j < k$. So, $\phi(j-i) = \phi(k-j) = \phi(k-i)$. Take $x = j - i$, $y = k - j$, and $z = k - i$. Then $\phi(x) = \phi(y) = \phi(z)$ and $x + y = z$, as desired. □

Now that we proved Schur's theorem, let us pause and think about what we gained by translating the problem to graph theory. We were able to apply Ramsey's theorem, whose proof considers restrictions to subgraphs, which would have been rather unnatural if we had worked exclusively in the integers. Graphs gave us greater flexibility.

Later in the book, we will see other more sophisticated examples of this idea. We will gain new perspectives by bringing number theory problems to graph theory.

Exercise 0.1.14 (Schur's lower bound). Let $N(r)$ denote the smallest positive integer in Schur's theorem (Theorem 0.1.2). Show that $N(r) \geq 3N(r-1) - 1$ for every r. Deduce that $N(r) \geq (3^r + 1)/2$ for every r. Also deduce that there exists a coloring of the edges of $K_{(3^r+1)/2}$ with r colors so that there are no monochromatic triangles.

Exercise 0.1.15 (Upper bound on Ramsey numbers). Let s and t be positive integers. Show that if the edges of a complete graph on $\binom{s+t-2}{s-1}$ vertices are colored with red and blue, then there must be either a red K_s or a blue K_t.

Exercise 0.1.16 (Monochromatic triangles compared to random coloring).

(a) True or false: If the edges of K_n are colored using 2 colors, then at least $1/4 - o(1)$ fraction of all triangles are monochromatic. (Note that $1/4$ is the fraction one expects if the edges were colored uniformly at random.)

(b) True or false: if the edges of K_n are colored using 3 colors, then at least $1/9 - o(1)$ fraction of all triangles are monochromatic.

(c*) True or false: if the edges of K_n are colored using 2 colors, then at least $1/32 - o(1)$ fraction of all copies of K_4's are monochromatic.

0.2 Progressions

Additive combinatorics describes a rapidly growing body of mathematics motivated by simple-to-state questions about addition and multiplication of integers. (The name "additive combinatorics" became popular in the 2000s, when the field witnessed a rapid explosion thanks to the groundbreaking works of Gowers, Green, Tao, and others; previously the area was more commonly known as "combinatorial number theory.") The problems and methods in additive combinatorics are deep and far-reaching, connecting many different areas of mathematics such as graph theory, harmonic analysis, ergodic theory, discrete geometry, and model theory.

Here we highlight some important developments in additive combinatorics, particularly concerning progressions. The ideas behind these developments form some of the core themes of this book.

Towards Szemerédi's Theorem

Schur's theorem is one of the earliest results in additive combinatorics. It has important variations and extensions, such as the following seminal result of van der Waerden (1927) on monochromatic arithmetic progressions.

Theorem 0.2.1 (van der Waerden's theorem)
If the integers are colored using finitely many colors, then there exist arbitrarily long monochromatic arithmetic progressions.

Note that having arbitrarily long arithmetic progressions is very different from having infinitely long arithmetic progressions, as seen in the next exercise.

Exercise 0.2.2. Show that $\mathbb{Z}$ may be colored using two colors so that it contains no infinitely long arithmetic progressions.

Erdős and Turán (1936) conjectured a stronger statement, that any subset of the integers with positive density contains arbitrarily long arithmetic progressions. To be precise, we say that $A \subseteq \mathbb{Z}$ has ***positive upper density*** if

$$\limsup_{N\to\infty} \frac{|A \cap \{-N, \ldots, N\}|}{2N + 1} > 0. \tag{0.2}$$

(There are several variations of definition of density – the exact formulation is not crucial here.) The Erdős and Turán conjecture speculates that the "true" reason for van der Waerden's theorem is not so much having finitely many colors (as in Ramsey's theorem), but rather that some color class necessarily has positive density. (The analogous claim is false for graphs, since a triangle-free graph can have edge-density up to $1/2$; we explore this topic further in the next chapter.)

Roth (1953) proved the Erdős and Turán conjecture for 3-term arithmetic progressions using Fourier analysis. It took another two decades before Szemerédi (1975) fully settled the conjecture in a combinatorial tour de force. These theorems by Roth and Szemerédi are landmark results in additive combinatorics. Much of what we will discuss in the book is motivated by these results and the developments around them.

Theorem 0.2.3 (Roth's theorem)
Every subset of the integers with positive upper density contains a 3-term arithmetic progression.

Theorem 0.2.4 (Szemerédi's theorem)
Every subset of the integers with positive upper density contains arbitrarily long arithmetic progressions.

Szemerédi's theorem is deep and intricate. This important work led to many subsequent developments in additive combinatorics. Several different proofs of Szemerédi's theorem have since been discovered, and some of them have blossomed into rich areas of mathematical research. Here are some of the most influential modern proofs of Szemerédi's theorem (in historical order):

- The ergodic theoretic approach by Furstenberg (1977);
- Higher-order Fourier analysis by Gowers (2001);
- Hypergraph regularity lemma independently by Rödl et al. (2005) and Gowers (2001).

Another modern proof of Szemerédi's theorem results from the **density Hales–Jewett theorem**, which was originally proved by Furstenberg and Katznelson (1978) using ergodic theory. Subsequently a new combinatorial proof was found in the first successful Polymath Project (Polymath 2012), an online collaborative project initiated by Gowers.

Each approaches has its own advantages and disadvantages. For example, the ergodic approach led to multidimensional and polynomial generalizations of Szemerédi's theorem, which we discuss in what follows. On the other hand, the ergodic approach does not give any concrete quantitative bounds. Fourier analysis and its generalizations produce the best quantitative bounds to Szemerédi's theorem. They also led to deep results about counting patterns in the prime numbers. However, there appear to be difficulties and obstructions extending Fourier analysis to higher dimensions.

The relationships between these different approaches to Szemerédi's theorem are not yet completely understood. A unifying theme underlying all known approaches to Szemerédi's theorem is

the dichotomy between structure and pseudorandomness.

This phrase was popularized by Tao (2007b) and others. It will be a theme throughout this book. We will see facets of this dichotomy in both graph theory and additive combinatorics.

Quantitative Bounds on Szemerédi's Theorem

There is much interest in obtaining better quantitative bounds on Szemerédi's theorem. Roth's initial proof showed that every subset of $[N]$ avoiding 3-term arithmetic progressions has size $O(N/\log\log N)$. (We will see this proof in Chapter 6.) Roth's upper bound has been improved steadily over time, all via refinement of his Fourier analytic technique. At the time of this writing, the current best upper bound is $N/(\log N)^{1+c}$ for some constant $c > 0$ (Bloom and Sisask 2020). For 4-term arithmetic progressions, the best known upper bound is $N/(\log N)^c$ (Green and Tao 2017). For k-term arithmetic progressions, with fixed $k \geq 5$, the best known upper bound is $N/(\log\log N)^{c_k}$ (Gowers 2001).

As for lower bounds, Behrend (1946) constructed a subset of $[N]$ of size $Ne^{-c\sqrt{\log N}}$ that avoids 3-term arithmetic progressions. This is an important construction that we will see in Section 2.5. Some researchers think that this lower bound is closer to the truth, since for a variant of Roth's theorem (namely avoiding solutions to $x + y + z = 3w$), Behrend's construction is quite close to the truth (Schoen and Shkredov 2014; Schoen and Sisask 2016).

Erdős famously conjectured the following.

Conjecture 0.2.5 (Erdős conjecture on arithmetic progressions)
Every subset A of integers with $\sum_{a \in A} 1/a = \infty$ contains arbitrarily long arithmetic progressions.

This is a strengthening of the Erdős–Turán conjecture (later Szemerédi's theorem), since every subset of integers with positive density necessarily has a divergent harmonic sum. Erdős' conjecture was motivated by the primes (see the Green–Tao theorem in what follows). It has an attractive statement and is widely publicized. The supposed connection between divergent harmonic series and arithmetic progressions seems magical. However, this connection is perhaps somewhat misleading. The hypothesis on divergent harmonic series implies that there are infinitely many N for which $|A \cap [N]| \geq N/(\log N(\log\log N)^2)$. So the Erdős conjecture is really about an upper bound on Szemerédi's theorem. As mentioned earlier, it is plausible that the true upper bound for Szemerédi may be much lower than $1/\log N$. Nevertheless, the "logarithmic barrier" proposed by the Erdős conjecture has a special symbolic and historical status. Erdős' conjecture for k-term arithmetic progressions is now proved for $k = 3$ thanks to the new $N/(\log N)^{1+c}$ upper bound (Bloom and Sisask 2020), but it remains very much open for all $k \geq 4$.

Improving quantitative bounds on Szemerédi's theorem remains an active area of research. Perhaps by the time you read this book (or when I update it to a future edition), these bounds will have been significantly improved.

Extensions of Szemerédi's Theorem

Instead of working over subsets of integers, what happens if we consider subsets of the lattice $\mathbb{Z}^d$? We say that $A \subseteq \mathbb{Z}^d$ has ***positive upper density*** if

$$\limsup_{N\to\infty} \frac{|A \cap [-N,N]^d|}{(2N+1)^d} > 0$$

(as before, other similar definitions are possible). How can we generalize the notion of a subset of $\mathbb{Z}$ containing arbitrarily long arithmetic progressions? We could desire A to contain $k \times k \times \cdots \times k$ cubical grids for arbitrarily large k. Equivalently, we say that $A \subseteq \mathbb{Z}^d$ ***contains arbitrary constellations*** if for every finite set $F \subseteq \mathbb{Z}^d$, there is some $a \in \mathbb{Z}^d$ and $t \in \mathbb{Z}_{>0}$ such that $a + t \cdot F = \{a + tx : x \in F\}$ is contained in A. In other words, A contains every finite pattern F (allowing dilation and translation, as captured by $a + t \cdot F$). The following multidimensional generalization of Szemerédi's theorem was proved by Furstenberg and Katznelson (1978) using ergodic theory, though a combinatorial proof was later discovered as a consequence of the hypergraph regularity method.

Theorem 0.2.6 (Multidimensional Szemerédi theorem)
Every subset of $\mathbb{Z}^d$ of positive upper density contains arbitrary constellations.

For example, the theorem implies that every subset of $\mathbb{Z}^2$ of positive upper density contains a $k \times k$ axis-aligned square grid for every k.

There is also a polynomial extension of Szemerédi's theorem. Let us first state a special case, originally conjectured by Lovász and proved independently by Furstenberg (1977) and Sárkőzy (1978).

Theorem 0.2.7 (Furstenberg–Sárkőzy theorem)
Any subset of the integers with positive upper density contains two numbers differing by a perfect square.

In other words, the set always contains $\{x, x + y^2\}$ for some $x \in \mathbb{Z}$ and $y \in \mathbb{Z}_{>0}$. What about other polynomial patterns? The following polynomial generalization was proved by Bergelson and Leibman (1996).

Theorem 0.2.8 (Polynomial Szemerédi theorem)
Suppose $A \subseteq \mathbb{Z}$ has positive upper density. If $P_1, \ldots, P_k \in \mathbb{Z}[X]$ are polynomials with $P_1(0) = \cdots = P_k(0) = 0$, then there exist $x \in \mathbb{Z}$ and $y \in \mathbb{Z}_{>0}$ such that $x + P_1(y), \ldots, x + P_k(y) \in A$.

In fact, Bergelson and Leibman proved a common generalization – a multidimensional polynomial Szemerédi theorem. (Can you guess what it says?)

We will not discuss the polynomial Szemerédi theorem in this book. Currently the only known proof of the most general form of the polynomial Szemerédi theorem uses ergodic theory, though quantitative bounds are known for certain patterns (e.g., Peluse 2020).

Building on Szemerédi's theorem as well as other important developments in number theory, Green and Tao (2008) proved their famous theorem that settled an old folklore conjecture about prime numbers. Their theorem is one of the most celebrated mathematical achievements of this century.

Theorem 0.2.9 (Green–Tao theorem)
The primes contain arbitrarily long arithmetic progressions.

We will discuss the Green–Tao theorem in Chapter 9. The theorem has been extended to polynomial progressions (Tao and Ziegler 2008) and to higher dimensions (Tao and Ziegler 2015; also see Fox and Zhao 2015).

0.3 What's Next in the Book?

One of our goals is to understand two different proofs of Roth's theorem, which has the following finitary statement. We say that a set is ***3-AP-free*** if it does not contain a 3-term arithmetic progression.

Theorem 0.3.1 (Roth's theorem)
Every 3-AP-free subset of $[N]$ has size $o(N)$.

Roth originally proved his result using **Fourier analysis** (also called the **Hardy–Littlewood circle method** in this context). We will see Roth's proof in Chapter 6.

In the 1970s, Szemerédi developed the **graph regularity method**. It is now a central technique in extremal graph theory. Ruzsa and Szemerédi (1978) used the graph regularity method to give a new graph theoretic proof of Roth's theorem. We will see this proof as well as other applications of the graph regularity method in Chapter 2.

Extremal graph theory, broadly speaking, concerns questions of the form: what is the maximum (or minimum) possible number of some structure in a graph with certain prescribed properties? A starting point (historically and also pedagogically) in extremal graph theory is the following question:

Question 0.3.2 (Triangle-free graphs)
What is the maximum number of edges in a triangle-free n-vertex graph?

This question has a relatively simple answer, and it will be the first topic in the next chapter. We will then explore related questions about the maximum number of edges in a graph without some given subgraph.

Although Question 0.3.2 above sounds similar to Roth's theorem, it does not actually allow us to deduce Roth's theorem. Instead, we need to consider the following question.

Question 0.3.3
What is the maximum number of edges in an n-vertex graph where every edge is contained in a unique triangle?

This innocent looking question turns out to be incredibly mysterious. In Chapter 2, we develop the graph regularity method and use it to prove that any such graph must have $o(n^2)$ edges. And we then deduce Roth's theorem from this graph theoretic claim.

The graph regularity method illustrates the dichotomy of structure and pseudorandomness in graph theory. Some of the later chapters dive further into related concepts. Chapter 3 explores **pseudorandom graphs** – what does it mean for a graph to look random? Chapter 4 concerns **graph limits**, a convenient analytic language for capturing many important concepts in earlier chapters. Chapter 5 explores **graph homomorphism inequalities**, revisiting questions from extremal graph theory with an analytic lens.

And then we switch gears (but not entirely) to some core topics in additive combinatorics. Chapter 6 contains the Fourier analytic proof of **Roth's theorem**. There will be many thematic similarities between elements of the Fourier analytic proof and earlier topics. Chapter 7 explores the structure of set addition. Here we prove **Freiman's theorem** on sets with small additive doubling, a cornerstone result in additive combinatorics. It also plays a key role in Gowers' proof of Szemerédi's theorem, generalizing Fourier analysis to higher-order Fourier analysis, although we will not go into the latter topic in this book (see Further Reading at the end of Chapter 7). In Chapter 8, we explore the **sum-product problem**, which is closely connected to incidence geometry (and we will see another graph theoretic proof there). In Chapter 9, we discuss the **Green–Tao theorem** and prove an extension of Szemerédi's theorem to sparse pseudorandom sets, which plays a central role in the proof of the Green–Tao theorem.

I hope that you will enjoy this book. I have been studying this subject since I began graduate school. I still think about these topics nearly every day. My goal is to organize and distill the beautiful mathematics in this field as a friendly introduction.

The chapters do have some logical dependencies, but not many. Each topic can be studied and enjoyed on its own, though you will gain a lot more by appreciating the overall themes and connections.

There is still a lot that we do not know. Perhaps you too will be intrigued by the boundless open questions that are still waiting to be explored.

Further Reading

The book *Ramsey Theory* by Graham, Rothschild, and Spencer (1990) is a wonderful introduction to the subject. It has beautiful accounts of theorems of Ramsey, van der Waerden, Hales–Jewett, Schur, Rado, and others, that form the foundation of Ramsey theory.

For a survey of modern developments in additive combinatorics, check out the book review by Green (2009a) of *Additive Combinatorics* by Tao and Vu (2006).

Chapter Summary

- **Schur's theorem.** Every coloring of $\mathbb{N}$ using finitely many colors contains a monochromatic solution to $x + y = z$.
  - Proof: set up a graph whose triangles correspond to solutions to $x + y = z$, and then apply **Ramsey's theorem**.
- **Szemerédi's theorem.** Every subset of $\mathbb{N}$ with positive density contains arbitrarily long arithmetic progressions.
  - A foundational result that led to important developments in **additive combinatorics**.
  - Several different proofs, each illustrating the **dichotomy of structure of pseudorandomness** in a different context.
  - Extensions: multidimensional, polynomial, primes (Green–Tao).

1

Forbidding a Subgraph

Chapter Highlights

- Turán problem: determine the maximum number of edges in an n-vertex H-free graph
- Mantel's and Turán's theorems: K_r-free
- Kővári–Sós–Turán theorem: $K_{s,t}$-free
- Erdős–Stone–Simonovits theorem: H-free for general H
- Dependent random choice technique: H-free for a bounded degree bipartite H
- Lower bound constructions of H-free graphs for bipartite H
- Algebraic constructions: matching lower bounds for $K_{2,2}$, $K_{3,3}$, and $K_{s,t}$ for t much larger than s, and also for C_4, C_6, C_{10}
- Randomized algebraic constructions

We begin by completely answering the following question.

Question 1.0.1 (Triangle-free graph)
What is the maximum number of edges in a triangle-free n-vertex graph?

We will see the answer shortly. More generally, we can ask about what happens if we replace "triangle" by an arbitrary subgraph. This is a foundational problem in extremal graph theory.

Definition 1.0.2 (Extremal number / Turán number)
We write $\mathbf{ex}(\boldsymbol{n}, \boldsymbol{H})$ for the maximum number of edges in an n-vertex H-free graph. Here an ***H*-free graph** is a graph that does not contain H as a subgraph.

In this book, by H-free we always mean forbidding H as a subgraph rather than as an induced subgraph. (See Notation and Conventions at the beginning of the book for the distinction.)

Question 1.0.3 (Turán problem)
Determine $\mathrm{ex}(n, H)$. Given a graph H, how does a graph H. How does $\mathrm{ex}(n, H)$ grow as $n \to \infty$?

The Turán problem is one of the most basic problems in extremal graph theory. It is named after Pál Turán for his fundamental work on the subject. Research on this problem has led to many important techniques. We will see a fairly satisfactory answer to the Turán problem for nonbipartite graphs H. We also know the answer for a small number of bipartite graphs H. However, for nearly all bipartite graphs H much mystery remains.

In the first part of the chapter, we focus on techniques for upper bounding $\mathrm{ex}(n, H)$. In the last few sections, we turn our attention to lower bounding $\mathrm{ex}(n, H)$ when H is a bipartite graph.

1.1 Forbidding a Triangle: Mantel's Theorem

We begin by answering Question 1.0.1: what is the maximum number of edges in an n-vertex triangle-free graph? This question was answered in the early 1900s by Willem Mantel, whose theorem is considered the starting point of extremal graph theory.

Let us partition the n vertices into two equal halves (differing by one if n is odd) and then put in all edges across the two parts. This is the complete bipartite graph $K_{\lfloor n/2 \rfloor, \lceil n/2 \rceil}$ and is triangle-free. For example,

$$K_{4,4} = \quad .$$

The graph $K_{\lfloor n/2 \rfloor, \lceil n/2 \rceil}$ has $\lfloor n/2 \rfloor \lceil n/2 \rceil = \lfloor n^2/4 \rfloor$ edges (one can check this equality by separately considering even and odd n).

Mantel (1907) proved that $K_{\lfloor n/2 \rfloor, \lceil n/2 \rceil}$ has the greatest number of edges among all triangle-free graphs.

Theorem 1.1.1 (Mantel's theorem)
Every n-vertex triangle-free graph has at most $\lfloor n^2/4 \rfloor$ edges.

Using the notation of Definition 1.0.2, Mantel's theorem says that

$$\mathrm{ex}(n, K_3) = \left\lfloor \frac{n^2}{4} \right\rfloor.$$

Moreover, we will see that $K_{\lfloor n/2 \rfloor, \lceil n/2 \rceil}$ is the unique maximizer of the number of edges among n-vertex triangle-free graphs.

We give two different proofs of Mantel's theorem, each illustrating a different technique.

First proof of Mantel's theorem. Let $G = (V, E)$ be a triangle-free graph with $|V| = n$ vertices and $|E| = m$ edges. For every edge xy of G, note that x and y have no common neighbors or else it would create a triangle.

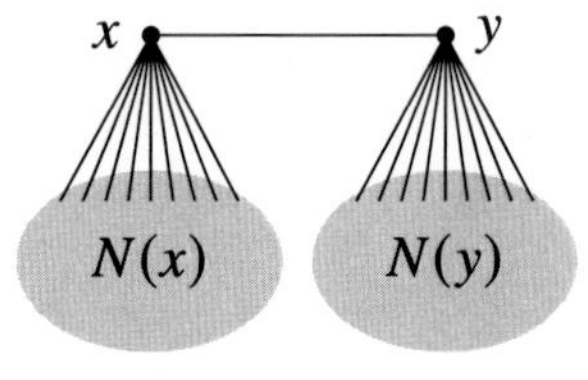

Therefore, $\deg x + \deg y \le n$, which implies that

$$\sum_{xy \in E} (\deg x + \deg y) \le mn.$$

On the other hand, note that for each vertex x, the term $\deg x$ appears once in the preceding sum for each edge incident to x, and so it appears a total of $\deg x$ times. We then apply the Cauchy–Schwarz inequality to get

$$\sum_{xy \in E} (\deg x + \deg y) = \sum_{x \in V} (\deg x)^2 \geq \frac{1}{n} \left(\sum_{x \in V} \deg x \right)^2 = \frac{(2m)^2}{n}.$$

Comparing the two inequalities, we obtain $(2m)^2/n \leq mn$, and hence $m \leq n^2/4$. Since m is an integer, we obtain $m \leq \lfloor n^2/4 \rfloor$, as claimed. □

Second proof of Mantel's theorem. Let $G = (V, E)$ be a triangle-free graph. Let v be a vertex of maximum degree in G. Since G is triangle-free, the neighborhood $N(v)$ of v is an independent set.

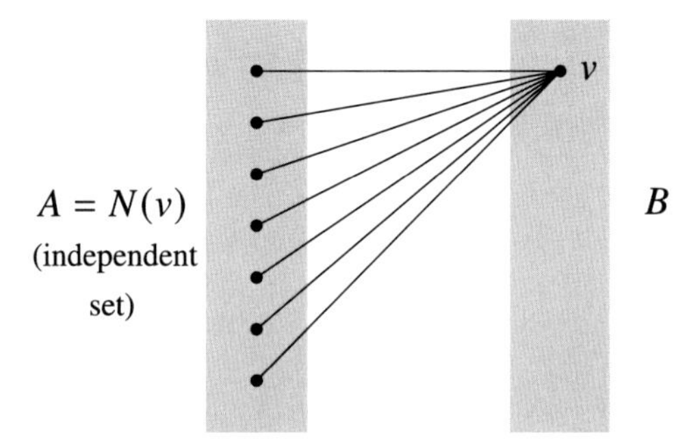

Partition $V = A \cup B$ where $A = N(v)$ and $B = V \setminus A$. Since v is a vertex of maximum degree, we have $\deg x \leq \deg v = |A|$ for all $x \in V$. Since A contains no edges, every edge of G has at least one endpoint in B. Therefore,

$$|E| \leq \sum_{x \in B} \deg x \leq |B| \max_{x \in B} \deg x \leq |A|\,|B| \leq \left(\frac{|A| + |B|}{2} \right)^2 = \frac{n^2}{4}, \tag{1.1}$$

as claimed. □

Remark 1.1.2 (The equality case in Mantel's theorem)**.** The second proof just presented shows that every n-vertex triangle-free graph with exactly $\lfloor n^2/4 \rfloor$ edges must be isomorphic to $K_{\lfloor n/2 \rfloor, \lceil n/2 \rceil}$. Indeed, in (1.1), the inequality $|E| \leq \sum_{x \in B} \deg x$ is tight only if B is an independent set, the inequality $\sum_{x \in B} \deg x \leq |A|\,|B|$ is tight if B is complete to A, and $|A|\,|B| < \lfloor n^2/4 \rfloor$ unless $|A| = |B|$ (if n is even) or $||A| - |B|| = 1$ (if n is odd).

(Exercise: also deduce the equality case from the first proof.)

In general, it is a good idea to keep the equality case in mind when following the proofs, or when coming up with your own proofs, to make sure you are not giving away too much at any step.

The next exercise can be solved by a neat application of Mantel's theorem.

Exercise 1.1.3. Let X and Y be independent and identically distributed random vectors in $\mathbb{R}^d$ according to some arbitrary probability distribution. Prove that

$$\mathbb{P}(|X+Y| \geq 1) \geq \frac{1}{2}\mathbb{P}(|X| \geq 1)^2.$$

Hint: Consider i.i.d. $X_1, \ldots, X_n$, and then let $n \to \infty$.

The next several exercises explore extensions of Mantel's theorem. It is useful to revisit the proof techniques.

Exercise 1.1.4 (Many triangles). Show that a graph with n vertices and m edges has at least

$$\frac{4m}{3n}\left(m - \frac{n^2}{4}\right) \text{ triangles.}$$

Exercise 1.1.5. Prove that every n-vertex nonbipartite triangle-free graph has at most $(n-1)^2/4 + 1$ edges.

Exercise 1.1.6 (Stability). Let G be an n-vertex triangle-free graph with at least $\lfloor n^2/4 \rfloor - k$ edges. Prove that G can be made bipartite by removing at most k edges.

Exercise 1.1.7. Show that every n-vertex triangle-free graph with minimum degree greater than $2n/5$ is bipartite.

Exercise 1.1.8*. Prove that every n-vertex graph with at least $\lfloor n^2/4 \rfloor + 1$ edges contains at least $\lfloor n/2 \rfloor$ triangles.

Exercise 1.1.9*. Let G be an n-vertex graph with $\lfloor n^2/4 \rfloor - k$ edges (here $k \in \mathbb{Z}$) and t triangles. Prove that G can be made bipartite by removing at most $k + 6t/n$ edges, and that this constant 6 is the best possible.

Exercise 1.1.10*. Prove that every n-vertex graph with at least $\lfloor n^2/4 \rfloor + 1$ edges contains some edge in at least $(1/6 - o(1))n$ triangles, and that this constant $1/6$ is the best possible.

1.2 Forbidding a Clique: Turán's Theorem

We generalize Mantel's theorem from triangles to cliques.

Question 1.2.1 (K_{r+1}-free graph)
What is the maximum number of edges in a K_{r+1}-free graph on n vertices?

Construction 1.2.2 (Turán graph)
The ***Turán graph*** $\boldsymbol{T_{n,r}}$ is defined to be the complete n-vertex r-partite graph with part sizes differing by at most 1 (so each part has size $\lfloor n/r \rfloor$ or $\lceil n/r \rceil$).

***Example* 1.2.3.** $T_{10,3} = K_{3,3,4}$:

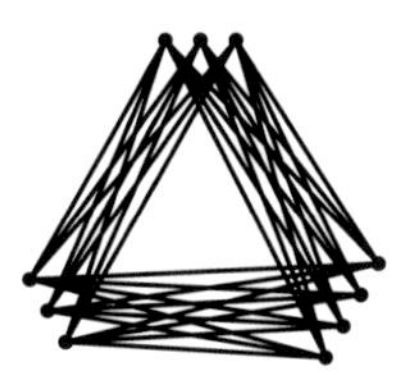

Turán (1941) proved the following fundamental result.

Theorem 1.2.4 (Turán's theorem)
The Turán graph $T_{n,r}$ maximizes the number of edges among all n-vertex K_{r+1}-free graphs. It is also the unique maximizer.

The first part of the theorem says that

$$\mathrm{ex}(n, K_{r+1}) = e(T_{n,r}).$$

It is not too hard to give a precise formula for $e(T_{n,r})$, though there is a small, annoying dependence on the residue class of n mod r. The following bound is good enough for most purposes.

Exercise 1.2.5. Show that

$$e(T_{n,r}) \leq \left(1 - \frac{1}{r}\right) \frac{n^2}{2},$$

with equality if and only if n is divisible by r.

Corollary 1.2.6 (Turán's theorem)

$$\mathrm{ex}(n, K_{r+1}) \leq \left(1 - \frac{1}{r}\right) \frac{n^2}{2}.$$

Even when n is not divisible by r, the difference between $e(T_{n,r})$ and $(1 - 1/r)n^2/2$ is $O(nr)$. As we are generally interested in the regime when r is fixed, this difference is a negligible lower-order contribution. That is,

$$\mathrm{ex}(n, K_{r+1}) = \left(1 - \frac{1}{r} - o(1)\right) \frac{n^2}{2}, \quad \text{for fixed } r \text{ as } n \to \infty.$$

Every r-partite graph is automatically K_{r+1}-free. Let us first consider an easy special case of the problem.

Lemma 1.2.7 (Maximum number of edges in an r-partite graph)
Among n-vertex r-partite graphs, $T_{n,r}$ is the unique graph with the maximum number of edges.

Proof. Suppose we have an n-vertex r-partite graph with the maximum possible number of edges. It should be a complete r-partite graph. If there were two vertex parts A and B with $|A| + 2 \leq |B|$, then moving a vertex from B (the larger part) to A (the smaller part) would increase the number of edges by $(|A| + 1)(|B| - 1) - |A|\,|B| = |B| - |A| - 1 > 0$. Thus, all the vertex parts must have sizes within one of each other. The Turán graph $T_{n,r}$ is the unique such graph. □

We will see three proofs of Turán's theorem. The first proof extends our second proof of Mantel's theorem.

First proof of Turán's theorem. We prove by induction on r. The case $r = 1$ is trivial, as a K_2-free graph is empty. Now assume $r > 1$ and that $\mathrm{ex}(n, K_r) = e(T_{n,r-1})$ for every n.

Let $G = (V, E)$ be a K_{r+1}-free graph. Let v be a vertex of maximum degree in G. Since G is K_{r+1}-free, the neighborhood $A = N(v)$ of v is K_r-free. So, by the induction hypothesis,

$$e(A) \leq \mathrm{ex}(|A|, K_r) = e(T_{|A|,r-1}).$$

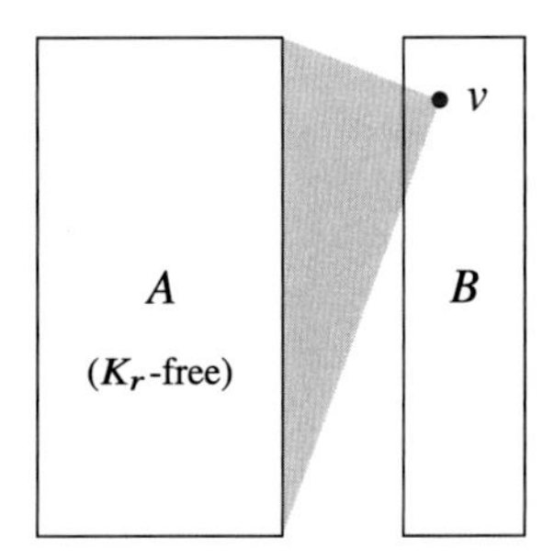

Let $B = V \setminus A$. Since v is a vertex of maximum degree, we have $\deg x \leq \deg v = |A|$ for all $x \in V$. So, the number of edges with at least one vertex in B is

$$e(A, B) + e(B) \leq \sum_{x \in B} \deg x \leq |B| \max_{x \in B} \deg x \leq |A|\,|B|\,.$$

Thus,

$$e(G) = e(A) + e(A, B) + e(B) \leq e(T_{|A|,r-1}) + |A|\,|B| \leq e(T_{n,r}),$$

where the final step follows from the observation that $e(T_{|A|,r-1}) + |A|\,|B|$ is the number of edges in an n-vertex r-partite graph (with part of size $|B|$ and the remaining vertices equitably partitioned into $r - 1$ parts) and Lemma 1.2.7.

To have equality in each of the preceding steps, B must be an independent set (or else $\sum_{y \in B} \deg(y) < |A|\,|B|$) and A must induce $T_{|A|,r-1}$, so that G is r-partite. We knew from Lemma 1.2.7 that the Turán graph $T_{n,r}$ uniquely maximizes the number of edges among r-partite graphs. □

The second proof starts out similarly to our first proof of Mantel's theorem. Recall that in Mantel's theorem, the initial observation was that in a triangle-free graph, given an edge, its two endpoints must have no common neighbors (or else they form a triangle). Generalizing, in a K_4-free graph, given a triangle, its three vertices have no common neighbor. The rest of the proof proceeds somewhat differently from earlier. Instead of summing over all edges as we did before, we remove the triangle and apply induction to the rest of the graph.

Second proof of Turán's theorem. We fix r and proceed by induction on n. The statement is trivial for $n \leq r$, as the Turán graph is the complete graph $K_n = T_{n,r}$ and thus maximizes the number of edges.

Now, assume that $n > r$ and that Turán's theorem holds for all graphs on fewer than n vertices. Let $G = (V, E)$ be an n-vertex K_{r+1}-free graph with the maximum possible number of edges. By the maximality assumption, G contains K_r as a subgraph, since otherwise we could add an edge to G and it would still be K_{r+1}-free. Let A be the vertex set of an r-clique in G, and let $B := V \setminus A$.

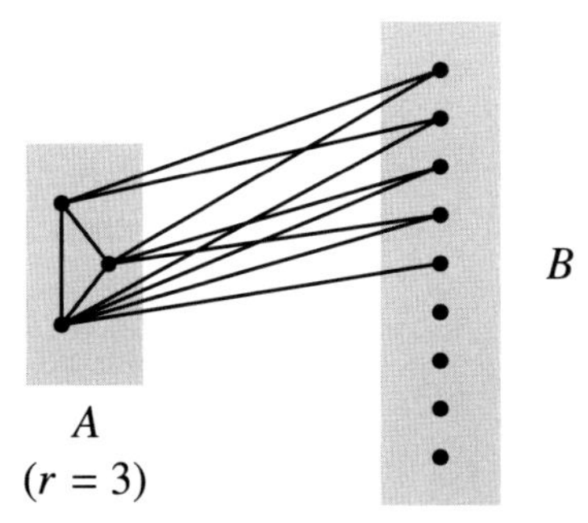

Since G is K_{r+1}-free, every $x \in B$ has at most $r - 1$ neighbors in A. So

$$e(A, B) = \sum_{y \in B} \deg(y, A) \leq \sum_{y \in B} (r - 1) = (r - 1)(n - r).$$

We have

$$\begin{aligned} e(G) &= e(A) + e(A, B) + e(B) \\ &\leq \binom{r}{2} + (r - 1)(n - r) + e(T_{n-r,r}) = e(T_{n,r}), \end{aligned}$$

where the inequality uses the induction hypothesis on $G[B]$, which is K_{r+1}-free, and the final equality can be seen by removing a K_r from $T_{n,r}$.

Finally, let us check when equality occurs. To have equality in each of the preceding steps, the subgraph induced on B must be $T_{n-r,r}$ by induction. To have $e(A) = \binom{r}{2}$, A must induce a clique. To have $e(A, B) = (r - 1)(n - r)$, every vertex of B must be adjacent to all but one vertex in A. Also, two vertices x, y lying in distinct parts of $G[B] \cong T_{n-r,r}$ cannot "miss" the same vertex v of A, or else $A \cup \{x, y\} \setminus \{v\}$ would be an K_{r+1}-clique. This then forces G to be $T_{n,r}$. □

The third proof uses a method known as **Zykov symmetrization**. The idea here is that if a K_{r+1}-free graph is not a Turán graph, then we should be able make some local modifications (namely replacing a vertex by a clone of another vertex) to get another K_{r+1}-free graph with strictly more edges.

Third proof of Turán's theorem. As before, let G be an n-vertex K_{r+1}-free graph with the maximum possible number of edges.

We claim that if x and y are nonadjacent vertices, then $\deg x = \deg y$. Indeed, suppose $\deg x > \deg y$. We can modify G by removing y and adding in a clone of x (a new vertex x' with the same neighborhood as x but not adjacent to x), as illustrated here.

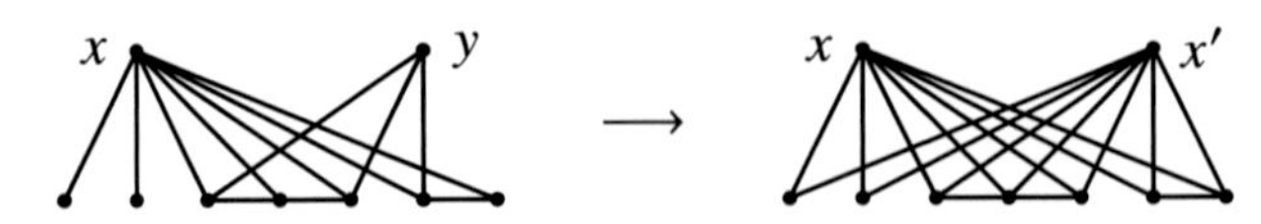

The resulting graph would still be K_{r+1}-free (since a clique cannot contain both x and its clone) and has strictly more edges than G, thereby contradicting the assumption that G has the maximum possible number of edges.

Suppose x is nonadjacent to both y and z in G. We claim that y and z must be nonadjacent. We just saw that $\deg x = \deg y = \deg z$. If yz is an edge, then by deleting y and z from G and adding two clones of x, we obtain a K_{r+1}-free graph with one more edge than G. This would contradict the maximality of G.

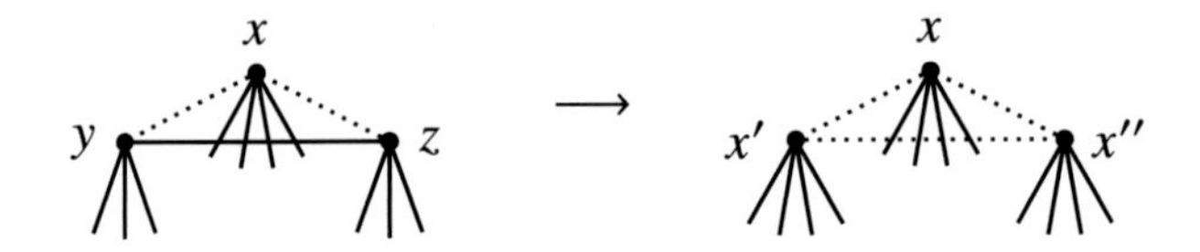

Therefore, nonadjacency is an equivalence relation among vertices of G. So the complement of G is a union of cliques. Hence G is a complete multipartite graph, which has at most r parts since G is K_{r+1}-free. Among all complete r-partite graphs, the Turán graph $T_{n,r}$ is the unique graph that maximizes the number of edges, by Lemma 1.2.7. Therefore, G is isomorphic to $T_{n,r}$. □

The last proof we give in this section uses the probabilistic method. This probabilistic proof was given in the book *The Probabilistic Method* by Alon and Spencer, though the key inequality is due earlier to Caro and Wei. In what follows, we prove Turán's theorem in the formulation of Corollary 1.2.6, that is, $\mathrm{ex}(n, K_{r+1}) \le (1 - 1/r)n^2/2$. A more careful analysis of the proof can yield the stronger statement of Theorem 1.2.4, which we omit.

Fourth proof of Turán's theorem (Corollary 1.2.6). Let $G = (V, E)$ be an n-vertex, K_{r+1}-free graph. Consider a uniform random ordering of the vertices. Let

$$X = \{v \in V : v \text{ is adjacent to all earlier vertices in the random ordering}\}.$$

Then X is a clique. Since the ordering was chosen uniformly at random,

$$\mathbb{P}(v \in X) = \mathbb{P}(v \text{ appears before all its nonneighbors}) = \frac{1}{n - \deg v}.$$

Since G is K_{r+1}-free, $|X| \le r$. So, by linearity of expectations

$$\begin{aligned} r \ge \mathbb{E}|X| &= \sum_{v \in V} \mathbb{P}(v \in X) \\ &= \sum_{v \in V} \frac{1}{n - \deg v} \ge \frac{n}{n - (\sum_{v \in V} \deg v)/n} = \frac{n}{n - 2m/n}. \end{aligned}$$

Rearranging gives

$$m \le \left(1 - \frac{1}{r}\right)\frac{n^2}{2}. \qquad \square$$

In Chapter 5, we will see another proof of Turán's theorem using a method known as **graph Lagrangians**.

Exercise 1.2.8. Let G be a K_{r+1}-free graph. Prove that there exists an r-partite graph H on the same vertex set as G such that $\deg_H(x) \ge \deg_G(x)$ for every vertex x (here $\deg_H(x)$ is the degree of x in H, and likewise with $\deg_G(x)$ for G). Give another proof of Turán's theorem from this fact.

The following exercise is an extension of Exercise 1.1.6.

Exercise 1.2.9* (Stability). Let G be an n-vertex K_{r+1}-free graph with at least $e(T_{n,r}) - k$ edges, where $T_{n,r}$ is the Turán graph. Prove that G can be made r-partite by removing at most k edges.

The next exercise is a neat geometric application of Turán's theorem.

Exercise 1.2.10. Let S be a set of n points in the plane, with the property that no two points are at distance greater than 1. Show that S has at most $\lfloor n^2/3 \rfloor$ pairs of points at distance greater than $1/\sqrt{2}$. Also, show that the bound $\lfloor n^2/3 \rfloor$ is tight (i.e., cannot be improved).

1.3 Turán Density and Supersaturation

Turán's theorem exactly determines $\mathrm{ex}(n, H)$ when H is a clique. Such precise answers are actually quite rare in extremal graph theory. We are often content with looser bounds and asymptotics.

We will go on to bound $\mathrm{ex}(n, H)$ for other values of H later. But for now, let us take a short detour and think about the structure of the problem.

Turán Density

In this chapter, we will define the ***edge density*** of a graph G to be

$$e(G) \Big/ \binom{v(G)}{2}.$$

So, the edge density of a clique is 1. Later in the book, we will consider a different normalization $2e(G)/v(G)^2$ for edge density, which is more convenient for other purposes. When $v(G)$ is large, there is no significant difference between the two choices.

Next, we use an averaging/sampling argument to show that $\mathrm{ex}(n, H)/\binom{n}{2}$ is nonincreasing in n.

Proposition 1.3.1 (Monotonicity of Turán numbers)

For every graph H and positive integer n,

$$\frac{\mathrm{ex}(n+1, H)}{\binom{n+1}{2}} \le \frac{\mathrm{ex}(n, H)}{\binom{n}{2}}.$$

Proof. Let G be an H-free graph on $n+1$ vertices. For each n-vertex subset S of $V(G)$, since $G[S]$ is also H-free, we have

$$\frac{e(G[S])}{\binom{n}{2}} \leq \frac{\mathrm{ex}(n,H)}{\binom{n}{2}}.$$

Varying S uniformly over all n-vertex subsets of $V(G)$, the left-hand side averages to the edge density of G by linearity of expectations. (Check this.) It follows that

$$\frac{e(G)}{\binom{n+1}{2}} \leq \frac{\mathrm{ex}(n,H)}{\binom{n}{2}}.$$

The claim then follows. □

For every fixed H, the sequence $\mathrm{ex}(n,H)/\binom{n}{2}$ is nonincreasing and bounded between 0 and 1. It follows that it approaches a limit.

Definition 1.3.2 (Turán density)

The ***Turán density*** of a graph H is defined to be

$$\boldsymbol{\pi(H)} := \lim_{n\to\infty} \frac{\mathrm{ex}(n,H)}{\binom{n}{2}}.$$

Here are some additional equivalent definitions of Turán density:

- $\pi(H)$ is the smallest real number so that for every $\varepsilon > 0$ there is some $n_0 = n_0(H,\varepsilon)$ so that for every $n \geq n_0$, every n-vertex graph with at least $(\pi(H)+\varepsilon)\binom{n}{2}$ edges contains H as a subgraph;
- $\pi(H)$ is the smallest real number so that every n-vertex H-free graph has edge density $\leq \pi(H) + o(1)$.

Recall, from Turán's theorem, that

$$\mathrm{ex}(n, K_{r+1}) = \left(1 - \frac{1}{r} - o(1)\right)\frac{n^2}{2}, \quad \text{for fixed } r \text{ as } n \to \infty,$$

which is equivalent to

$$\pi(K_{r+1}) = 1 - \frac{1}{r}.$$

In the next couple of sections we will prove the **Erdős–Stone–Simonovits theorem**, which determines the Turán density for every graph H:

$$\pi(H) = 1 - \frac{1}{\chi(H) - 1},$$

where $\chi(H)$ is the chromatic number of H. It should be surprising that the Turán density of H depends only on the chromatic number of H.

With the Erdős–Stone–Simonovits theorem, it may seem as if the Turán problem is essentially understood, but actually this would be very far from the truth. We will see in the next section that $\pi(H) = 0$ for every bipartite graph H. In other words $\mathrm{ex}(n,H) = o(n^2)$. Actual asymptotics growth rate of $\mathrm{ex}(n,H)$ is often unknown.

In a different direction, the generalization to hypergraphs, while looking deceptively similar, turns out to be much more difficult, and very little is known here.

***Remark* 1.3.3** (Hypergraph Turán problem). Generalizing from graphs to hypergraphs, given an r-uniform hypergraph H, we write $\mathrm{ex}(n, H)$ for the maximum number of edges in an n-vertex r-uniform hypergraph that does not contain H as a subgraph. A straightforward extension of Proposition 1.3.1 gives that $\mathrm{ex}(n, H)/\binom{n}{r}$ is a nonincreasing function of n, for each fixed H. So we can similarly define the hypergraph Turán density:

$$\pi(H) := \lim_{n\to\infty} \frac{\mathrm{ex}(n, H)}{\binom{n}{r}}.$$

The exact value of $\pi(H)$ is known in very few cases. It is a major open problem to determine $\pi(H)$ when H is the complete 3-uniform hypergraph on four vertices (also known as a tetrahedron), and more generally when H is a complete hypergraph.

Supersaturation

We know from Mantel's theorem that any n-vertex graph G with $> n^2/4$ edges must contain a triangle. What if G has a lot more edges? It turns out that G must have a lot of triangles. In particular, an n-vertex graph with $> (1/4 + \varepsilon)n^2$ edges must have at least δn^3 triangles for some constant $\delta > 0$ depending on $\varepsilon > 0$. This is indeed a lot of triangles, since there are could only be at most $O(n^3)$ triangles no matter what. (Exercise 1.1.4 asks you to give a more precise quantitative lower bound on the number of triangles. The optimal dependence of δ on ε is a difficult problem that we will discuss in Chapter 5.)

It turns out there is a general phenomenon in combinatorics where, once some density crosses an existence threshold (e.g., the Turán density is the threshold for H-freeness), it will be possible to find not just one copy of the desired object, but in fact lots and lots of copies. This fundamental principle, called **supersaturation**, is useful for many applications, including in our upcoming determination of $\pi(H)$ for general H.

Theorem 1.3.4 (Supersaturation)
For every $\varepsilon > 0$ and graph H there exist some $\delta > 0$ and n_0 such that every graph on $n \geq n_0$ vertices with at least $(\pi(H) + \varepsilon)\binom{n}{2}$ edges contains at least $\delta n^{v(H)}$ copies of H as a subgraph.

Equivalently: every n-vertex graph with $o(n^{v(H)})$ copies of H has edge density $\leq \pi(H) + o(1)$ (here H is fixed). The **sampling argument** in the following proof is useful in many applications.

Proof. By the definition of the Turán density, there exists some n_0 (depending on H and ε) such that every n_0-vertex graph with at least $(\pi(H)+\varepsilon/2)\binom{n_0}{2}$ edges contains H as a subgraph.

Let $n \geq n_0$ and G be an n-vertex graph with at least $(\pi(H) + \varepsilon)\binom{n}{2}$ edges. Let S be an n_0-element subset of $V(G)$, chosen uniformly at random. Let X denote the edge density of $G[S]$. By averaging, $\mathbb{E}X$ equals the edge density of G, and so $\mathbb{E}X \geq \pi(H) + \varepsilon$. Then $X \geq \pi(H) + \varepsilon/2$ with probability $\geq \varepsilon/2$ (or else $\mathbb{E}X$ could not be as large as $\pi(H) + \varepsilon$). So, from the previous paragraph, we know that with probability $\geq \varepsilon/2$, $G[S]$ contains a copy of H. This gives us $\geq (\varepsilon/2)\binom{n}{n_0}$ copies of H, but each copy of H may be counted up to $\binom{n-v(H)}{n_0-v(H)}$ times. Thus, the number of copies of H in G is

$$\geq \frac{(\varepsilon/2)\binom{n}{n_0}}{\binom{n-v(H)}{n_0-v(H)}} = \Omega_{H,\varepsilon}(n^{v(H)}). \qquad \square$$

Exercise 1.3.5 (Supersaturation for hypergraphs). Let H be an r-uniform hypergraph with hypergraph Turán density $\pi(H)$. Prove that every n-vertex r-uniform hypergraph with $o(n^{v(H)})$ copies of H has at most $(\pi(H) + o(1))\binom{n}{r}$ edges.

Exercise 1.3.6 (Density Ramsey). Prove that for every s and r, there is some constant $c > 0$ so that for every sufficiently large n, if the edges of K_n are colored using r colors, then at least c fraction of all copies of K_s are monochromatic.

Exercise 1.3.7 (Density Szemerédi). Let $k \geq 3$. Assuming Szemerédi's theorem for k-term arithmetic progressions (i.e., every subset of $[N]$ without a k-term arithmetic progression has size $o(N)$), prove the following density version of Szemerédi's theorem:

For every $\delta > 0$ there exist c and N_0 (both depending only on k and δ) such that for every $A \subseteq [N]$ with $|A| \geq \delta N$ and $N \geq N_0$, the number of k-term arithmetic progressions in A is at least cN^2.

1.4 Forbidding a Complete Bipartite Graph: Kővári–Sós–Turán Theorem

In this section, we provide an upper bound on $\mathrm{ex}(n, K_{s,t})$, the maximum number of edges in an n-vertex $K_{s,t}$-free graph. It is a major open problem to determine the asymptotic growth of $\mathrm{ex}(n, K_{s,t})$. For certain pairs (s, t) the answer is known, as we will discuss later in the chapter.

Problem 1.4.1 (Zarankiewicz problem)
Determine $\mathrm{ex}(n, K_{s,t})$, the maximum number of edges in an n-vertex $K_{s,t}$-free graph.

Zarankiewicz (1951) originally asked a related problem: determine the maximum number of 1's in an $m \times n$ matrix without an $s \times t$ submatrix with all entries 1.

The main theorem of this section is the fundamental result due to Kővári, Sós, and Turán (1954). We will refer to it as the **KST theorem**, which stands both for its discoverers, as well as for the forbidden subgraph $K_{s,t}$.

Theorem 1.4.2 (Kővári–Sós–Turán theorem – "KST theorem")
For positive integers $s \leq t$, there exists some constant $C = C(s, t)$, such that, for all n,

$$\mathrm{ex}(n, K_{s,t}) \leq Cn^{2-1/s}.$$

The proof proceeds by double counting.

Proof. Let G be an n-vertex $K_{s,t}$-free graph with m edges. Let

$$X = \text{number of copies of } K_{s,1} \text{ in } G.$$

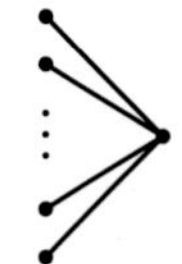

(When $s = 1$, we set $X = 2e(G)$.) The strategy is to count X in two ways. First we count $K_{s,1}$ by first embedding the "left" s vertices of $K_{s,1}$. Then we count $K_{s,1}$ by first embedding the "right" single vertex of $K_{s,1}$.

Upper bound on X. Since G is $K_{s,t}$-free, every s-vertex subset of G has $\le t-1$ common neighbors. Therefore,

$$X \le \binom{n}{s}(t-1).$$

Lower bound on X. For each vertex v of G, there are exactly $\binom{\deg v}{s}$ ways to pick s of its neighbors to form a $K_{s,1}$ as a subgraph. Therefore,

$$X = \sum_{v \in V(G)} \binom{\deg v}{s}.$$

To obtain a lower bound on this quantity in terms of the number of edges m of G, we use a standard trick of viewing $\binom{x}{s}$ as a convex function on the reals, namely, letting

$$f_s(x) = \begin{cases} x(x-1)\cdots(x-s+1)/s! & \text{if } x \ge s-1. \\ 0 & x < s-1. \end{cases}$$

Then $f(x) = \binom{x}{s}$ for all nonnegative integers x. Furthermore f_s is a convex function. Since the average degree of G is $2m/n$, it follows by convexity that

$$X = \sum_{v \in V(G)} f_s(\deg v) \ge n f_s\left(\frac{2m}{n}\right).$$

(It would be a sloppy mistake to lower bound X by $n\binom{2m/n}{s}$.)

Combining the upper bound and the lower bound. We find that

$$n f_s\left(\frac{2m}{n}\right) \le X \le \binom{n}{s}(t-1).$$

Since $f_s(x) = (1+o(1))x^s/s!$ for $x \to \infty$ and fixed s, we find that, as $n \to \infty$,

$$\frac{n}{s!}\left(\frac{2m}{n}\right)^s \le (1+o(1))\frac{n^s}{s!}(t-1).$$

Therefore,

$$m \le \left(\frac{(t-1)^{1/s}}{2} + o(1)\right) n^{2-1/s}.$$ □

The final bound in the proof gives us a somewhat more precise estimate than stated in Theorem 1.4.2. Let us record it here for future reference.

Theorem 1.4.3 (KST theorem)
Fix positive integers $s \le t$. Then, as $n \to \infty$,

$$\mathrm{ex}(n, K_{s,t}) \le \left(\frac{(t-1)^{1/s}}{2} + o(1)\right) n^{2-1/s}.$$

It has been long conjectured that the KST theorem is tight up to a constant factor.

Conjecture 1.4.4 (Tightness of KST bound)
For positive integers $s \le t$, there exists a constant $c = c(s,t) > 0$ such that for all $n \ge 2$,

$$\mathrm{ex}(n, K_{s,t}) \ge cn^{2-1/s}.$$

(In other words, $\mathrm{ex}(n, K_{s,t}) = \Theta_{s,t}(n^{2-1/s})$.)

In the final sections of this chapter, we will produce some constructions showing that Conjecture 1.4.4 is true for $K_{2,t}$ and $K_{3,t}$. We also know that the conjecture is true if t is much larger than s. The first open case of the conjecture is $K_{4,4}$.

Here is an easy consequence of the KST theorem.

Corollary 1.4.5
For every bipartite graph H, there exists some constant $c > 0$ so that $\mathrm{ex}(n, H) = O_H(n^{2-c})$.

Proof. Suppose the two vertex parts of H have sizes s and t, with $s \le t$. Then $H \subseteq K_{s,t}$. And thus every n-vertex H-free graph is also $K_{s,t}$-free, and thus has $O_{s,t}(n^{2-1/s})$ edges. □

In particular, the Turán density $\pi(H)$ of every bipartite graph H is zero.

The KST theorem gives a constant c in the preceding corollary that depends on the number of vertices on the smaller part of H. In Section 1.7, we will use the dependent random choice technique to give a proof of the corollary showing that c only has to depend on the maximum degree of H.

Geometric Applications of the KST Theorem

The following famous problem was posed by Erdős (1946).

Question 1.4.6 (Erdős unit distance problem)
What is the maximum number of unit distances formed by a set of n points in $\mathbb{R}^2$?

In other words, given n distinct points in the plane, at most how many pairs of these points can be exactly distance 1 apart? We can draw a graph with these n points as vertices, with edges joining points exactly a unit distance apart.

To get a feeling for the problem, let us play with some constructions. For small values of n, it is not hard to check by hand that the following configurations are optimal.

$n =$ 3 4 5 6 7

What about for larger values of n? If we line up the n points equally spaced on a line, we get $n - 1$ unit distances.

We can be a bit more efficient by chaining up triangles. The following construction gives us $2n - 3$ unit distances.

The construction for $n = 6$ looks like it was obtained by copying and translating a unit triangle. We can generalize this idea to obtain a recursive construction. Let $f(n)$ denote the maximum number of unit distances formed by n points in the plane. Given a configuration P with $\lfloor n/2 \rfloor$ points that has $f(\lfloor n/2 \rfloor)$ unit distances, we can copy P and translate it by a generic unit vector to get P'. The configuration $P \cup P'$ has at least $2f(\lfloor n/2 \rfloor) + \lfloor n/2 \rfloor$ unit distances. We can solve the recursion to get $f(n) \gtrsim n \log n$.

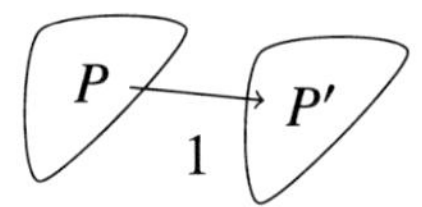

Now we take a different approach to obtain an even better construction. Take a square grid with $\lfloor \sqrt{n} \rfloor \times \lfloor \sqrt{n} \rfloor$ vertices. Instead of choosing the distance between adjacent points as the unit distance, we can scale the configuration so that $\sqrt{r}$ becomes the "unit" distance for some integer r. As an illustration, here is an example of a 5×5 grid with $r = 10$.

It turns out that by choosing the optimal r as a function of n, we can get at least

$$n^{1+c/\log \log n}$$

unit distances, where $c > 0$ is some absolute constant. The proof uses analytic number theory, which we omit as it would take us too far afield. The basic idea is to choose r to be a product of many distinct primes that are congruent to 1 modulo 4, so that r can be represented as a sum of two squares in many different ways, and then estimate the number of such ways.

It is conjectured that the last construction just discussed is close to optimal.

Conjecture 1.4.7 (Erdős unit distance conjecture)
Every set of n points in $\mathbb{R}^2$ has at most $n^{1+o(1)}$ unit distances.

The KST theorem can be used to prove the following upper bound on the number of unit distances.

Theorem 1.4.8 (Upper bound on the unit distance problem)
Every set of n points in $\mathbb{R}^2$ has $O(n^{3/2})$ unit distances.

Proof. Every unit distance graph is $K_{2,3}$-free. Indeed, for every pair of distinct points, there are at most two other points that are at a unit distance from both points.

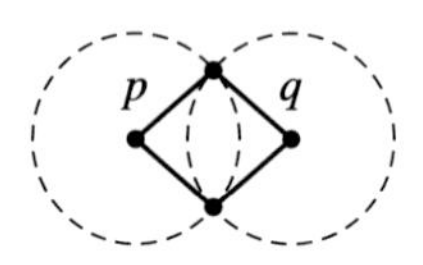

So, the number of edges is at most $\mathrm{ex}(n, K_{2,3}) = O(n^{3/2})$ by Theorem 1.4.2. □

There is a short proof of a better bound of $O(n^{4/3})$ using the crossing number inequality (see Section 8.2), and this is the best known upper bound to date.

Erdős (1946) also asked the following related question.

Question 1.4.9 (Erdős distinct distances problem)
What is the minimum number of distinct distances formed by n points in $\mathbb{R}^2$?

Let $g(n)$ denote the answer. The asymptotically best construction for the minimum number of distinct distances is also a square grid, same as earlier. It can be shown that a square grid with $\lfloor\sqrt{n}\rfloor \times \lfloor\sqrt{n}\rfloor$ points has on the order of $n/\sqrt{\log n}$ distinct distances. This is conjectured to be optimal (i.e., $g(n) \lesssim n/\sqrt{\log n}$).

Let $f(n)$ denote the maximum number of unit distances among n points in the answer. We have $f(n)g(n) \geq \binom{n}{2}$, since each distance occurs at most $f(n)$ times. So an upper bound on $f(n)$ gives a lower bound on $g(n)$ (but not conversely).

A breakthrough on the distinct distances problem was obtained by Guth and Katz (2015).

Theorem 1.4.10 (Guth–Katz distinct distances theorem)
A set of n points in $\mathbb{R}^2$ form $\Omega(n/\log n)$ distinct distances.

In other words, $g(n) \gtrsim n/\log n$, thereby matching the upper bound example up to a factor of $O(\sqrt{\log n})$. The Guth–Katz proof is quite sophisticated. It uses tools ranging from the polynomial method to algebraic geometry.

Exercises

Exercise 1.4.11. Show that a C_4-free bipartite graph between two vertex parts of sizes a and b has at most $ab^{1/2} + b$ edges.

Exercise 1.4.12 (Density KST). Prove that for every pair of positive integers $s \leq t$, there are constants $C, c > 0$ such that every n-vertex graph with $p\binom{n}{2}$ edges contains at least $cp^{st}n^{s+t}$ copies of $K_{s,t}$, provided that $p \geq Cn^{-1/s}$.

The next exercise asks you to think about the quantitative dependencies in the proof of the KST theorem.

Exercise 1.4.13. Show that for every $\varepsilon > 0$, there exists $\delta > 0$ such that every graph with n vertices and at least εn^2 edges contains a copy of $K_{s,t}$ where $s \geq \delta \log n$ and $t \geq n^{0.99}$.

The next exercise illustrates a bad definition of density of a subset of $\mathbb{Z}^2$ (it always ends up being either 0 or 1).

Exercise 1.4.14 (How not to define density). Let $S \subseteq \mathbb{Z}^2$. Define

$$d_k(S) = \max_{\substack{A,B\subseteq \mathbb{Z}\\ |A|=|B|=k}} \frac{|S \cap (A \times B)|}{|A|\,|B|}.$$

Show that $\lim_{k\to\infty} d_k(S)$ exists and is always either 0 or 1.

1.5 Forbidding a General Subgraph: Erdős–Stone–Simonovits Theorem

Turán's theorem tells us that

$$\mathrm{ex}(n, K_{r+1}) = \left(1 - \frac{1}{r} - o(1)\right)\frac{n^2}{2} \quad \text{for fixed } r.$$

The KST theorem implies that

$$\mathrm{ex}(n, H) = o(n^2) \quad \text{for any fixed bipartite graph } H.$$

In this section, we extend these results and determine $\mathrm{ex}(n, H)$, up to an $o(n^2)$ error term, for every graph H. In other words, we will compute the Turán density $\pi(H)$.

Initially it seems possible that the Turán density $\pi(H)$ might depend on H in some complicated way. It turns out that it only depends on the ***chromatic number*** $\boldsymbol{\chi(H)}$ of H, which is the smallest number of colors needed to color the vertices of H such that no two adjacent vertices receive the same color.

Suppose $\chi(H) = r$. Then H cannot be a subgraph of any $(r-1)$-partite graph. In particular, the Turán graph $T_{n,r-1}$ is H-free. (Recall from Construction 1.2.2 that $T_{n,r-1}$ is the complete $(r-1)$-partite graph with n vertices divided into nearly equal parts.) Therefore,

$$\mathrm{ex}(n, H) \geq e(T_{n,r-1}) = \left(1 - \frac{1}{r-1} + o(1)\right)\frac{n^2}{2}.$$

The main theorem of this section, which follows, is a matching upper bound.

Theorem 1.5.1 (Erdős–Stone–Simonovits theorem)
For every graph H, as $n \to \infty$,

$$\mathrm{ex}(n, H) = \left(1 - \frac{1}{\chi(H) - 1} + o(1)\right)\frac{n^2}{2}.$$

In other words, the Turán density of H is

$$\pi(H) = 1 - \frac{1}{\chi(H) - 1}.$$

Remark **1.5.2** (History). Erdős and Stone (1946) proved this result when H is a complete multipartite graph. Erdős and Simonovits (1966) observed that the general case follows as a quick corollary. The proof given here is due to Erdős (1971).

Example **1.5.3.** When $H = K_{r+1}$, $\chi(H) = r + 1$, and so Theorem 1.5.1 agrees with Turán's theorem.

***Example* 1.5.4.** When H is the Petersen graph (shown in what follows), which has chromatic number 3, Theorem 1.5.1 tells us that $\operatorname{ex}(n, H) = (1/4 + o(1))n^2$. The Turán density of the Petersen graph is the same as that of a triangle, which may be somewhat surprising since the Petersen graph seems more complicated than the triangle.

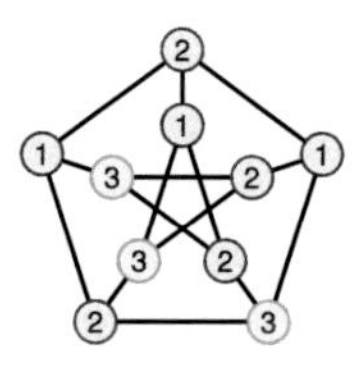

It suffices to establish the Erdős–Stone–Simonovits theorem for complete r-partite graphs H, since every H with $\chi(H) = r$ is a subgraph of some complete r-partite graph.

Theorem 1.5.5 (Erdős–Stone theorem)
Fix $r \geq 2$ and $s \geq 1$. Let $H = K_{s,\ldots,s}$ be the complete r-partite graph with s vertices in each part. Then

$$\operatorname{ex}(n, H) = \left(1 - \frac{1}{r-1} + o(1)\right)\frac{n^2}{2}.$$

In other words, using the notation $K_r[s]$ for ***s*-blow-up** of K_r, obtained by replacing each vertex of K_r by s duplicates of itself (so that $K_r[s] = H$ in the preceding theorem statement), the Erdős–Stone theorem says that

$$\pi(K_r[s]) = \pi(K_r) = 1 - \frac{1}{r-1}.$$

In Section 1.3, we saw supersaturation (Theorem 1.3.4): when the edge density is significantly above the Turán density threshold $\pi(H)$, one finds not just a single copy of H but actually many copies. The Erdős–Stone theorem can be viewed in this light: above edge density $\pi(H)$, we find a large blow-up of H.

The proof uses the following hypergraph extension of the KST theorem, which we will prove later in the section.

Recall the hypergraph Turán problem (Remark 1.3.3). Given an r-uniform hypergraph H (also known as an r-graph), we write $\operatorname{ex}(n, H)$ to be the maximum number of edges in an H-free r-graph.

The analogue of a complete bipartite graph for an r-graph is a complete r-partite r-graph $\boldsymbol{K^{(r)}_{s_1,\ldots,s_r}}$. Its vertex set consists of disjoint vertex parts $V_1, \ldots, V_r$ with $|V_i| = s_i$ for each i. Every r-tuple in $V_1 \times \cdots \times V_r$ is an edge.

Theorem 1.5.6 (Hypergraph KST)
For every fixed positive integers $r \geq 2$ and s,

$$\operatorname{ex}(n, K^{(r)}_{s,\ldots,s}) = o(n^r).$$

Proof of the Erdős–Stone theorem (Theorem 1.5.5). We already saw the lower bound to $\operatorname{ex}(n, H)$ using a Turán graph. It remains to prove an upper bound.

Let G be an H-free graph (where $H = K_{s,\ldots,s}$ is the complete r-partite graph in the theorem). Let $G^{(r)}$ be the r-graph with the same vertex set as G and whose edges are the r-cliques in G. Note that $G^{(r)}$ is $K^{(r)}_{s,\ldots,s}$-free, or else a copy of $K^{(r)}_{s,\ldots,s}$ in $G^{(r)}$ would be supported by a copy of H in G. Thus, by the hypergraph KST theorem (Theorem 1.5.6), $G^{(r)}$ has $o(n^r)$ edges. So, G has $o(n^r)$ copies of K_r, and thus by the supersaturation theorem quoted earlier, the edge density of G is at most $\pi(K_r) + o(1)$, which equals $1 - 1/(r-1) + o(1)$ by Turán's theorem. □

In Section 2.6, we will give another proof of the Erdős–Stone–Simonovits theorem using the graph regularity method.

Hypergraph KST

To help keep notation simple, we first consider what happens for 3-uniform hypergraphs.

Theorem 1.5.7 (KST for 3-graphs)
For every s, there is some C such that

$$\mathrm{ex}(n, K^{(3)}_{s,s,s}) \le Cn^{3-1/s^2}.$$

Recall that the KST theorem (Theorem 1.4.2) was proved by counting the number of copies of $K_{s,1}$ in the graph in two different ways. For 3-graphs, we instead count the number of copies of $K^{(3)}_{s,1,1}$ in two different ways, one of which uses the KST theorem for $K_{s,s}$-free graphs.

Proof. Let G be a $K^{(3)}_{s,s,s}$-free 3-graph with n vertices and m edges. Let X denote the number of copies of $K^{(3)}_{s,1,1}$ in G. (When $s = 1$, we count each copy three times.)

Upper bound on X. Given a set S of s vertices, consider the set T of all unordered pairs of distinct vertices that would form a $K^{(3)}_{s,1,1}$ with S (i.e., every triple formed by combining a pair in T and a vertex of S is an edge of G). Note that T is the edge-set of a graph on the same n vertices. If T contains a $K_{s,s}$, then together with S we would have a $K^{(3)}_{s,s,s}$. Thus, T is $K_{s,s}$-free, and hence by Theorem 1.4.2, $|T| = O_s(n^{2-1/s})$. Therefore,

$$X \lesssim_s \binom{n}{s} n^{2-1/s} \lesssim_s n^{s+2-1/s}.$$

Lower bound on X. We write $\deg(u, v)$ for the number of edges in G containing both u and v. Then, summing over all unordered pairs of distinct vertices u, v in G, we have

$$X = \sum_{u,v} \binom{\deg(u,v)}{s}.$$

As in the proof of Theorem 1.4.2, let

$$f_s(x) = \begin{cases} x(x-1)\cdots(x-s+1)/s! & \text{if } x \ge s-1, \\ 0 & x < s-1. \end{cases}$$

Then, f_s is convex and $f_s(x) = \binom{x}{s}$ for all nonnegative integers x. Since the average of $\deg(u, v)$ is $3m/\binom{n}{2}$,

$$X = \sum_{u,v} f_s(\deg(u,v)) \geq \binom{n}{2} f_s\left(\frac{3m}{\binom{n}{2}}\right).$$

Combining the upper and lower bounds, we have

$$\binom{n}{2}\left(\frac{3m}{\binom{n}{2}}\right)^s \lesssim_s n^{s+2-1/s}$$

and hence

$$m = O_s(n^{3-1/s^2}). \qquad \square$$

Exercise 1.5.8. Prove that $\mathrm{ex}(n, K^{(3)}_{r,s,t}) = O_{r,s,t}(n^{3-1/(rs)})$.

We can iterate further, using the same technique to prove an analogous result for every uniformity, thereby giving us the statement (Theorem 1.5.6) used in our proof of the Erdős–Stone–Simonovits theorem earlier. Feel free to skip reading the next proof if you feel comfortable with generalizing the preceding proof to r-graphs.

Theorem 1.5.9 (Hypergraph KST)
For every $r \geq 2$ and $s \geq 1$, there is some C such that

$$\mathrm{ex}(n, K^{(r)}_{s,\dots,s}) \leq C n^{r-s^{-r+1}},$$

where $K^{(r)}_{s,\dots,s}$ is the complete r-partite r-graph with s vertices in each of the r parts.

Proof. We prove by induction on r. The cases $r = 2$ and $r = 3$ were covered previously in Theorem 1.4.2 and Theorem 1.5.7. Assume that $r \geq 3$ and that the theorem has already been established for smaller values of r. (Actually, we could have started at $r = 1$ if we adjust the definitions appropriately.)

Let G be a $K^{(r)}_{s,\dots,s}$-free r-graph with n vertices and m edges. Let X denote the number of copies of $K^{(r)}_{s,1,\dots,1}$ in G. (When $s = 1$, we count each copy r times.)

Upper bound on X. Given a set S of s vertices, consider the set T of all unordered $(r-1)$-tuples of vertices that would form a $K^{(r)}_{s,1,\dots,1}$ with S (where S is in one part and the $r-1$ new vertices are each in its own part). Note that T is the edge-set of an $(r-1)$ graph on the same n vertices. If T contains a $K^{(r-1)}_{s,\dots,s}$, then together with S we would have a $K^{(r)}_{s,\dots,s}$. Thus, T is $K^{(r-1)}_{s,\dots,s}$-free, and by the induction hypothesis, $|T| = O_{r,s}(n^{r-1-s^{-r+2}})$. Hence

$$X \lesssim_{r,s} \binom{n}{s} n^{r-1-s^{-r+2}} \lesssim_{r,s} n^{r+s-1-s^{-r+2}}.$$

Lower bound on X. Given a set U of vertices, we write $\deg U$ for the number of edges containing all vertices in U. Then

$$X = \sum_{U \in \binom{V(G)}{r-1}} \binom{\deg U}{s}.$$

Let $f_s(x)$ be defined as in the previous proof. Since the average of $\deg U$ over all $(r-1)$-element subsets U is $rm/\binom{n}{r-1}$, we have

$$X = \sum_{U \in \binom{V(G)}{r-1}} f_s(\deg U) \ge \binom{n}{r-1} f_s\left(\frac{rm}{\binom{n}{r-1}}\right).$$

Combining the upper and lower bounds, we have

$$\binom{n}{r-1} f_s\left(\frac{rm}{\binom{n}{r-1}}\right) \lesssim_{r,s} n^{s+r-1-s^{-r+2}}$$

and hence

$$m = O_{r,s}(n^{r-s^{-r+1}}). \qquad \square$$

Exercise 1.5.10 (Forbidding a multipartite complete hypergraph with unbalanced parts). Prove that, for every sequence of positive integers $s_1, \ldots, s_r$, there exists C so that

$$\mathrm{ex}(n, K^{(r)}_{s_1,\ldots,s_r}) \le C n^{r-1/(s_1 \cdots s_{r-1})}.$$

Exercise 1.5.11 (Erdős–Stone for hypergraphs). Let H be an r-graph. Show that $\pi(H[s]) = \pi(H)$, where $H[s]$, the s-blow-up of H, is obtained by replacing every vertex of H by s duplicates of itself.

1.6 Forbidding a Cycle

In this section, we consider the problem of determining $\mathrm{ex}(n, C_\ell)$, the maximum number of edges in an n-vertex graph without an ℓ-cycle.

Odd Cycles

First let us consider forbidding odd cycles. Let k be a positive integer. Then C_{2k+1} has chromatic number 3, and so the Erdős–Stone–Simonovits theorem (Theorem 1.5.1) tells us that

$$\mathrm{ex}(n, C_{2k+1}) = (1 + o(1))\frac{n^2}{4}.$$

In fact, an even stronger statement is true. If n is large enough (as a function of k), then the complete bipartite graph $K_{\lfloor n/2 \rfloor, \lceil n/2 \rceil}$ is always the extremal graph, just like in the triangle case.

Theorem 1.6.1 (Exact Turán number of an odd cycle)
Let k be a positive integer. Then, for all sufficiently large integer n (i.e., $n \ge n_0(k)$ for some $n_0(k)$), one has

$$\mathrm{ex}(n, C_{2k+1}) = \left\lfloor \frac{n^2}{4} \right\rfloor.$$

We will not prove this theorem. See Füredi and Gunderson (2015) for a more recent proof.

More generally, Simonovits (1974) developed a stability method for exactly determining the Turán number of nonbipartite color-critical graphs.

Theorem 1.6.2 (Exact Turán number of a color-critical graph)
Let F be a graph with chromatic number $r + 1 \geq 3$ and such that one can remove some edge from F to reduce its chromatic number to r. Then for all sufficiently large n (i.e., $n \geq n_0(F)$ for some $n_0(F)$), the Turán graph $T_{n,r}$ uniquely maximizes the number of edges among all n-vertex F-free graphs.

Forbidding Even Cycles

Let us now turn to forbidding even cycles. Since C_{2k} is bipartite, we know from the KST theorem that $\mathrm{ex}(n, C_{2k}) = o(n^2)$. The following upper bound was determined by Bondy and Simonovits (1974).

Theorem 1.6.3 (Even cycles)
For every integer $k \geq 2$, there exists a constant C so that

$$\mathrm{ex}(n, C_{2k}) \leq Cn^{1+1/k}.$$

***Remark* 1.6.4** (Tightness). We will see in Section 1.10 a matching lower bound construction (up to constant factors) for $k = 2, 3, 5$. For all other values of k, it is an open question whether a matching lower bound construction exists.

Instead of proving the preceding theorem, we will prove a weaker result, stated in what follows. This weaker result has a short and neat proof, which hopefully gives some intuition as to why the above theorem should be true.

Theorem 1.6.5 (Short even cycles)
For any integer $k \geq 2$, there exists a constant C so that every graph G with n vertices and at least $Cn^{1+1/k}$ edges contains an even cycle of length at most $2k$.

In other words, Theorem 1.6.5 says that

$$\mathrm{ex}(n, \{C_4, C_6, \ldots, C_{2k}\}) = O_k(n^{1+1/k}).$$

Here, given a set $\mathcal{F}$ of graphs, $\mathrm{ex}(n, \mathcal{F})$ denotes the maximum number of edges in an n-vertex graph that does not contain any graph in $\mathcal{F}$ as a subgraph.

To prove this theorem, we first clean up the graph by removing some edges and vertices to get a bipartite subgraph with large minimum degree.

Lemma 1.6.6 (Large bipartite subgraph)
Every G has a bipartite subgraph with at least $e(G)/2$ edges.

Proof. Color every vertex with red or blue independently and uniformly at random. Then the expected number of nonmonochromatic edges is $e(G)/2$. Hence there exists a coloring that has at least $e(G)/2$ nonmonochromatic edges, and these edges form the desired bipartite subgraph. □

Lemma 1.6.7 (Large average degree implies subgraph with large minimum degree) Let $t > 0$. Every graph with average degree $2t$ has a nonempty subgraph with minimum degree greater than t.

Proof. Let G be a graph with average degree $2t$. Removing a vertex of degree at most t cannot decrease the average degree, since the total degree goes down by at most $2t$ and so the postdeletion graph has average degree of at least $(2e(G) - 2t)/(v(G) - 1)$, which is at least $2e(G)/v(G)$ since $2e(G)/v(G) \geq 2t$. Let us repeatedly delete vertices of degree at most t in the remaining graph until every vertex has degree more than t. This algorithm must terminate with a nonempty graph since we cannot ever drop below $2t$ vertices in this process (as such a graph would have average degree less than $2t$). □

Proof of Theorem 1.6.5. The idea is to use a **breath-first search**. Suppose G contains no even cycles of length at most $2k$. Applying Lemma 1.6.6 followed by Lemma 1.6.7, we find a bipartite subgraph G' of G with minimum degree $> t := e(G)/(2v(G))$. Let u be an arbitrary vertex of G'. For each $i = 0, 1, \ldots, k$, let A_i denote the set of vertices at distance exactly i from u.

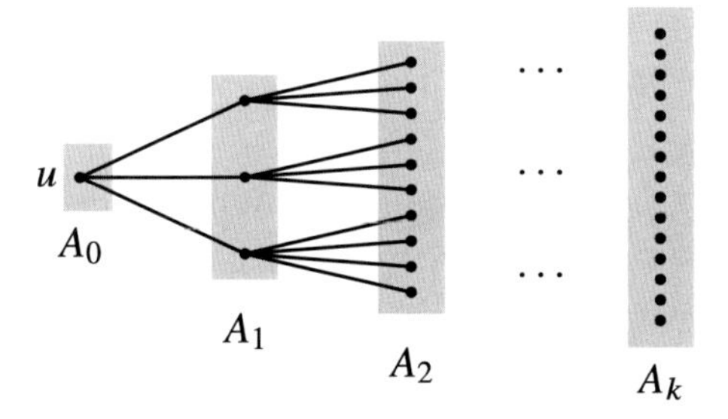

For each $i = 1, \ldots, k - 1$, every vertex of A_i has

- no neighbors inside A_i (or else G' would not be bipartite),
- exactly one neighbor in A_{i-1} (else we can backtrace through two neighbors, which must converge at some point to form an even cycle of length at most $2k$),
- and thus $> t - 1$ neighbors in A_{i+1} (by the minimum degree assumption on G').

Therefore, each layer A_i expands to the next by a factor of at least $t - 1$. Hence

$$v(G) \geq |A_k| \geq (t-1)^k \geq \left(\frac{e(G)}{2v(G)} - 1\right)^k$$

and thus

$$e(G) \leq 2v(G)^{1+1/k} + 2v(G).$$ □

Exercise 1.6.8 (Extremal number of trees). Let T be a tree with k edges. Show that $\mathrm{ex}(n, T) \leq kn$.

1.7 Forbidding a Sparse Bipartite Graph: Dependent Random Choice

Every bipartite graph H is contained in some $K_{s,t}$, and thus by the KST theorem (Theorem 1.4.2), $\mathrm{ex}(n, H) \leq \mathrm{ex}(n, K_{s,t}) = O_{s,t}(n^{2-1/s})$. The main result of this section gives a significant improvement when the maximum degree of H is small. The proof introduces an important probabilistic technique known as **dependent random choice**.

Theorem 1.7.1 (Bounded degree bipartite graph: Turán number upper bound)
Let H be a bipartite graph with vertex bipartition $A \cup B$ such that every vertex in A has degree at most r. Then there exists a constant $C = C_H$ such that for all n,

$$\mathrm{ex}(n, H) \le C n^{2-1/r}.$$

***Remark* 1.7.2** (History). The result was first proved by Füredi (1991). The proof given here is due to Alon, Krivelevich, and Sudakov (2003a). For more applications of the dependent random choice technique see the survey by Fox and Sudakov (2011).

***Remark* 1.7.3** (Tightness). The exponent $2-1/r$ is best possible as a function of r. Indeed, we will see in the following section that for every r there exists some s so that $\mathrm{ex}(n, K_{r,s}) \ge cn^{2-1/r}$ for some $c = c(r, s) > 0$.

On the other hand, for specific graphs G, Theorem 1.7.1 may not be tight. For example, $\mathrm{ex}(n, C_6) = \Theta(n^{4/3})$, whereas Theorem 1.7.1 only tells us that $\mathrm{ex}(n, C_6) = O(n^{3/2})$.

Given a graph G with many edges, we wish to find a large subset U of vertices such that every r-vertex subset of U has many common neighbors in G. (Even the case $r = 2$ is interesting.) Once such a U is found, we can then embed the B-vertices of H into U. It will then be easy to embed the vertices of A. The tricky part is to find such a U.

***Remark* 1.7.4** (Intuition). We want to host a party so that each pair of partygoers has many common friends. (Here G is the friendship graph.) Whom should we invite? Inviting people uniformly at random is not a good idea. (Why?) Perhaps we can pick some random individual (Alice) to host a party inviting all her friends. Alice's friends are expected to share some common friends – at least they all know Alice.

We can take a step further, and pick a few people at random (Alice, Bob, Carol, David) and have them host a party and invite all their common friends. This will likely be an even more sociable crowd. At least all the partygoers will know all the hosts, and likely even more. As long as the social network is not too sparse, there should be lots of invitees.

Some invitees (e.g., Zack) might feel a bit out of place at the party – maybe they don't have many common friends with other partygoers. (They all know the hosts but maybe Zack doesn't know many others.) To prevent such awkwardness, the hosts will cancel Zack's invitation. There shouldn't be too many people like Zack. The party must go on.

Here is the technical statement that we will prove. While there are many parameters, the specific details are less important compared to the proof technique. This is quite a tricky proof.

Theorem 1.7.5 (Dependent random choice)
Let n, r, m, t be positive integers and $\alpha > 0$. Then every graph G with n vertices and at least $\alpha n^2/2$ edges contains a vertex subset U with

$$|U| \ge n\alpha^t - \binom{n}{r}\left(\frac{m}{n}\right)^t$$

such that every r-element subset S of U has more than m common neighbors in G.

***Remark* 1.7.6** (Parameters). In the theorem statement, t is an auxiliary parameter that does not appear in the conclusion. While one can optimize for t, it is instructive and convenient to leave it as is. The theorem is generally applied to graphs with at least n^{2-c} edges, for some small $c > 0$, and we can play with the parameters to get $|U|$ and m both large as desired.

Proof. We say that an r-element subset of $V(G)$ is "bad" if it has at most m common neighbors in G.

Let $u_1, \ldots, u_t$ be vertices chosen uniformly and independently at random from $V(G)$. (These vertices are chosen "with replacement," i.e., they can repeat.) Let A be their common neighborhood. (Keep in mind that $u_1, \ldots, u_t, A$ are random. It may be a bit confusing in this proof to see what is random and what is not.)

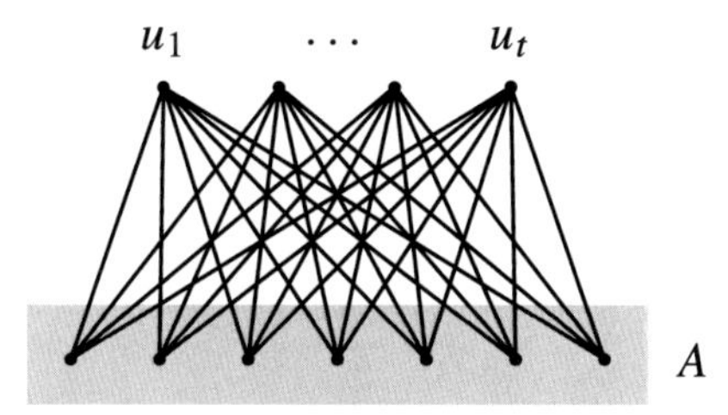

Each fixed vertex $v \in V(G)$ has probability $(\deg(v)/n)^t$ of being adjacent to all of $u_1, \ldots, u_t$, and so, by linearity of expectations and convexity,

$$\mathbb{E}\,|A| = \sum_{v \in V(G)} \mathbb{P}(v \in A) = \sum_{v \in V(G)} \left(\frac{\deg(v)}{n}\right)^t \geq n\left(\frac{1}{n}\sum_{v \in V}\frac{\deg(v)}{n}\right)^t \geq n\alpha^t.$$

For any fixed $R \subseteq V(G)$,

$$\mathbb{P}(R \subseteq A) = \mathbb{P}(R \text{ is complete to } u_1, \ldots, u_t) = \left(\frac{\#\text{ common neighbors of } R}{n}\right)^t.$$

If R is a bad r-vertex subset, then it has at most m common neighbors, and so

$$\mathbb{P}(R \subseteq A) \leq \left(\frac{m}{n}\right)^t.$$

Therefore, summing over all $\binom{n}{r}$ possible r-vertex subsets $R \subseteq V(G)$, by linearity of expectation,

$$\mathbb{E}[\text{the number of bad } r\text{-vertex subsets of } A] \leq \binom{n}{r}\left(\frac{m}{n}\right)^t.$$

Let U be obtained from A by deleting an element from each bad r-vertex subset. So, U has no bad r-vertex subsets. Also,

$$\begin{aligned}\mathbb{E}\,|U| &\geq \mathbb{E}\,|A| - \mathbb{E}[\text{the number of bad } r\text{-vertex subsets of } A] \\ &\geq n\alpha^t - \binom{n}{r}\left(\frac{m}{n}\right)^t.\end{aligned}$$

Thus, there exists some U with at least this size, with the property that all its r-vertex subsets have more than m common neighbors. □

Now we are ready to show Theorem 1.7.1, which we recall says that, for a bipartite graph H with vertex bipartition $A \cup B$ such that every vertex in A has degree at most r, one has $\operatorname{ex}(n, H) = O_H(n^{2-1/r})$.

Proof of Theorem 1.7.1. Let G be a graph with n vertices and at least $Cn^{2-\frac{1}{r}}$ edges. By choosing C large enough (depending only on $|A| + |B|$), we have

$$n\left(2Cn^{-\frac{1}{r}}\right)^r - \binom{n}{r}\left(\frac{|A| + |B|}{n}\right)^r \geq |B|.$$

We want to show that G contains H as a subgraph. By dependent random choice (Theorem 1.7.5 applied with $t = r$), we can embed the B-vertices of H into G so that every r-vertex subset of B (now viewed as a subset of $V(G)$) has $> |A| + |B|$ common neighbors.

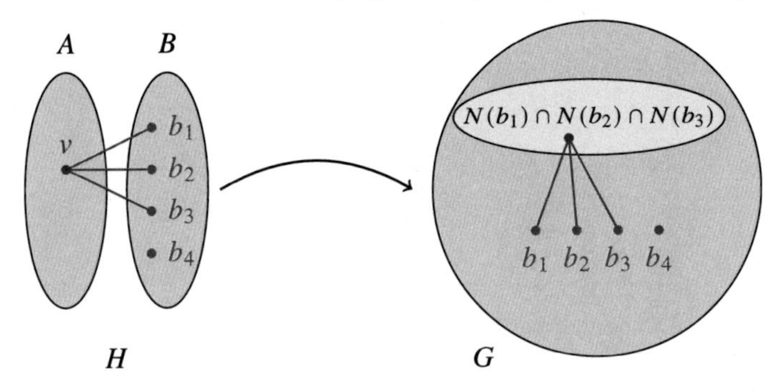

Next, we embed the vertices of A one at a time. Suppose we need to embed $v \in A$ (some previous vertices of A may have already been embedded at this point). Note that v has $\leq r$ neighbors in B, and these $\leq r$ vertices in B have $> |A| + |B|$ common neighbors in G. While some of these common neighbors may have already been used up in earlier steps to embed vertices of H, there are enough of them that they cannot all be used up, and thus we can embed v to some remaining common neighbor. This process ends with an embedding of H into G. □

Exercise 1.7.7. Let H be a bipartite graph with vertex bipartition $A \cup B$, such that r vertices in A are complete to B, and all remaining vertices in A have degree at most r. Prove that there is some constant $C = C_H$ such that $\operatorname{ex}(n, H) \leq Cn^{2-1/r}$ for all n.

Exercise 1.7.8. Let $\varepsilon > 0$. Show that, for sufficiently large n, every K_4-free graph with n vertices and at least εn^2 edges contains an independent set of size at least $n^{1-\varepsilon}$.

Exercise 1.7.9 (Extremal numbers of degenerate graphs).

(a*) Prove that there is some absolute constant $c > 0$ so that for every positive integer r, every n-vertex graph with at least $n^{2-c/r}$ edges contains disjoint nonempty vertex subsets A and B such that every subset of at most r vertices in A has at least n^c common neighbors in B and every subset of at most r vertices in B has at least n^c neighbors in A.

Hint: Apply the technique from the dependent random choice proof repeatedly back and forth to the two vertex parts.

(b) We say that a graph H is ***r*-degenerate** if its vertices can be ordered so that every vertex has at most r neighbors that appear before it in the ordering. Show that for every r-degenerate bipartite graph H there is some constant $C > 0$ so that $\mathrm{ex}(n, H) \leq Cn^{2-c/r}$, where c is the same absolute constant from part (a). (Here c should not depend on H or r.)

1.8 Lower Bound Constructions: Overview

We proved various upper bounds on $\mathrm{ex}(n, H)$ in earlier sections. When H is nonbipartite, the Turán graph construction (Construction 1.2.2) shows that the upper bound in the Erdős–Stone–Simonovits theorem (Theorem 1.5.1) is tight up to lower-order terms. However, when H is bipartite, so that $\mathrm{ex}(n, H) = o(n^2)$, we have not seen any nontrivial lower bound constructions. In the remainder of this chapter, we will see some methods for constructing H-free graphs for bipartite H. In some cases, these constructions will have enough edges to match the upper bounds on $\mathrm{ex}(n, H)$ from earlier sections. However, for most bipartite graphs H, there is a gap in known upper and lower bounds on $\mathrm{ex}(n, H)$. It is a central problem in extremal graph theory to close this gap.

We will see three methods for constructing H-free graphs.

Randomized Constructions

The idea is to take a random graph at a density that gives a small number of copies of H, and then destroy these copies of H by removing some edges from the random graph. The resulting graph is then H-free. This method is easy to implement and applies quite generally to all H. For example, it will be shown that

$$\mathrm{ex}(n, H) = \Omega_H\left(n^{2-\frac{v(H)-2}{e(H)-1}}\right).$$

However, bounds arising from this method are usually not tight.

Algebraic Constructions

The idea is to use algebraic geometry over a finite field to construct a graph. Its vertices correspond to geometric objects such as points or lines. Its edges correspond to incidences or other algebraic relations. These constructions sometimes give tight bounds. They work for a small number of graphs H, and usually require a different ad hoc idea for each H. They work rarely, but when they do, they can appear quite mysterious, or even magical. Many important tight lower bounds on bipartite extremal numbers arise this way. In particular, it will be shown that

$$\mathrm{ex}(n, K_{s,t}) = \Omega_{s,t}\left(n^{2-1/s}\right) \qquad \text{whenever } t \geq (s-1)! + 1,$$

thereby matching the KST theorem (Theorem 1.4.2) for such s, t. Also, it will be shown that

$$\mathrm{ex}(n, C_{2k}) = \Omega_k\left(n^{1+1/k}\right) \qquad \text{whenever } k \in \{2, 3, 5\},$$

thereby matching Theorem 1.6.3 for these values of k.

Randomized Algebraic Constructions

In algebraic constructions, usually we specify the edges using some specific well-chosen polynomials. A powerful recent idea is to choose the edge-defining polynomials at random.

1.9 Randomized Constructions

We use the probabilistic method to construct an H-free graph. The Erdős–Rényi random graph $\mathbf{G}(n,p)$ is the random graph on n vertices where every pair of vertices forms an edge independently with probability p. We first take a $\mathbf{G}(n,p)$ with an appropriately chosen p. The number of copies of H in $\mathbf{G}(n,p)$ is expected to be small, and we can destroy all such copies of H from the random graph by removing some edges. The remaining graph will then be H-free.

The method of starting with a simple random object and then modifying it is sometimes called **alteration method** or the **deletion method**.

Theorem 1.9.1 (Randomized lower bound)
Let H be a graph with at least two edges. Then there exists a constant $c = c_H > 0$, so that for all $n \geq 2$, there exists an H-free graph on n vertices with at least $cn^{2-\frac{v(H)-2}{e(H)-1}}$ edges. In other words,

$$\mathrm{ex}(n,H) \geq cn^{2-\frac{v(H)-2}{e(H)-1}}.$$

Proof. Let G be an instance of the Erdős–Rényi random graph $\mathbf{G}(n,p)$, with

$$p = \frac{1}{4}n^{-\frac{v(H)-2}{e(H)-1}}$$

(chosen with hindsight). We have $\mathbb{E}\, e(G) = p\binom{n}{2}$. Let X denote the number of copies of H in G. Then, our choice of p ensures that

$$\mathbb{E}X \leq p^{e(H)}n^{v(H)} \leq \frac{p}{2}\binom{n}{2} = \frac{1}{2}\mathbb{E}\, e(G).$$

Thus,

$$\mathbb{E}[e(G) - X] \geq \frac{p}{2}\binom{n}{2} \gtrsim n^{2-\frac{v(H)-2}{e(H)-1}}.$$

Take a graph G such that $e(G) - X$ is at least its expectation. Remove one edge from each copy of H in G, and we get an H-free graph with at least $e(G) - X \gtrsim n^{2-\frac{v(H)-2}{e(H)-1}}$ edges. □

For some graphs H, we can bootstrap Theorem 1.9.1 to give an even better lower bound. For example, if

$H =$ [graph], ,

then $v(H) = 10$ and $e(H) = 20$, so applying Theorem 1.9.1 directly gives

$$\mathrm{ex}(n,H) \gtrsim n^{2-8/19}.$$

On the other hand, any $K_{4,4}$-free graph is automatically H-free. Applying Theorem 1.9.1 to $K_{4,4}$ (8-vertex 16-edge) actually gives a better lower bound ($2 - 6/15 > 2 - 8/19$):

$$\mathrm{ex}(n, H) \geq \mathrm{ex}(n, K_{4,4}) \gtrsim n^{2-6/15}.$$

In general, given H, we should apply Theorem 1.9.1 to the subgraph of H with the maximum $(e(H) - 1)/(v(H) - 2)$ ratio. This gives the following corollary, which sometimes gives a better lower bound than directly applying Theorem 1.9.1.

Definition 1.9.2 (2-density)
The ***2-density*** of a graph H is defined by

$$\boldsymbol{m_2(H)} := \max_{\substack{H' \subseteq H \\ e(H') \geq 2}} \frac{e(H') - 1}{v(H') - 2}.$$

Corollary 1.9.3 (Randomized lower bound)
For any graph H with at least two edges, there exists constant $c = c_H > 0$ such that

$$\mathrm{ex}(n, H) \geq cn^{2-1/m_2(H)}.$$

Proof. Let H' be the subgraph of H with $m_2(H) = \frac{e(H')-1}{v(H')-2}$. Then, $\mathrm{ex}(n, H) \geq \mathrm{ex}(n, H')$, and we can apply Theorem 1.9.1 to get $\mathrm{ex}(n, H) \geq cn^{2-1/m_2(H)}$. □

Example **1.9.4.** Theorem 1.9.1 combined with the upper bound from the KST theorem (Theorem 1.4.2) gives that for every fixed $2 \leq s \leq t$,

$$n^{2-\frac{s+t-2}{st-1}} \lesssim \mathrm{ex}(n, K_{s,t}) \lesssim n^{2-\frac{1}{s}}.$$

When t is large compared to s, the exponents in preceding two bounds are close to each other (but never equal). When $t = s$, the preceding bounds specialize to

$$n^{2-\frac{2}{s+1}} \lesssim \mathrm{ex}(n, K_{s,s}) \lesssim n^{2-\frac{1}{s}}.$$

In particular, for $s = 2$,

$$n^{4/3} \lesssim \mathrm{ex}(n, K_{2,2}) \lesssim n^{3/2}.$$

It turns out that the upper bound is tight. We will show this in the next section using an algebraic construction.

Exercise 1.9.5. Show that if H is a bipartite graph containing a cycle of length $2k$, then $\mathrm{ex}(n, H) \gtrsim_H n^{1+1/(2k-1)}$.

Exercise 1.9.6. Find a graph H with $\chi(H) = 3$ and $\mathrm{ex}(n, H) > \frac{1}{4}n^2 + n^{1.99}$ for all sufficiently large n.

1.10 Algebraic Constructions

In this section, we use algebraic methods to construct $K_{s,t}$-free graphs for certain values of (s, t), as well as C_{2k}-free graphs for certain values of k. In both cases, the constructions are optimal in that they match the upper bounds up to a constant factor.

$K_{2,2}$-*free*

We begin by constructing $K_{2,2}$-free graphs with the number of edges matching the KST theorem. The construction is due to Erdős, Rényi, and Sós (1966) and Brown (1966) independently.

Theorem 1.10.1 (Construction of $K_{2,2}$-free graphs)

$$\operatorname{ex}(n, K_{2,2}) \geq \left(\frac{1}{2} - o(1)\right) n^{3/2}.$$

Combining with the KST theorem, we obtain the corollary.

Corollary 1.10.2 (Turán number of $K_{2,2}$)

$$\operatorname{ex}(n, K_{2,2}) = \left(\frac{1}{2} - o(1)\right) n^{3/2}.$$

Before giving the proof of Theorem 1.10.1, let us first sketch the geometric intuition. Given a set of points $\mathcal{P}$ and a set of lines $\mathcal{L}$, the ***point-line incidence graph*** is the bipartite graph with two vertex parts $\mathcal{P}$ and $\mathcal{L}$, where $p \in \mathcal{P}$ and $\ell \in \mathcal{L}$ are adjacent if $p \in \ell$.

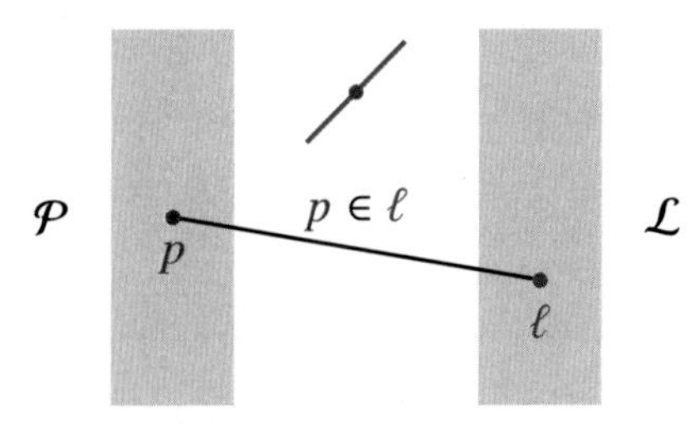

A point-line incidence graph is C_4-free. Indeed, a C_4 would correspond to two lines both passing through two distinct points, which is impossible.

We want to construct a set of points and a set of lines so that there are many incidences. To do this, we take all points and all lines in a finite field plane $\mathbb{F}_p^2$. There are p^2 points and $p^2 + p$ lines. Since every line contains p points, the graph has around p^3 edges, and so $\operatorname{ex}(2p^2 + p, K_{2,2}) \geq p^3$. By rounding down an integer n to the closest number of the form $2p^2 + p$ for a prime p, we already see that $\operatorname{ex}(n, K_{2,2}) \gtrsim n^{3/2}$ for all n. Here we use a theorem from number theory regarding large gaps in primes, which we quote in what follows without proof. (This strategy does not work if we instead take points and lines in $\mathbb{R}^2$; see the Szemerédi–Trotter theorem in Section 8.2.)

Theorem 1.10.3 (Large gaps between primes)
The largest prime below N has size $N - o(N)$.

***Remark* 1.10.4** (Large gaps between primes). The preceding result already follows from the prime number theorem, which says that the number of primes up to N is $(1 + o(1))N/\log N$. The best quantitative result, due to Baker, Harman, and Pintz (2001), says that there exists a prime in $[N - N^{0.525}, N]$ for all sufficiently large N. Cramer's conjecture, which is wide open and based on a random model of the primes, speculates that the $o(N)$ in Theorem 1.10.3 may be replaced by $O((\log N)^2)$. An easier claim is Bertrand's postulate, which says that there is a prime between N and $2N$ for every N, and this already suffices for proving $\mathrm{ex}(n, K_{2,2}) \gtrsim n^{3/2}$.

To get a better constant in the preceding construction, we optimize somewhat by using the same vertices to represent both points and lines. This pairing of points and lines is known as ***polarity*** in projective geometry, and this construction is known as the ***polarity graph***. (This usually refers to the projective plane version of the construction.)

Proof of Theorem 1.10.1. Let p denote the largest prime such that $p^2 - 1 \le n$. Then $p = (1 - o(1))\sqrt{n}$ by Theorem 1.10.3. Let G be a graph with vertex set $V(G) = \mathbb{F}_p^2 \setminus \{(0,0)\}$ and an edge between (x, y) and (a, b) if and only if $ax + by = 1$ in $\mathbb{F}_p$.

For any two distinct vertices (a, b) and (a', b') in $V(G)$, they have at most one common neighbor since there is at most one solution to the system $ax + by = 1$ and $a'x + b'y = 1$. Therefore, G is $K_{2,2}$-free. (This is where we use the fact that two lines intersect in at most one point.)

For every $(a, b) \in V(G)$, there are exactly p vertices (x, y) satisfying $ax+by = 1$. However, one of those vertices could be (a, b) itself. So every vertex in G has degree p or $p - 1$. Hence G has at least $(p^2 - 1)(p - 1)/2 = (1/2 - o(1))n^{3/2}$ edges. □

$K_{3,3}$-*free*

Next, we construct $K_{3,3}$-free graphs with the number of edges matching the KST theorem. This construction is due to Brown (1966).

Theorem 1.10.5 (Construction of $K_{3,3}$-free graphs)
For every n,

$$\mathrm{ex}(n, K_{3,3}) \ge \left(\frac{1}{2} - o(1)\right) n^{5/3}.$$

Consider the incidences between points in $\mathbb{R}^3$ and unit spheres. This graph is $K_{3,3}$-free since no three unit spheres can share three distinct common points. Again, one needs to do this over a finite field to attain the desired bounds, but it is easier to visualize the setup in Euclidean space, where it is clearly true.

Proof sketch. Let p be the largest prime less than $n^{1/3}$. Fix a nonzero element $d \in \mathbb{F}_p$, which we take to be a quadratic residue if $p \equiv 3 \pmod 4$ and a quadratic nonresidue if $p \not\equiv 3 \pmod 4$. Construct a graph G with vertex set $V(G) = \mathbb{F}_p^3$, and an edge between (x, y, z) and $(a, b, c) \in V(G)$ if and only if

$$(a - x)^2 + (b - y)^2 + (c - z)^2 = d.$$

It turns out that each vertex has $(1 - o(1))p^2$ neighbors. (The intuition here is that, for a fixed (a, b, c), if we choose $x, y, z \in \mathbb{F}_p$ independently and uniformly at random, then the resulting sum $(a - x)^2 + (b - y)^2 + (c - z)^2$ is roughly uniformly distributed, and hence equals d with probability close to $1/p$.) It remains to show that the graph is $K_{3,3}$-free. To see this, think about how one might prove this claim in $\mathbb{R}^3$ via algebraic manipulations. We compute the radical planes between pairs of spheres as well as the intersections of these radical planes (i.e., the radical axis). The claim boils down to the fact that no sphere has three collinear points, which is true due to the quadratic (non)residue hypothesis on d. The details are omitted.

Thus G is a $K_{3,3}$-free graph on $p^3 \leq n$ vertices and with at least $(1/2 - o(1))p^5 = (1/2 - o(1))n^{5/3}$ edges. □

It is unknown if the preceding ideas can be extended to construct $K_{4,4}$-free graphs with $\Omega(n^{7/4})$ edges. It is a major open problem to determine the asymptotics of $\text{ex}(n, K_{4,4})$.

Conjecture 1.10.6 (KST theorem is tight)
For every fixed $s \geq 4$, one has

$$\text{ex}(n, K_{s,s}) = \Theta_s(n^{2-1/s}).$$

$K_{s,t}$*-free*

Now we present a substantial generalization of the above constructions, due to Kollár, Rónyai, and Szabó (1996) and Alon, Rónyai, and Szabó (1999). It gives a matching lower bound (up to a constant factor) to the KST theorem for $K_{s,t}$ whenever t is sufficiently large compared to s.

Theorem 1.10.7 (Tightness of KST bound when $t > (s - 1)!$)
Fix a positive integer $s \geq 2$. Then

$$\text{ex}(n, K_{s,(s-1)!+1}) \geq \left(\frac{1}{2} - o(1)\right) n^{2-1/s}.$$

Corollary 1.10.8 (Tightness of KST bound when $t > (s - 1)!$)
If $t > (s - 1)!$, then

$$\text{ex}(n, K_{s,t}) = \Theta_{s,t}(n^{2-1/s}).$$

We first prove a slightly weaker version of Theorem 1.10.7, namely that

$$\text{ex}(n, K_{s,s!+1}) \geq \left(\frac{1}{2} - o(1)\right) n^{2-1/s}$$

(Kollár, Rónyai, and Szabó 1996). Afterwards, we will modify the construction to prove Theorem 1.10.7.

Let p be a prime. Recall that the ***norm map*** $N \colon \mathbb{F}_{p^s} \to \mathbb{F}_p$ is defined by

$$N(x) := x \cdot x^p \cdot x^{p^2} \cdots x^{p^{s-1}} = x^{\frac{p^s - 1}{p-1}}.$$

Note that $N(x) \in \mathbb{F}_p$ for all $x \in \mathbb{F}_{p^s}$ since $N(x)^p = N(x)$ and $\mathbb{F}_p$ is the set of elements in $\mathbb{F}_{p^s}$ invariant under the automorphism $x \mapsto x^p$. Furthermore, since $\mathbb{F}_{p^s}^\times$ is a cyclic group of order $p^s - 1$, we know that

$$\left|\{x \in \mathbb{F}_{p^s} : N(x) = 1\}\right| = \frac{p^s - 1}{p - 1}. \tag{1.2}$$

Construction 1.10.9 (Norm graph)
NormGraph$_{p,s}$ is defined to be the graph with vertex set $\mathbb{F}_{p^s}$ and an edge between distinct $a, b \in \mathbb{F}_{p^s}$ if $N(a + b) = 1$.

By (1.2), every vertex in NormGraph$_{p,s}$ has degree at least

$$\frac{p^s - 1}{p - 1} - 1 \geq p^{s-1}.$$

(We had to subtract 1 in case $N(x + x) = 1$.) And thus the number of edges is at least $p^{2s-1}/2$. It remains to establish that NormGraph$_{p,s}$ is $K_{s,s!+1}$-free. Once this is done, we can take p to be the largest prime at most $n^{1/s}$, and then

$$\mathrm{ex}(n, K_{s,s!+1}) \geq \mathrm{ex}(p^s, K_{s,s!+1}) \geq \frac{p^{2s-1}}{2} \geq \left(\frac{1}{2} - o(1)\right) n^{2-1/s}.$$

Proposition 1.10.10
NormGraph$_{p,s}$ is $K_{s,s!+1}$-free for all $s \geq 2$.

We wish to upper bound the number of common neighbors to a set of s vertices. This amount to showing that a certain system of algebraic equations cannot have too many solutions. We quote without proof the following key algebraic result from Kollár, Rónyai, and Szabó (1996), which can be proved using algebraic geometry.

Theorem 1.10.11
Let $\mathbb{F}$ be any field and $a_{ij}, b_i \in \mathbb{F}$ such that $a_{ij} \neq a_{i'j}$ for all $i \neq i'$. Then the system of equations

$$\begin{aligned}
(x_1 - a_{11})(x_2 - a_{12}) \cdots (x_s - a_{1s}) &= b_1 \\
(x_1 - a_{21})(x_2 - a_{22}) \cdots (x_s - a_{2s}) &= b_2 \\
&\vdots \\
(x_1 - a_{s1})(x_2 - a_{s2}) \cdots (x_s - a_{ss}) &= b_s
\end{aligned}$$

has at most $s!$ solutions $(x_1, \ldots, x_s) \in \mathbb{F}^s$.

***Remark* 1.10.12** (Special base $b = 0$). Consider the special case when all the b_i are 0. In this case, since the a_{ij} are distinct for each fixed j, every solution to the system corresponds to a permutation $\pi \colon [s] \to [s]$, setting $x_i = a_{i\pi(i)}$. So there are exactly $s!$ solutions in this special case. The difficult part of the theorem says that the number of solutions cannot increase if we move b away from the origin.

Proof of Proposition 1.10.10. Consider distinct $y_1, y_2, \dots, y_s \in \mathbb{F}_{p^s}$. We wish to bound the number of common neighbors x. Recall that in a field with characteristic p, we have the identity $(x+y)^p = x^p + y^p$ for all x, y. So,

$$\begin{aligned} 1 = N(x+y_i) &= (x+y_i)(x+y_i)^p \dots (x+y_i)^{p^{s-1}} \\ &= (x+y_i)(x^p+y_i^p)\dots(x^{p^{s-1}}+y_i^{p^{s-1}}) \end{aligned}$$

for all $1 \le i \le s$. By Theorem 1.10.11, these s equations (as i ranges over $[s]$) have at most $s!$ solutions in x. Note the hypothesis of Theorem 1.10.11 is satisfied since $y_i^p = y_j^p$ if and only if $y_i = y_j$ in $\mathbb{F}_{p^s}$. □

Now we modify the norm graph construction to forbid $K_{s,(s-1)!+1}$, thereby yielding Theorem 1.10.7.

Construction 1.10.13 (Projective norm graph)
Let $\mathbf{ProjNormGraph}_{p,s}$ be the graph with vertex set $\mathbb{F}_{p^{s-1}} \times \mathbb{F}_p^\times$, where two vertices $(X,x),(Y,y) \in \mathbb{F}_{p^{s-1}} \times \mathbb{F}_p^\times$ are adjacent if and only if

$$N(X+Y) = xy.$$

In $\mathrm{ProjNormGraph}_{p,s}$, every vertex (X,x) has degree $p^{s-1}-1$ since its neighbors are $(Y, N(X+Y)/x)$ for all $Y \neq -X$. There are $(p^{s-1}-1)p^{s-1}(p-1)/2$ edges. As earlier, it remains to show that this graph is $K_{s,(s-1)!+1}$-free. Once we know this, by taking p to be the largest prime satisfying $p^{s-1}(p-1) \le n$, we obtain the desired lower bound

$$\mathrm{ex}(n, K_{s,(s-1)!+1}) \ge \frac{1}{2}(p^{s-1}-1)p^{s-1}(p-1) \ge \left(\frac{1}{2} - o(1)\right) n^{2-1/s}.$$

Proposition 1.10.14
$\mathrm{ProjNormGraph}_{p,s}$ is $K_{s,(s-1)!+1}$-free.

Proof. Fix distinct $(Y_1,y_1),\dots,(Y_s,y_s) \in \mathbb{F}_{p^{s-1}} \times \mathbb{F}_p^\times$. We wish to show that there are at most $(s-1)!$ solutions $(X,x) \in \mathbb{F}_{p^{s-1}} \times \mathbb{F}_p^\times$ to the system of equations

$$N(X+Y_i) = xy_i, \qquad i = 1,\dots,s.$$

Assume this system has at least one solution. Then if $Y_i = Y_j$ with $i \neq j$ we must have that $y_i = y_j$. Therefore all the Y_i are distinct. For each $i < s$, dividing $N(X+Y_i) = xy_i$ by $N(X+Y_s) = xy_s$ gives

$$N\left(\frac{X+Y_i}{X+Y_s}\right) = \frac{y_i}{y_s}, \qquad i = 1,\dots,s-1.$$

Dividing both sides by $N(Y_i - Y_s)$ gives

$$N\left(\frac{1}{X+Y_s} + \frac{1}{Y_i - Y_s}\right) = \frac{y_i}{N(Y_i-Y_s)y_s}, \qquad i = 1,\dots,s-1.$$

Now apply Theorem 1.10.11 (same as in the proof of Proposition 1.10.10). We deduce that there are at most $(s-1)!$ choices for X, and each such X automatically determines $x = N(X+Y_1)/y_1$. Thus, there are at most $(s-1)!$ solutions (X,x). □

C_4, C_6, C_{10}*-free*

Finally, let us turn to constructions of C_{2k}**-free graphs**. We had mentioned in Section 1.6 that $\mathrm{ex}(C_{2k}, n) = O_k(n^{1+1/k})$. We saw a matching lower bound construction for 4-cycles. Now we give matching constructions for 6-cycles and 10-cycles. (It remains an open problem for other cycle lengths.)

Theorem 1.10.15 (Tight lower bound for avoiding C_{2k} for $k \in \{2,3,5\}$)
Let $k \in \{2,3,5\}$. Then there is a constant $c > 0$ such that for every n,

$$\mathrm{ex}(n, C_{2k}) \geq cn^{1+1/k}.$$

***Remark* 1.10.16** (History). The existence of such C_{2k}-free graphs for $k \in \{3,5\}$ is due to Benson (1966) and Singleton (1966). The construction given here is due to Wenger (1991), with a simplified description due to Conlon (2021).

The following construction generalizes the point-line incidence graph construction earlier for the C_4-free graph in Theorem 1.10.1. Here we consider a special set of lines in $\mathbb{F}_q^k$, whereas previously for C_4 we took all lines in $\mathbb{F}_q^2$.

Construction 1.10.17 (C_{2k}-free construction for $k \in \{2,3,5\}$)
Let q be a prime power. Let $\mathcal{L}$ denote the set of all lines in $\mathbb{F}_q^k$ whose direction can be written as $(1, t, \ldots, t^{k-1})$ for some $t \in \mathbb{F}_q$. Let $G_{q,k}$ denote the bipartite point-line incidence graph with vertex sets $\mathbb{F}_q^k$ and $\mathcal{L}$. That is, $(p, \ell) \in \mathbb{F}_q^k \times \mathcal{L}$ is an edge if and only if $p \in \ell$.

We have $|\mathcal{L}| = q^k$, since to specify a line in $\mathcal{L}$ we can provide a point with first coordinate equal to zero, along with a choice of $t \in \mathbb{F}_q$ giving the direction of the line. So the graph $G_{q,k}$ has $n = 2q^k$ vertices. Since each line contains exactly q points, there are exactly $q^{k+1} \asymp n^{1+1/k}$ edges in the graph. It remains to show that this graph is C_{2k}-free whenever $k \in \{2,3,5\}$. Then Theorem 1.10.15 would follow after the usual trick of taking q to be the largest prime with $2q^k < n$.

Proposition 1.10.18
Let $k \in \{2,3,5\}$. The graph $G_{q,k}$ from Construction 1.10.17 is C_{2k}-free.

Proof. A $2k$-cycle in $G_{q,k}$ would correspond to p_1, ℓ_1, ..., p_k, ℓ_k with distinct p_1, ..., $p_k \in \mathbb{F}_q^k$ and distinct ℓ_1, ..., $\ell_k \in \mathcal{L}$, and p_i, $p_{i+1} \in \ell_i$ for all i (indices taken mod k). Let $(1, t_i, \ldots, t_i^{k-1})$ denote the direction of ℓ_i.

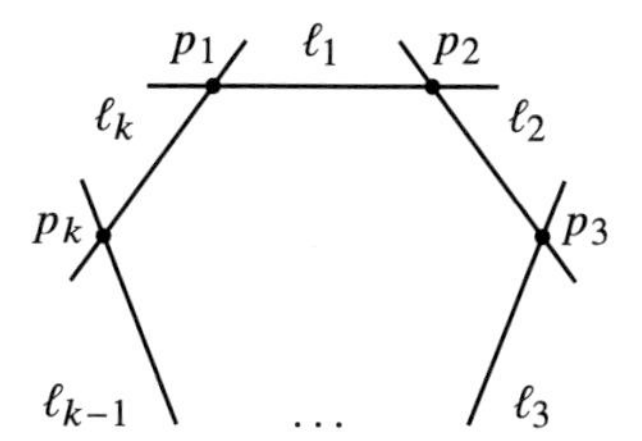

Then

$$p_{i+1} - p_i = a_i(1, t_i, \ldots, t_i^{k-1})$$

for some $a_i \in \mathbb{F}_q \setminus \{0\}$. Thus (recall that $p_{k+1} = p_1$)

$$\sum_{i=1}^{k} a_i(1, t_i, \ldots, t_i^{k-1}) = \sum_{i=1}^{k} (p_{i+1} - p_i) = 0. \tag{1.3}$$

The vectors $(1, t_i, \ldots, t_i^{k-1})$, $i = 1, \ldots, k$, after deleting duplicates, are linearly independent. One way to see this is via the Vandermonde determinant

$$\begin{vmatrix} 1 & x_1 & x_1^2 & \cdots & x_1^{k-1} \\ 1 & x_2 & x_2^2 & \cdots & x_2^{k-1} \\ \vdots & \vdots & \vdots & \ddots & \vdots \\ 1 & x_k & x_k^2 & \cdots & x_k^{k-1} \end{vmatrix} = \prod_{i<j} (x_j - x_i).$$

For (1.3) to hold, each vector $(1, t_i, \ldots, t_i^{k-1})$ must appear at least twice in the sum, with their coefficients a_i adding up to zero.

Since the lines $\ell_1, \ldots, \ell_k$ are distinct, for each $i = 1, \ldots, k$ (indices taken mod k), the lines ℓ_i and ℓ_{i+1} cannot be parallel. So, $t_i \neq t_{i+1}$. When $k \in \{2, 3, 5\}$ it is impossible to select $t_1, \ldots, t_k$ with no equal consecutive terms (including wraparound) and so that each value is repeated at least twice. Therefore the $2k$-cycle cannot exist. (Why does the argument fail for C_8-freeness?) □

1.11 Randomized Algebraic Constructions

In this section, we show how to add randomness to algebraic constructions, thereby combining the power of both approaches. This idea is due to Bukh (2015).

The algebraic constructions in the previous section can be abstractly described as follows. Take a graph whose vertices are points in some algebraic set (e.g., some finite field geometry), with two vertices x and y being adjacent if some algebraic relationship such as $f(x, y) = 0$ is satisfied. Previously, this f was carefully chosen by hand. The new idea is to take f to be a random polynomial.

We illustrate this technique by giving another proof of the tightness of the KST bound on extremal numbers for $K_{s,t}$ when t is large compared to s.

Theorem 1.11.1 (Tightness of KST bound for large t)
For every $s \geq 2$, there exists some t so that

$$\mathrm{ex}(n, K_{s,t}) \geq \left(\frac{1}{2} - o(1)\right) n^{2-1/s}.$$

The construction we present here has a worse dependence of t on s than in Theorem 1.10.7. The main purpose of this section is to illustrate the technique of randomized algebraic constructions. Bukh (2021) later gave a significant extension of this technique which shows that $\mathrm{ex}(n, K_{s,t}) = \Omega_s(n^{2-1/s})$ for some t close to 9^s, improving on Theorem 1.10.7, which required $t > (s-1)!$.

Proof idea. Take a random polynomial $f(X_1,\ldots,X_s,Y_1,\ldots,Y_s)$ symmetric in the X and Y variables (i.e., $f(X,Y) = f(Y,X)$), but otherwise uniformly chosen among all polynomials with degree up to d with coefficients in $\mathbb{F}_q$. Consider a graph with vertex set $\mathbb{F}_q^s$ and where X and Y are adjacent if $f(X,Y) = 0$.

Given an s-vertex set U, let Z_U denote the set of common neighbors of U. It is an algebraic set: the common zeros of the polynomials $f(X,y)$, $y \in U$. Due to the Lang–Weil bound from algebraic geometry, Z_U is either bounded in size, $|Z_U| \le C$ (the zero-dimensional case), or it must be quite large, say, $|Z_U| > q/2$ (the positive-dimensional case). This is unlike an Erdős–Rényi random graph.

One can then deduce, using Markov's inequality, that

$$\mathbb{P}(|Z_U| > C) = \mathbb{P}\left(|Z_U| > \frac{q}{2}\right) \le \frac{\mathbb{E}[|Z_U|^k]}{(q/2)^k} = \frac{O_k(1)}{(q/2)^k},$$

which is quite small (much smaller compared to an Erdős–Rényi random graph). So, typically very few sets U have size $> C$. By deleting these bad U's from the vertex set of the graph, we obtain a $K_{s,C+1}$-free graph with around q^s vertices and on the order of q^{2s-1} edges. ■

Now we begin the actual proof. Let q be the largest prime power satisfying $q^s \le n$. Due to prime gaps (Theorem 1.10.3), we have $q = (1 - o(1))n^{1/s}$. So it suffices to construct a $K_{s,t}$-free graph on q^s vertices with $(1/2 - o(1))q^{2s-1}$ edges.

Let $d = s^2 + s$. (The reason for this choice will come up later.) Let

$$f \in \mathbb{F}_q[X_1, X_2, \ldots, X_s, Y_1, Y_2, \ldots, Y_s]_{\le d}$$

be a polynomial chosen uniformly at random among all polynomials with degree at most d in each of $X = (X_1, X_2, \ldots, X_s)$ and $Y = (Y_1, Y_2, \ldots, Y_s)$ and furthermore satisfying $f(X,Y) = f(Y,X)$. In other words,

$$f = \sum_{\substack{i_1+\cdots+i_s \le d \\ j_1+\cdots+j_s \le d}} a_{i_1,\ldots,i_s,j_1,\ldots,j_s} X_1^{i_1} \cdots X_s^{i_s} Y_1^{j_1} \cdots Y_s^{j_s}$$

where the coefficients $a_{i_1,\ldots,i_s,j_1,\ldots,j_s} \in \mathbb{F}_q$ are chosen subject to $a_{i_1,\ldots,i_s,j_1,\ldots,j_s} = a_{j_1,\ldots,j_s,i_1,\ldots,i_s}$ but otherwise independently and uniformly at random.

Let G be the graph with vertex set $\mathbb{F}_q^s$, with distinct $x, y \in \mathbb{F}_q^s$ adjacent if and only if $f(x,y) = 0$.

Then G is a random graph. The next two lemmas show that G behaves in some ways like a random graph with edges independently appearing with probability $1/q$. Indeed, the next lemma shows that every pair of vertices form an edge with probability $1/q$.

Lemma 1.11.2 (Random polynomial)
Suppose f is randomly chosen as earlier. For all $u, v \in \mathbb{F}_q^s$,

$$\mathbb{P}[f(u,v) = 0] = \frac{1}{q}.$$

Proof. Note that resampling the constant term of f does not change its distribution. Thus, $f(u,v)$ is uniformly distributed in $\mathbb{F}_q$ for a fixed (u,v). Hence $f(u,v)$ takes each value with probability $1/q$. □

More generally, we show in what follows that the expected occurrence of small subgraphs mirrors that of the usual random graph with independent edges. We write $\binom{U}{2}$ for the set of unordered pairs of element from U.

Lemma 1.11.3 (Random polynomial)
Suppose f is randomly chosen as previously. Let $W \subseteq \mathbb{F}_q^s$ with $|W| \leq d+1$. Then the vector $(f(u,v))_{\{u,v\}\in\binom{W}{2}}$ is uniformly distributed in $\mathbb{F}_q^{\binom{W}{2}}$. In particular, for any $E \subseteq \binom{W}{2}$, one has

$$\mathbb{P}[f(u,v) = 0 \text{ for all } \{u,v\} \in E] = q^{-|E|}.$$

Proof. We first perform multivariate Lagrange interpolation to show that $(f(u,v))_{\{u,v\}}$ can take all possible values. For each pair $u, v \in W$ with $u \neq v$, we can find some polynomial $\ell_{u,v} \in \mathbb{F}[X_1, \ldots, X_s]$ of degree 1 such that $\ell_{u,v}(u) = 1$ and $\ell_{u,v}(v) = 0$. For each $u \in W$, let

$$q_u(X) = \prod_{v \in W\setminus\{u\}} \ell_{u,v}(X) \in \mathbb{F}[X_1, \ldots, X_s],$$

which has degree $|W| - 1 \leq d$. It satisfies $q_u(u) = 1$, and $q_u(v) = 0$ for all $v \in W \setminus \{u\}$.

Let

$$p(X,Y) = \sum_{\{u,v\}\in\binom{W}{2}} c_{u,v}(q_u(X)q_v(Y) + q_v(X)q_u(Y)),$$

with $c_{u,v} \in \mathbb{F}_q$. Note that $p(X,Y) = p(Y,X)$. Also, $p(u,v) = c_{u,v}$ for all distinct $u, v \in W$.

Now let each $c_{u,v} \in \mathbb{F}_q$ be chosen independently and uniformly at random so that $p(X,Y)$ is a random polynomial. Note that $f(X,Y)$ and $p(X,Y)$ are independent random polynomials both with degree at most d in each of X and Y. Since f is chosen uniformly at random, it has the same distribution as $f + p$. Since $(p(u,v))_{u,v} = (c_{u,v})_{u,v} \in \mathbb{F}_q^{\binom{|W|}{2}}$ is uniformly distributed, the same must be true for $(f(u,v))_{u,v}$ as well. □

Now fix $U \subseteq \mathbb{F}_q^s$ with $|U| = s$. We want to show that it is rare for U to have many common neighbors. We will use the method of moments. Let

$$\begin{aligned} Z_U &= \text{the set of common neighbors of } U \\ &= \{x \in \mathbb{F}_q^s \setminus U : f(x,u) = 0 \text{ for all } u \in U\}. \end{aligned}$$

Then using Lemma 1.11.3, for any $k \leq s^2 + 1$,

$$\begin{aligned} \mathbb{E}[|Z_U|^k] &= \mathbb{E}\Big[\Big(\sum_{v\in\mathbb{F}_q^s\setminus U} 1\{v \in Z_U\}\Big)^k\Big] \\ &= \sum_{v^{(1)},\ldots,v^{(k)}\in\mathbb{F}_q^s\setminus U} \mathbb{E}[1\{v^{(1)}, \ldots, v^{(k)} \in Z_U\}] \\ &= \sum_{v^{(1)},\ldots,v^{(k)}\in\mathbb{F}_q^s\setminus U} \mathbb{P}[f(u,v) = 0 \text{ for all } u \in U \text{ and } v \in \{v^{(1)}, \ldots, v^{(k)}\}] \\ &= \sum_{v^{(1)},\ldots,v^{(k)}\in\mathbb{F}_q^s\setminus U} q^{-|U|\#\{v^{(1)},\ldots,v^{(k)}\}}, \end{aligned}$$

with the final step due to Lemma 1.11.3 applied with $W = U \cup \{v^{(1)}, \ldots, v^{(k)}\}$, which has cardinality $\le |U| + k \le s + s^2 + 1 = d + 1$. Note that $\#\{v^{(1)}, \ldots, v^{(k)}\}$ counts distinct elements in the set. Thus, continuing the preceding calculation,

$$\begin{aligned} &= \sum_{r \le k} \binom{q^s - |U|}{r} q^{-rs} \#\{\text{surjections } [k] \to [r]\} \\ &\le \sum_{r \le k} \#\{\text{surjections } [k] \to [r]\} \\ &= O_k(1). \end{aligned}$$

Applying the preceding with $k = s^2 + 1$ and using Markov's inequality, we get

$$\mathbb{P}(|Z_U| \ge \lambda) = \mathbb{P}(|Z_U|^{s^2+1} \ge \lambda^{s^2+1}) \le \frac{\mathbb{E}\big[|Z_U|^{s^2+1}\big]}{\lambda^{s^2+1}} \le \frac{O_s(1)}{\lambda^{s^2+1}}. \tag{1.4}$$

***Remark* 1.11.4.** All the probabilistic arguments up to this point would be identical had we used a random graph with independent edges appearing with probability p. In both settings, $|Z_U|$ is a random variable with constant order expectation. However, their distributions are extremely different, as we will soon see. For a random graph with independent edges, $|Z_U|$ behaves like a Poisson random variable, and consequently, for any constant t, $\mathbb{P}(|Z_U| \ge t)$ is bounded from below by a constant. Consequently, many s-element sets of vertices are expected to have at least t common neighbors, and so this method will not work. However, this is not the case with the random algebraic construction. It is impossible for $|Z_U|$ to take on certain ranges of values. If $|Z_U|$ is somewhat large, then it must be very large.

Note that Z_U is defined by s polynomial equations. The next result tells us that the number of points on such an algebraic variety must be either bounded or at least around q.

Lemma 1.11.5 (Dichotomy: number of common zeros)
For all s, d there exists a constant C such that if $f_1(X), \ldots, f_s(X)$ are polynomials on $\mathbb{F}_q^s$ of degree at most d, then

$$\{x \in \mathbb{F}_q^s : f_1(x) = \ldots f_s(x) = 0\}$$

has size either at most C or at least $q - C\sqrt{q}$.

The lemma can be deduced from the following important result from algebraic geometry due to Lang and Weil (1954), which says that the number of points of an r-dimensional algebraic variety in $\mathbb{F}_q^s$ is roughly q^r, as long as certain irreducibility hypotheses are satisfied. We include here the statement of the Lang–Weil bound. Here $\overline{\mathbb{F}}_q$ denotes the algebraic closure of $\mathbb{F}_q$.

Theorem 1.11.6 (Lang–Weil bound)
Let $g_1, \ldots, g_m \in \mathbb{F}_q[X]$ be polynomials of degree at most d. Let

$$V = \left\{x \in \overline{\mathbb{F}}_q^s : g_1(x) = g_2(x) = \ldots = g_m(x)\right\}.$$

Suppose V is an irreducible variety. Then

$$\left|V \cap \mathbb{F}_q^s\right| = q^{\dim V}(1 + O_{s,m,d}(q^{-1/2})).$$

The two cases in Lemma 1.11.5 then correspond to the zero-dimensional case and the positive-dimensional case, though some care is needed to deal with what happens if the variety is reducible in the field closure. We refer the reader to Bukh (2015) for details on how to deduce Lemma 1.11.5 from the Lang–Weil bound.

Now, continuing our proof of Theorem 1.11.1. Recall $Z_U = \{x \in \mathbb{F}_q^s \setminus U : f(x,u) = 0 \text{ for all } u \in U\}$. Apply Lemma 1.11.5 to the polynomials $f(X,u)$, $u \in U$. Then for large enough q there exists a constant C from Lemma 1.11.5 such that either $|Z_U| \le C$ (bounded) or $|Z_U| > q/2$ (very large). Thus, by (1.4),

$$\mathbb{P}(|Z_U| > C) = \mathbb{P}\left(|Z_U| > \frac{q}{2}\right) \le \frac{O_s(1)}{(q/2)^{s^2+1}}.$$

So, the expected number of s-element subset U with $|Z_U| > C$ is

$$\le \binom{q^s}{s} \frac{O_s(1)}{(q/2)^{s^2+1}} = O_s(1/q).$$

Remove from G a vertex from every s-element U with $|Z_U| > C$. Then the resulting graph is $K_{s,\lceil C \rceil+1}$-free. Since we remove at most q^s edges for each deleted vertex, the expected number of remaining edges is at least

$$\frac{1}{q}\binom{q^s}{2} - O_s(q^{s-1}) = \left(\frac{1}{2} - o(1)\right) q^{2s-1}.$$

Finally, given n, we can take the largest prime q satisfying $q^s \le n$ to finish the proof of Theorem 1.11.1.

Further Reading

Graph theory is a huge subject. There are many important topics that are quite far from the main theme of this book. For a standard introduction to the subject (especially on more classical aspects), several excellent graph theory textbooks are available: Bollobás (1998), Bondy and Murty (2008), Diestel (2017), and West (1996). The three-volume *Combinatorial Optimization* by Schrijver (2003) is also an excellent reference for graph theory, with a focus on combinatorial algorithms.

The following surveys discuss in more depth various topics encountered in this chapter:

- *The History of Degenerate (Bipartite) Extremal Graph Problems* by Füredi and Simonovits (2013);
- *Hypergraph Turán Problems* by Keevash (2011);
- *Dependent Random Choice* by Fox and Sudakov (2011).

Chapter Summary

- **Turán number** $\mathrm{ex}(n, H)$ = the maximum number of edges in an n-vertex H-free graph.
- **Turán's theorem.** Among all n-vertex K_{r+1}-free graphs, the Turán graph $T_{n,r}$ (a complete r-partite graph with nearly equal sized parts) uniquely maximizes the number of edges.
- **Erdős–Stone–Simonovits Theorem.** For any fixed graph H,

$$\mathrm{ex}(n, H) = \left(1 - \frac{1}{\chi(H) - 1} + o(1)\right)\frac{n^2}{2}.$$

- **Supersaturation** (from one copy to many copies): an n-vertex graph with $\geq \mathrm{ex}(n, H) + \varepsilon n^2$ edges has $\geq \delta n^{v(H)}$ copies of H, for some constant $\delta > 0$ depending on $\varepsilon > 0$ and H, and provided that n is sufficiently large.
- **Kővári–Sós–Turán theorem.** For fixed $s \leq t$,

$$\mathrm{ex}(n, K_{s,t}) = O_{s,t}(n^{2-1/s}).$$

  - Tight for $K_{2,2}$, $K_{3,3}$, and more generally, for $K_{s,t}$ with t much larger than s (algebraic constructions).
  - Conjectured to be tight in general.
- **Even cycles.** For any integer $k \geq 2$,

$$\mathrm{ex}(n, C_{2k}) = O_k(n^{1+1/k}).$$

  - Tight for $k \in \{2, 3, 5\}$ (algebraic constructions).
  - Conjectured to be tight in general.
- **Randomized constructions** for constructing H-free graphs: destroying all copies of H from a random graph.
- **Algebraic construction**: define edges using polynomials over $\mathbb{F}_q^n$.
- **Randomized algebraic constructions**: randomly select the polynomials.

2

Graph Regularity Method

Chapter Highlights

- Szemerédi's graph regularity lemma: partitioning an arbitrary graph into a bounded number of parts with random-like edges between parts
- Graph regularity method: recipe and applications
- Graph removal lemma
- Roth's theorem: a graph theoretic proof using the triangle removal lemma
- Strong regularity and induced graph removal lemma
- Graph property testing
- Hypergraph removal lemma and Szemerédi's theorem

In this chapter, we discuss a powerful technique in extremal graph theory, developed in the 1970s, known as Szemerédi's graph regularity lemma. The graph regularity method has wide ranging applications, and is now considered a central technique in the field. The regularity lemma produces a "rough structural" decomposition of an arbitrary graph (though it is mainly useful for graphs with quadratically many edges). It then allows us to model an arbitrary graph by a random graph.

The regularity method introduces us to a central theme of the book: **the dichotomy of structure and pseudorandomness**. This dichotomy is analogous to the more familiar concept of "signal and noise," namely that a complex system can be decomposed into a structural piece with plenty of information content (the signal) as well as a random-like residue (the noise). This idea will show up again later in Chapter 6 when we discuss Fourier analysis in additive combinatorics.

In general, we face two related challenges:

- How to decompose an object into a structured piece and a random-like piece?
- How to analyze the resulting components and their interactions?

We begin the chapter with the statement and the proof of the graph regularity lemma. We then prove Roth's theorem using the regularity method. This proof, due to Ruzsa and Szemerédi (1978), is not the original proof by Roth (1953), whose original Fourier analytic proof we will see in Chapter 6. Nevertheless, it is important for being historically one of the first major applications of the graph regularity method. Similar to the proof of Schur's theorem in Chapter 0, this graph theoretic proof of Roth's theorem demonstrates a fruitful connection between graph theory and additive combinatorics.

By **the regularity method**, we mean both the graph regularity lemma as well as methods for applying it. Rather than some specific set of theorems, graph regularity should be viewed as a general technique malleable to adaptations. Do not get bogged down by specific choices of parameters in the statements and proofs in what follows; rather, focus on the main ideas and techniques.

Many students experience a steep learning curve when studying the regularity method. The technical details can obscure the underlying intuition. Also, the style of arguments may be quite different from the type of combinatorial proofs they encountered earlier in their studies (e.g., the type of proofs from earlier in this book). Section 2.7 contains important exercises on applying the graph regularity method, which are essential for understanding the material.

2.1 Szemerédi's Graph Regularity Lemma

In this section, we state and prove the graph regularity lemma. Let us first give an informal statement.

Graph Regularity Lemma (Informal) The vertex set of every graph can be partitioned into a bounded number of parts so that the graph looks random-like between most pairs of parts.

Following is an illustration of what the outcome of the partition looks like. Here the vertex set of a graph is partitioned into five parts. Between a pair of parts (including between a part and itself) is a random-like graph with a certain edge-density (e.g., 0.4 between the first and second parts, 0.7 between the first and third parts, . . .).

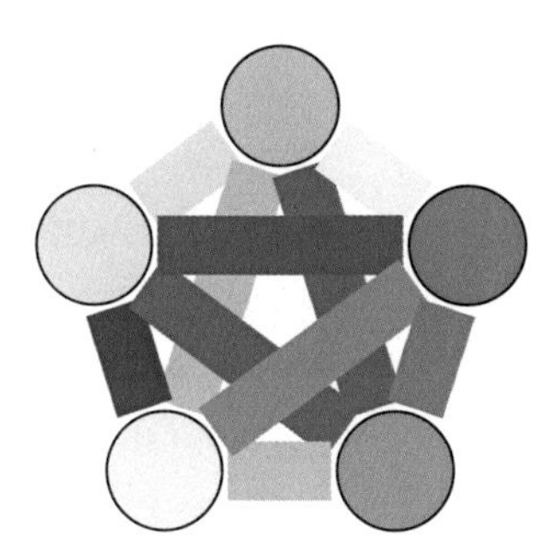

Definition 2.1.1 (Edge density)
Let X and Y be sets of vertices in a graph G. Let $e_G(X,Y)$ be the number of edges between X and Y; that is,

$$\boldsymbol{e_G(X,Y)} := |\{(x,y) \in X \times Y : xy \in E(G)\}|.$$

Define the ***edge density*** between X and Y in G by

$$\boldsymbol{d_G(X,Y)} := \frac{e_G(X,Y)}{|X|\,|Y|}.$$

We drop the subscript G if the context is clear.

We allow X and Y to overlap in the preceding definition. For intuition, it is mostly fine to picture the bipartite setting, where X and Y are automatically disjoint.

What should it mean for a graph to be "random-like"? We will explore the concept of **pseudorandom graphs** in depth in Chapter 3. Given vertex sets X and Y, we would like the edge density between them to not change much even if we restrict X and Y to smaller subsets. Intuitively, this says that the edges are somewhat evenly distributed.

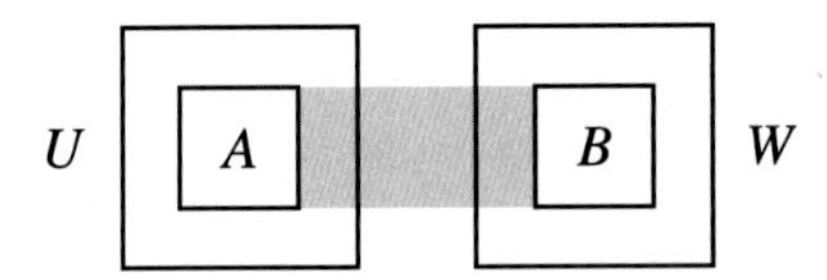

Definition 2.1.2 (ε-regular pair)
Let G be a graph and $U, W \subseteq V(G)$. We call (U, W) an ***ε-regular pair*** in G if, for all $A \subseteq U$ and $B \subseteq W$ with $|A| \geq \varepsilon |U|$ and $|B| \geq \varepsilon |W|$, one has

$$|d(A, B) - d(U, W)| \leq \varepsilon.$$

If (U, W) is not ε-regular, then we say that their irregularity is ***witnessed*** by some $A \subseteq U$ and $B \subseteq W$ satisfying $|A| \geq \varepsilon |U|$, $|B| \geq \varepsilon |W|$, and $|d(A, B) - d(U, W)| > \varepsilon$.

We need the hypotheses $|A| \geq \varepsilon |U|$ and $|B| \geq \varepsilon |W|$ since the definition would be too restrictive otherwise. For example, by taking $A = \{x\}$ and $B = \{y\}$, $d(A, B)$ could end up being both 0 (if $xy \notin E$) and 1 (if $xy \in E$).

Remark **2.1.3** (Different roles of ε). The ε in $|A| \geq \varepsilon |U|$ and $|B| \geq \varepsilon |W|$ plays a different role from the ε in $|d(A, B) - d(U, W)| \leq \varepsilon$. However, it is usually not important to distinguish these ε's. So we use only one ε for convenience of notation.

The "random-like" intuition is justified as random graphs indeed satisfy the above property. (This can be proved by the Chernoff bound; more on this in the next chapter.)

The following exercises can help you check your understanding of ε-regularity.

Exercise 2.1.4 (Basic inheritance of regularity). Let G be a graph and $X, Y \subseteq V(G)$. If (X, Y) is an $\varepsilon\eta$-regular pair, then (X', Y') is ε-regular for all $X' \subseteq X$ with $|X'| \geq \eta |X|$ and $Y' \subseteq Y$ with $|Y'| \geq \eta |Y|$.

Exercise 2.1.5 (An alternate definition of regular pairs). Let G be a graph and $X, Y \subseteq V(G)$. Say that (X, Y) is ***ε-homogeneous*** if, for all $A \subseteq X$ and $B \subseteq Y$, one has

$$|e(A, B) - |A|\,|B|\,d(X, Y)| \leq \varepsilon\,|X|\,|Y|\,.$$

Show that if (X, Y) is ε-regular, then it is ε-homogeneous. Also, show that, if (X, Y) is ε^3-homogeneous, then it is ε-regular.

Exercise 2.1.6 (Robustness of regularity). Prove that for every $\varepsilon' > \varepsilon > 0$, there exists $\delta > 0$ so that given an ε-regular pair (X, Y) in some graph, if we modify the graph by adding/deleting $\leq \delta |X|$ vertices to/from X, adding/deleting $\leq \delta |Y|$ vertices to/from Y, and adding/deleting $\leq \delta |X|\,|Y|$ edges, then the resulting new (X, Y) is still ε'-regular.

Next, let us define what it means for a vertex partition to be ε-regular.

Definition 2.1.7 (ε-regular partition)
Given a graph G, a partition $\mathcal{P} = \{V_1, \dots, V_k\}$ of its vertex set is an ***ε-regular partition*** if

$$\sum_{\substack{(i,j)\in[k]^2 \\ (V_i,V_j) \text{ not } \varepsilon\text{-regular}}} |V_i||V_j| \le \varepsilon |V(G)|^2.$$

In other words, all but at most an ε-fraction of pairs of vertices of G lie between ε-regular parts.

Remark **2.1.8.** When $|V_1| = \cdots = |V_k|$, the inequality says that at most εk^2 of pairs (V_i, V_j) are not ε-regular.

Also, note that the summation includes $i = j$. If none of the V_i's are too large, say $|V_i| \le \varepsilon n$ for each i, then the terms with $i = j$ contribute $\le \sum_i |V_i|^2 \le \varepsilon n \sum_i |V_i| = \varepsilon n^2$, which is neglible.

We are now ready to state Szemerédi's graph regularity lemma.

Theorem 2.1.9 (Szemerédi's graph regularity lemma)
For every $\varepsilon > 0$, there exists a constant M such that every graph has an ε-regular partition into at most M parts.

Proof of the Graph Regularity Lemma

Proof idea. We will generate the desired vertex partition according to the following algorithm:

(1) Start with the trivial partition of $V(G)$. (The trivial partition has a single part consisting of the whole set.)
(2) While the current partition $\mathcal{P}$ is not ε-regular:
    (a) For each (V_i, V_j) that is not ε-regular, find a witnessing pair in V_i and V_j.
    (b) Refine $\mathcal{P}$ using all the witnessing pairs. (Here, given two partitions $\mathcal{P}$ and Q of the same set, we say that ***Q refines $\mathcal{P}$*** if each part of Q is contained in a part of $\mathcal{P}$. In other words, we divide each part of $\mathcal{P}$ further to obtain Q.)

We repeat step (2) until the partition is ε-regular, at which point the algorithm terminates. The resulting partition is always ε-regular by design. It remains to show that the number of iterations is bounded as a function of ε. To see this, we keep track of a quantity that necessarily increases at each iteration of the procedure. This is called an **energy increment argument**. (The reason that we call it an "energy" is because it is the L^2 norm of a vector of edge-densities, and the kinetic energy in physics is also an L^2 norm.) ■

Definition 2.1.10 (Energy)
Let G be an n-vertex graph (whose dependence we drop from the notation). Let $U, W \subseteq V(G)$. Define

$$\boldsymbol{q(U,W)} := \frac{|U|\,|W|}{n^2} d(U,W)^2.$$

For partitions $\mathcal{P}_U = \{U_1, \ldots, U_k\}$ of U and $\mathcal{P}_W = \{W_1, \ldots, W_l\}$ of W, define

$$\boldsymbol{q(\mathcal{P}_U, \mathcal{P}_W)} := \sum_{i=1}^{k} \sum_{j=1}^{l} q(U_i, W_j).$$

Finally, for a partition $\mathcal{P} = \{V_1, \ldots, V_k\}$ of $V(G)$, define its ***energy*** to be

$$\boldsymbol{q(\mathcal{P})} := q(\mathcal{P}, \mathcal{P}) = \sum_{i=1}^{k} \sum_{j=1}^{k} q(V_i, V_j) = \sum_{i=1}^{k} \sum_{j=1}^{k} \frac{|V_i|\,|V_j|}{n^2} d(V_i, V_j)^2.$$

Since the edge density is always between 0 and 1, we have $0 \leq q(\mathcal{P}) \leq 1$ for all partitions $\mathcal{P}$. The following lemmas show that the energy cannot decrease upon refinement, and furthermore, it must increase substantially at each step of the preceding algorithm.

Lemma 2.1.11 (Energy never decreases under refinement)
Let G be a graph, $U, W \subseteq V(G)$, $\mathcal{P}_U$ a partition of U, and $\mathcal{P}_W$ a partition of W. Then $q(\mathcal{P}_U, \mathcal{P}_W) \geq q(U, W)$.

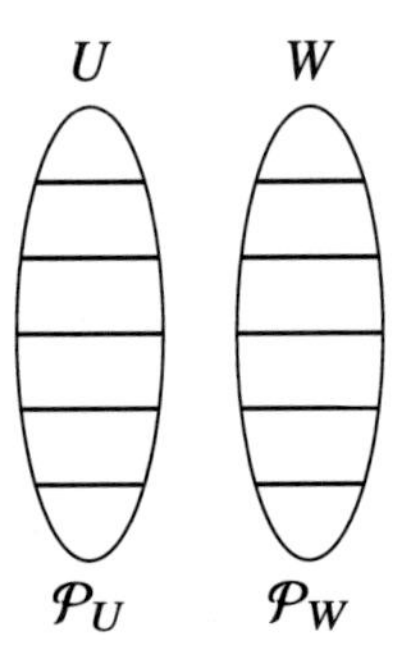

Proof. Let $n = v(G)$. Let $\mathcal{P}_U = \{U_1, \ldots, U_k\}$ and $\mathcal{P}_W = \{W_1, \ldots, W_l\}$. Choose $x \in U$ and $y \in W$ uniformly and independently at random. Let U_i be the part of $\mathcal{P}_U$ that contains x and W_j be the part of $\mathcal{P}_W$ that contains y. Define the random variable $Z := d(U_i, W_j)$. We have

$$\mathbb{E}[Z] = \sum_{i=1}^{k} \sum_{j=1}^{l} \frac{|U_i|}{|U|} \frac{|W_j|}{|W|} d(U_i, W_j) = d(U, W) = \sqrt{\frac{n^2}{|U||W|} q(U, W)}.$$

We have

$$\mathbb{E}[Z^2] = \sum_{i=1}^{k} \sum_{j=1}^{l} \frac{|U_i|}{|U|} \frac{|W_j|}{|W|} d(U_i, W_j)^2 = \frac{n^2}{|U||W|} q(\mathcal{P}_U, \mathcal{P}_W).$$

By convexity, $\mathbb{E}[Z^2] \geq \mathbb{E}[Z]^2$, which implies $q(\mathcal{P}_U, \mathcal{P}_W) \geq q(U, W)$. □

Lemma 2.1.12 (Energy never decreases under refinement)
Given two vertex partitions $\mathcal{P}$ and $\mathcal{P}'$ of some graph, if $\mathcal{P}'$ refines $\mathcal{P}$, then $q(\mathcal{P}) \leq q(\mathcal{P}')$.

Proof. The conclusion follows by applying Lemma 2.1.11 to each pair of parts of $\mathcal{P}$. In more detail, letting $\mathcal{P} = \{V_1, \ldots, V_m\}$, and supposing $\mathcal{P}'$ refines each V_i into a partition $\mathcal{P}'_{V_i} = \{V'_{i1}, \ldots, V'_{ik_i}\}$ of V_i, so that $\mathcal{P}' = \mathcal{P}'_{V_1} \cup \cdots \cup \mathcal{P}'_{V_m}$, we have

$$q(\mathcal{P}) = \sum_{i,j} q(V_i, V_j) \le \sum_{i,j} q(\mathcal{P}'_{V_i}, \mathcal{P}'_{V_j}) = q(\mathcal{P}'). \qquad \square$$

Lemma 2.1.13 (Energy boost for an irregular pair)
Let G be an n-vertex graph. If (U, W) is not ε-regular, as witnessed by $A \subseteq U$ and $B \subseteq W$, then

$$q(\{A, U \setminus A\}, \{B, W \setminus B\}) > q(U, W) + \varepsilon^4 \frac{|U|\,|W|}{n^2}.$$

This is the "red bull lemma," giving an energy boost when feeling irregular.

Proof. Define Z as in the proof of Lemma 2.1.11 for $\mathcal{P}_U = \{A, U \setminus A\}$ and $\mathcal{P}_W = \{B, W \setminus B\}$. Then

$$\mathrm{Var}(Z) = \mathbb{E}[Z^2] - \mathbb{E}[Z]^2 = \frac{n^2}{|U|\,|W|} \left(q(\mathcal{P}_U, \mathcal{P}_W) - q(U, W) \right).$$

We have $Z = d(A, B)$ with probability $\ge |A|\,|B| / (|U|\,|W|)$ (corresponding to the event $x \in A$ and $y \in B$). So,

$$\begin{aligned} \mathrm{Var}(Z) &= \mathbb{E}[(Z - \mathbb{E}[Z])^2] \\ &\ge \frac{|A|}{|U|} \frac{|B|}{|W|} (d(A, B) - d(U, W))^2 \\ &> \varepsilon \cdot \varepsilon \cdot \varepsilon^2. \end{aligned}$$

Putting the two inequalities together gives the claim. $\square$

The next lemma, corresponding to step (2)(b) of the preceding algorithm, shows that we can put all the witnessing pairs together to obtain an energy increment.

Lemma 2.1.14 (Energy boost for an irregular partition)
If a partition $\mathcal{P} = \{V_1, \ldots, V_k\}$ of $V(G)$ is not ε-regular, then there exists a refinement $\mathcal{Q}$ of $\mathcal{P}$ where every V_i is partitioned into at most 2^{k+1} parts, and such that

$$q(\mathcal{Q}) > q(\mathcal{P}) + \varepsilon^5.$$

Proof. Let

$$R = \{(i, j) \in [k]^2 : (V_i, V_j) \text{ is } \varepsilon\text{-regular}\} \qquad \text{and} \qquad \overline{R} = [k]^2 \setminus R.$$

For each pair (V_i, V_j) that is not ε-regular, find a pair $A^{i,j} \subseteq V_i$ and $B^{i,j} \subseteq V_j$ that witnesses the irregularity. Do this simultaneously for all $(i, j) \in \overline{R}$. Note that for $i \ne j$ we can take $A^{i,j} = B^{j,i}$ due to symmetry. When $i = j$, we should allow for the possibility of $A^{i,i}$ and $B^{i,i}$ to be distinct.

Let $\mathcal{Q}$ be a common refinement of $\mathcal{P}$ by all the $A^{i,j}$ and $B^{i,j}$ (i.e., the parts of $\mathcal{Q}$ are maximal subsets that are not "cut up" into small pieces by any element of $\mathcal{P}$ or by the $A^{i,j}$

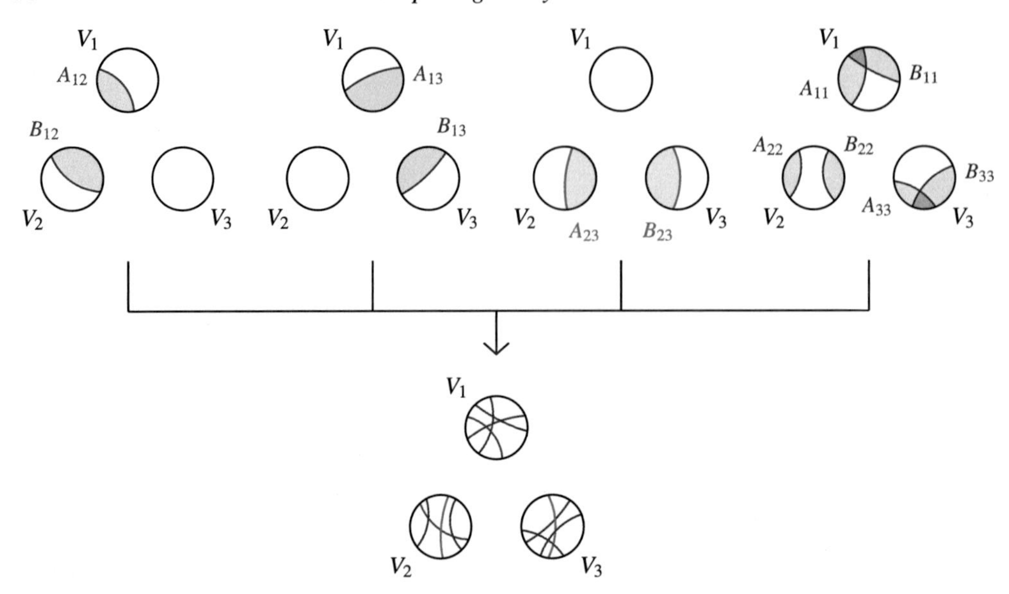

Figure 2.1 In the proof of Lemma 2.1.14, we refine the partition by taking a common refinement using witnesses of irregular pairs.

and $B^{i,j}$; intuitively, imagine regions of a Venn diagram). See Figure 2.1 for an illustration. There are $\leq k+1$ such distinct nonempty sets inside each V_i. So $\mathcal{Q}$ refines each V_i into at most 2^{k+1} parts. Let $\mathcal{Q}_i$ be the partition of V_i given by $\mathcal{Q}$. Then, using the monotonicity of energy under refinements (Lemma 2.1.11),

$$
\begin{aligned}
q(\mathcal{Q}) &= \sum_{(i,j)\in[k]^2} q(\mathcal{Q}_i,\mathcal{Q}_j) \\
&= \sum_{(i,j)\in R} q(\mathcal{Q}_i,\mathcal{Q}_j) + \sum_{(i,j)\in \overline{R}} q(\mathcal{Q}_i,\mathcal{Q}_j) \\
&\geq \sum_{(i,j)\in R} q(V_i,V_j) + \sum_{(i,j)\in \overline{R}} q(\{A^{i,j}, V_i\setminus A^{i,j}\},\{B^{i,j}, V_j\setminus B^{i,j}\}).
\end{aligned}
$$

By Lemma 2.1.13, the energy boost lemma, the preceding sum is

$$
> \sum_{(i,j)\in[k]^2} q(V_i,V_j) + \sum_{(i,j)\in \overline{R}} \varepsilon^4 \frac{|V_i|\,|V_j|}{n^2}.
$$

The first sum equals $q(\mathcal{P})$, and the second sum is $> \varepsilon^5$ by Lemma 2.1.13 since $\mathcal{P}$ is not ε-regular. This gives the desired inequality. □

***Remark* 2.1.15** (Refinements should be done simultaneously). Here is a subtle point in the preceding proof. The refinement $\mathcal{Q}$ must be obtained in a single step by refining $\mathcal{P}$ using all the witnessing sets $A^{i,j}$ *simultaneously*. If instead we pick out a pair $A^{i,j} \subseteq V_i$ and $A^{j,i} \subseteq V_j$, refine the partition using just this pair, and then iterate using another irregular pair $(V_{i'}, V_{j'})$, the energy boost step would not work. This is because ε-regularity (or lack thereof) is not well-preserved under taking refinements.

Proof of the graph regularity lemma (Theorem 2.1.9). Start with a trivial partition of the vertex set of the graph. Repeatedly apply Lemma 2.1.14 whenever the current partition is not ε-regular. By Lemma 2.1.14, the energy of the partition increases by more than ε^5 at each iteration. Since the energy of the partition is ≤ 1, we must stop after $< \varepsilon^{-5}$ iterations, terminating in an ε-regular partition.

If a partition has k parts, then Lemma 2.1.14 produces a refinement with $\leq k2^{k+1}$ parts. We start with a trivial partition with one part, and then refine $< \varepsilon^{-5}$ times. Observe the crude bound $k2^{k+1} \leq 2^{2^k}$. So, the total number of parts at the end is $\leq \text{tower}(\lceil 2\varepsilon^{-5} \rceil)$, where

$$\textbf{tower}(k) := \left. 2^{2^{\cdot^{\cdot^{2}}}} \right\} \text{height } k. \qquad \square$$

***Remark* 2.1.16** (The proof does not guarantee that the partition becomes "more regular" after each step.). Let us stress what the proof is *not* saying. It is *not* saying that the partition gets more and more regular under each refinement. Also, it is *not* saying that partition gets more regular as the energy gets higher. Rather, the energy simply bounds the number of iterations.

The bound on the number of parts guaranteed by the proof is a constant for each fixed $\varepsilon > 0$, but it grows extremely quickly as ε gets smaller. Is the poor quantitative dependence somehow due to a suboptimal proof strategy? Surprisingly, the tower-type bound is necessary, as shown by Gowers (1997).

Theorem 2.1.17 (Lower bound on the number of parts in a regularity partition)
There exists a constant $c > 0$ such that for all sufficiently small $\varepsilon > 0$, there exists a graph with no ε-regular partition into fewer than $\text{tower}(\lceil \varepsilon^{-c} \rceil)$ parts.

We do not include the proof here. See Moshkovitz and Shapira (2016) for a short proof. The general idea is to construct a graph that roughly reverse engineers the proof of the regularity lemma. So, there is essentially a unique ε-regular partition, which must have many parts.

***Remark* 2.1.18** (Irregular pairs are necessary in the regularity lemma.). Recall that in Definition 2.1.7 of an ε-regular partition, we are allowed to have some irregular pairs. Are irregular pairs necessary? It turns that we must permit them. Exercise 2.1.24 gives an example of a canonical example (a "half graph") where every regularity partition has irregular pairs.

The regularity lemma is quite flexible. For example, we can start with an arbitrary partition of $V(G)$ instead of the trivial partition in the proof in order to obtain a partition that is a refinement of a given partition. The exact same proof with this modification yields the following.

Theorem 2.1.19 (Regularity starting with an arbitrary initial partition)
For every $\varepsilon > 0$ and k, there exists a constant M such that for every graph G and a partition $\mathcal{P}_0$ of $V(G)$ at most k parts, there exists an ε-regular partition $\mathcal{P}$ of $V(G)$ that is a refinement of $\mathcal{P}_0$, and such that each part of $\mathcal{P}_0$ is refined into at most M parts.

Here is another strengthening of the regularity lemma. We impose the additional requirement that vertex parts should be as equal in size as possible. We say that a partition is ***equitable*** if all part sizes are within one of each other; that is, $\left||V_i| - |V_j|\right| \leq 1$. In other words, a partition of a set of size n into k parts is equitable if every part has size $\lfloor n/k \rfloor$ or $\lceil n/k \rceil$.

Theorem 2.1.20 (Equitable regularity lemma)
For all $\varepsilon > 0$ and m_0, there exists a constant M such that every graph has an ε-regular equitable partition of its vertex set into k parts with $m_0 \leq k \leq M$.

***Remark* 2.1.21.** The lower bound m_0 requirement on the number of parts is somewhat superficial. The reason for including it here is that it is often convenient to discard all the edges that lie within individual parts of the partition, and since there are most n^2/k such edges, they contribute negligibly if the number of parts k is not too small, which is true if we require $m_0 \geq 1/\varepsilon$ in the equitable regularity lemma statement.

There are several ways to guarantee equitability. One method is sketched in what follows. We equitize the partition at every step of the refinement iteration, so that at each step in the proof, we both obtain an energy increment and also end up with an equitable partition.

Proof sketch of the equitable regularity lemma (Theorem 2.1.20). Here is a modified algorithm:

(1) Start with an arbitrary equitable partition of the graph into m_0 parts.
(2) While the current equitable partition $\mathcal{P}$ is not ε-regular:
    (a) (Refinement/energy boost) Refine the partition using pairs that witness irregularity (as in the earlier proof). The new partition $\mathcal{P}'$ divides each part of $\mathcal{P}$ into $\leq 2^{|\mathcal{P}|}$ parts.
    (b) (Equitization) Modify $\mathcal{P}'$ into an equitable partition by arbitrarily chopping each part of $\mathcal{P}'$ into parts of size $|V(G)|/m$ (for some appropriately chosen $m = m(|\mathcal{P}'|, \varepsilon)$) plus some leftover pieces, which are then combined together and then divided into parts of size $|V(G)|/m$.

The refinement step (2)(a) increases energy by $\geq \varepsilon^5$ as before. The energy might go down in the equitization step (2)(b), but it should not decrease by much, provided that the m chosen in that step is large enough (say, $m = \lfloor 100\,|\mathcal{P}'|\,\varepsilon^{-5} \rfloor$). So overall, we still have an energy increment of $\geq \varepsilon^5/2$ at each step, and hence the process still terminates after $O(\varepsilon^{-5})$ steps. The total number of parts at the end is bounded. □

Exercise 2.1.22. Complete the details in the preceding proof sketch.

Exercise 2.1.23 (Making each part ε-regular to nearly all other parts)**.** Prove that for all $\varepsilon > 0$ and m_0, there exists a constant M so that every graph has an equitable vertex partition into k parts, with $m_0 \leq k \leq M$, such that each part is ε-regular with all but at most εk other parts.

The important example in the next exercise shows why we must allow irregular pairs in the graph regularity lemma.

Exercise 2.1.24 (Unavoidability of irregular pairs). Let the ***half-graph*** H_n be the bipartite graph on $2n$ vertices $\{a_1, \ldots, a_n, b_1, \ldots, b_n\}$ with edges $\{a_i b_j : i \le j\}$.

(a) For every $\varepsilon > 0$, explicitly construct an ε-regular partition of H_n into $O(1/\varepsilon)$ parts.
(b) Show that there is some $c > 0$ such that for every $\varepsilon \in (0, c)$, every positive integer k and sufficiently large multiple n of k, every partition of the vertices of H_n into k equal-sized parts contains at least ck pairs of parts that are not ε-regular.

The next exercise should remind you of the iteration technique from the proof of the graph regularity lemma.

Exercise 2.1.25 (Existence of a regular pair of subsets). Show that there is some absolute constant $C > 0$ such that for every $0 < \varepsilon < 1/2$, every graph on n vertices contains an ε-regular pair of vertex subsets each with size at least δn, where $\delta = 2^{-\varepsilon^{-C}}$.

Hint: Density increment (don't use energy).

This exercise asks for two different proofs of the following theorem.

Given a graph G, we say that $X \subseteq V(G)$ is ***ε-regular*** if the pair (X, X) is ε-regular; that is, for all $A, B \subseteq X$ with $|A|, |B| \ge \varepsilon |X|$, one has $|d(A, B) - d(X, X)| \le \varepsilon$.

Theorem 2.1.26 (ε-regular subset)
For every $\varepsilon > 0$, there exists $\delta > 0$ such that every graph contains an ε-regular subset of vertices that is an $\ge \delta$ fraction of the vertex set.

Exercise 2.1.27 (ε-regular subset).

(a) Prove Theorem 2.1.26 using Szemerédi's regularity lemma, showing that one can obtain the ε-regular subset by combining a suitable subcollection of parts from some regularity partition.
(b*) Give an alternative proof of the theorem with $\delta = \exp(-\exp(\varepsilon^{-C}))$ for some constant C.

Exercise 2.1.28* (Regularity partition into regular sets). Show that for every $\varepsilon > 0$ there exists M so that every graph has an ε-regular partition into at most M parts, with every part being ε-regular with itself.

2.2 Triangle Counting Lemma

Szemerédi's regularity lemma gave us a vertex partition of a graph. How can we use this partition?

In this section, we begin by establishing the **triangle counting lemma**. Given three vertex sets X, Y, Z, pairwise ε-regular in G, we can approximate it by a random tripartite graph on X, Y, Z with the same edge densities between parts. By comparing G to its random model approximation, we expect the number of triples $(x, y, z) \in X \times Y \times Z$ forming a triangle in G to be roughly

$$d(X, Y) d(X, Z) d(Y, Z) |X||Y||Z|.$$

The triangle counting lemma makes this intuition precise.

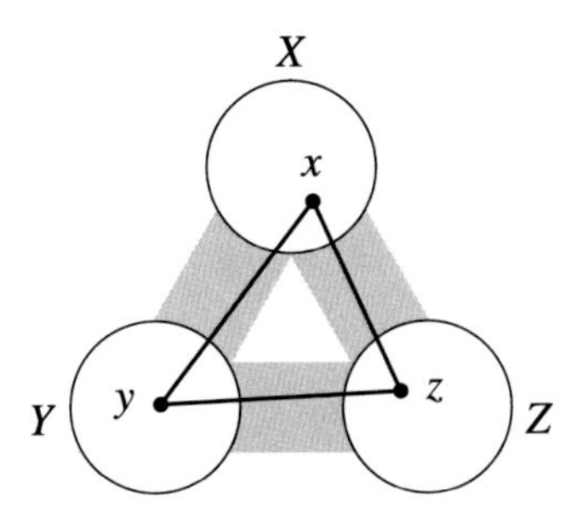

Theorem 2.2.1 (Triangle counting lemma)
Let G be a graph and X, Y, Z be subsets of the vertices of G such that (X,Y), (Y,Z), (Z,X) are all ε-regular pairs for some $\varepsilon > 0$. If $d(X,Y), d(X,Z), d(Y,Z) \geq 2\varepsilon$, then

$$\begin{aligned}&|\{(x,y,z) \in X \times Y \times Z : xyz \text{ is a triangle in } G\}| \\ &\qquad \geq (1-2\varepsilon)(d(X,Y)-\varepsilon)(d(X,Z)-\varepsilon)(d(Y,Z)-\varepsilon)\,|X|\,|Y|\,|Z|\,.\end{aligned}$$

***Remark* 2.2.2.** The vertex sets X, Y, Z do not have to be disjoint, but one does not lose any generality by assuming that they are disjoint in this statement. Indeed, starting with $X, Y, Z \subseteq V(G)$, one can always create an auxiliary tripartite graph G' with vertex parts being disjoint replicas of X, Y, Z and the edge relations in $X \times Y$ being the same for G and G', and likewise for $X \times Z$ and $Y \times Z$. Under this auxiliary construction, a triple in $X \times Y \times Z$ forms a triangle in G if and only if it forms a triangle in G'.

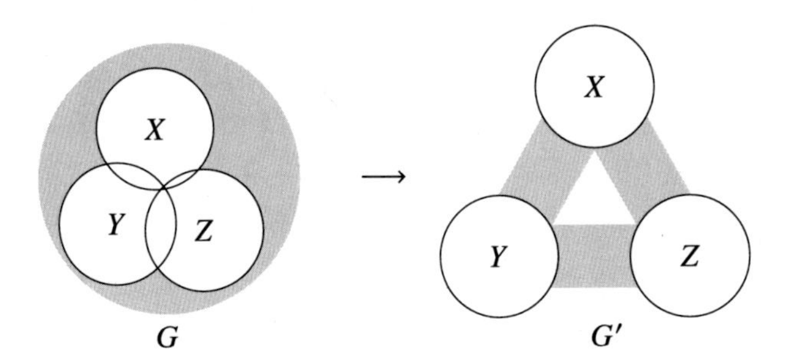

Now we show that in an ε-regular pair (X,Y), almost all vertices of X have roughly the same number of neighbors in Y. (The next lemma only states a lower bound on degree, but the same argument also gives an analogous upper bound.)

Lemma 2.2.3 (Most vertices have roughly the same degree)
Let (X,Y) be an ε-regular pair. Then fewer than $\varepsilon\,|X|$ vertices in X have fewer than $(d(X,Y)-\varepsilon)\,|Y|$ neighbors in Y. Likewise, fewer than $\varepsilon\,|Y|$ vertices in Y have fewer than $(d(X,Y)-\varepsilon)\,|X|$ neighbors in X.

Proof. Let A be the subset of vertices in X with $< (d(X,Y)-\varepsilon)\,|Y|$ neighbors in Y. Then $d(A,Y) < d(X,Y) - \varepsilon$, and thus $|A| < \varepsilon\,|X|$ by Definition 2.1.2, as (X,Y) is an ε-regular pair. The other claim is similar. $\square$

Proof of Theorem 2.2.1. By Lemma 2.2.3, we can find $X' \subseteq X$ with $|X'| \geq (1 - 2\varepsilon)|X|$ such that every vertex $x \in X'$ has $\geq (d(X,Y) - \varepsilon)|Y|$ neighbors in Y and $\geq (d(X,Z) - \varepsilon)|Z|$ neighbors in Z. Write $N_Y(x) = N(x) \cap Y$ and $N_Z(x) = N(x) \cap Z$.

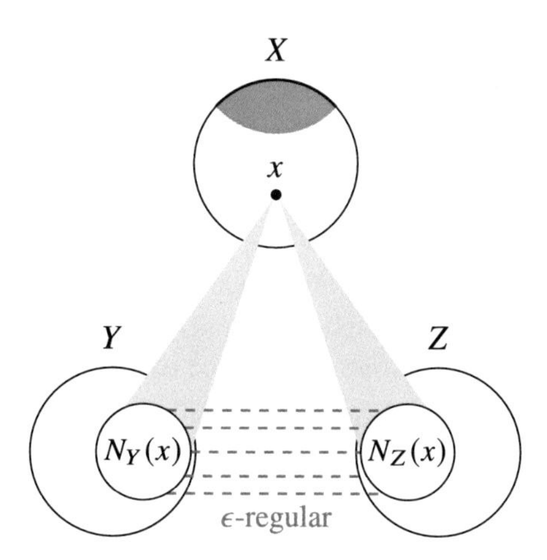

For each such $x \in X'$, we have $|N_Y(x)| \geq (d(X,Y) - \varepsilon)|Y| \geq \varepsilon|Y|$. Likewise, $|N_Z(x)| \geq \varepsilon|Z|$. Since (Y,Z) is ε-regular, the edge density between $N_Y(x)$ and $N_Z(x)$ is $\geq d(Y,Z) - \varepsilon$. So, for each $x \in X'$, the number of edges between $N_Y(x)$ and $N_Z(x)$ is

$$\geq (d(Y,Z) - \varepsilon)|N_Y(x)||N_Z(x)| \geq (d(X,Y) - \varepsilon)(d(X,Z) - \varepsilon)(d(Y,Z) - \varepsilon)|Y||Z|.$$

Multiplying by $|X'| \geq (1 - 2\varepsilon)|X|$, we obtain the desired lower bound on the number of triangles. □

***Remark* 2.2.4.** We only need the lower bound on the triangle count for our applications in this chapter, but the same proof can also be modified to give an upper bound, which we leave as an exercise.

2.3 Triangle Removal Lemma

The triangle removal lemma (Ruzsa and Szemerédi 1978) is one of the first major applications of the regularity method. Informally, the triangle removal lemma says that a graph with few triangles can be made triangle-free by removing a few edges. Here, "few triangles" means a subcubic number of triangles (i.e., asymptotically less than the maximum possible number) and "few edges" means a subquadratic number of edges.

> **Theorem 2.3.1** (Triangle removal lemma)
> For all $\varepsilon > 0$, there exists $\delta > 0$ such that any graph on n vertices with fewer than δn^3 triangles can be made triangle-free by removing fewer than εn^2 edges.

The triangle removal lemma can be equivalently stated as:

An n-vertex graph with $o(n^3)$ triangles can be made triangle-free by removing $o(n^2)$ edges.

Our proof of Theorem 2.3.1 demonstrates how to apply the graph regularity lemma. Here is a representative "recipe" for the regularity method.

***Remark* 2.3.2** (Regularity method recipe). Typical applications of the regularity method proceed in the following steps:

(1) ***Partition*** the vertex set of a graph using the regularity lemma.
(2) ***Clean*** the graph by removing edges that behave poorly in the regularity partition. Most commonly, we remove edges that lie between pairs of parts with
   (a) irregularity, or
   (b) low-density, or
   (c) one of the parts too small.
   This ends up removing a negligible number of edges.
(3) ***Count*** a certain pattern in the cleaned graph using a counting lemma.

To prove the triangle removal lemma, after cleaning the graph (which removes few edges), we claim that the resulting cleaned graph must be triangle-free, or else the triangle counting lemma would find many triangles, contradicting the hypothesis.

Proof of the triangle removal lemma (Theorem 2.3.1). Suppose we are given a graph on n vertices with $< \delta n^3$ triangles, for some parameter δ we will choose later. Apply the graph regularity lemma, Theorem 2.1.9, to obtain an $\varepsilon/4$-regular partition of the graph with parts $V_1, V_2, \cdots, V_m$. Next, for each $(i, j) \in [m]^2$, remove all edges between V_i and V_j if

(a) (V_i, V_j) is not $\varepsilon/4$-regular, or
(b) $d(V_i, V_j) < \varepsilon/2$, or
(c) $\min\{|V_i|, |V_j|\} < \varepsilon n/(4m)$.

Since the partition is $\varepsilon/4$-regular (recall Definition 2.1.7), the number of edges removed in (a) from irregular pairs is

$$\le \sum_{\substack{i,j \\ (V_i,V_j) \text{ not } (\varepsilon/4)\text{-regular}}} |V_i||V_j| \le \frac{\varepsilon}{4} n^2.$$

The number of edges removed in (b) from low-density pairs is

$$\le \sum_{\substack{i,j \\ d(V_i,V_j)<\varepsilon/2}} d(V_i, V_j)|V_i||V_j| \le \frac{\varepsilon}{2} \sum_{i,j} |V_i||V_j| = \frac{\varepsilon}{2} n^2.$$

The number of edges removed in (c) with an endpoint in a small part is

$$< m \cdot \frac{\varepsilon n}{4m} \cdot n = \frac{\varepsilon}{4} n^2.$$

In total, we removed $< \varepsilon n^2$ edges from the graph.

We claim that the remaining graph is triangle-free, provided that δ was chosen appropriately small. Indeed, suppose there remains a triangle whose three vertices lie in V_i, V_j, V_k (not necessarily distinct parts).

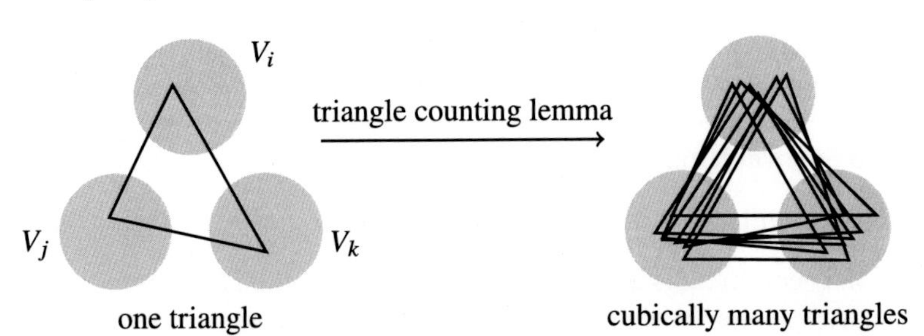

Because edges between the pairs described in (a) and (b) were removed, V_i, V_j, V_k satisfy the hypotheses of the triangle counting lemma (Theorem 2.2.1),

$$\begin{aligned}\#\{\text{triangles in } V_i \times V_j \times V_k\} &\geq \left(1 - \frac{\varepsilon}{2}\right)\left(\frac{\varepsilon}{4}\right)^3 |V_i|\,|V_j|\,|V_k| \\ &\geq \left(1 - \frac{\varepsilon}{2}\right)\left(\frac{\varepsilon}{4}\right)^3 \left(\frac{\varepsilon n}{4m}\right)^3,\end{aligned}$$

where the final step uses (c). Then, as long as

$$\delta < \frac{1}{6}\left(1 - \frac{\varepsilon}{2}\right)\left(\frac{\varepsilon}{4}\right)^3 \left(\frac{\varepsilon}{4m}\right)^3,$$

we would contradict the hypothesis that the original graph has $< \delta n^3$ triangles. (The extra factor of 6 in the preceding inequality is there to account for the possibility that $V_i = V_j = V_k$.) Since m is bounded for each fixed ε, we see that δ can be chosen to depend only on ε. □

The next corollary of the triangle removal lemma will soon be used to prove Roth's theorem. Here "diamond" refers to the following graph, consisting of two triangles sharing an edge.

Corollary 2.3.3 (Diamond-free lemma)
Let G be an n-vertex graph where every edge lies in a unique triangle. Then G has $o(n^2)$ edges.

Proof. Let G have m edges. Because each edge lies in exactly one triangle, the number of triangles in G is $m/3 = O(n^2) = o(n^3)$. By the triangle removal lemma (see the statement after Theorem 2.3.1), we can remove $o(n^2)$ edges to make G triangle-free. However, deleting an edge removes at most one triangle from the graph by assumption, so $m/3$ edges need to be removed to make G triangle-free. Thus, $m = o(n^2)$. □

***Remark* 2.3.4** (Quantitative dependencies in the triangle removal lemma)**.** Since the preceding proof of the triangle removal lemma applies the graph regularity lemma, the resulting bounds from the proof are quite poor: it shows that one can pick $\delta = 1/\text{tower}(\varepsilon^{-O(1)})$. Using a different but related method, Fox (2011) proved the triangle removal lemma with a slightly better dependence $\delta = 1/\text{tower}(O(\log(1/\varepsilon)))$. In the other direction, we know that the triangle removal lemma does not hold with $\delta = \varepsilon^{c\log(1/\varepsilon)}$ for a sufficiently small constant $c > 0$. The construction comes from the Behrend construction of large 3-AP-free sets that we will soon see in Section 2.5. Our knowledge of the quantitative dependence in Corollary 2.3.3 comes from the same source; specifically, we know that the $o(n^2)$ can be sharpened to $n^2/e^{\Omega(\log^*(1/\varepsilon))}$ (where $\log^*$, the iterated logarithm function, is the number of iterations of log that one needs to take to bring a number to at most 1). But the statement is false if the $o(n^2)$ is replaced by $n^2e^{-C\sqrt{\log n}}$ for some sufficiently large constant C. It is a major open problem to close the gap between the upper and lower bounds in these problems.

The triangle removal lemma was historically first considered in the following equivalent formulation.

Theorem 2.3.5 ((6, 3)-theorem)
Let H be an n-vertex 3-uniform hypergraph without a subgraph having six vertices and three edges. Then H has $o(n^2)$ edges.

Exercise 2.3.6. Deduce the (6, 3)-theorem from Corollary 2.3.3, and vice versa.

The following conjectural extension of the (6, 3)-theorem is a major open problem in extremal combinatorics. The conjecture is attributed to Brown, Erdős, and Sós (1973).

Conjecture 2.3.7 ((7, 4)-conjecture)
Let H be an n-vertex 3-uniform hypergraph without a subgraph having seven vertices and four edges. Then H has $o(n^2)$ edges.

2.4 Graph Theoretic Proof of Roth's Theorem

We will now prove Roth's theorem, which we saw in Chapter 0 and is restated below. The proof below, due to Ruzsa and Szemerédi (1978) connects graph theory and additive combinatorics, akin to the proof of Schur's theorem in Chapter 0.

We write ***3-AP*** for "3-term arithmetic progression." We say that A is ***3-AP-free*** if there are no $x, x + y, x + 2y \in A$ with $y \neq 0$.

Theorem 2.4.1 (Roth's theorem)
Let $A \subseteq [N]$ be 3-AP-free. Then $|A| = o(N)$.

Proof. Embed $A \subseteq \mathbb{Z}/M\mathbb{Z}$ with $M = 2N + 1$ (to avoid wraparounds). Since A is 3-AP-free in $\mathbb{Z}$, it is 3-AP-free in $\mathbb{Z}/M\mathbb{Z}$ as well.

Now, we construct a tripartite graph G whose parts X, Y, Z are all copies of $\mathbb{Z}/M\mathbb{Z}$. The edges of the graph are (since M is odd, we are allowed to divide by 2 in $\mathbb{Z}/M\mathbb{Z}$):

- $(x, y) \in X \times Y$ whenever $y - x \in A$;
- $(y, z) \in Y \times Z$ whenever $z - y \in A$;
- $(x, z) \in X \times Z$ whenever $(z - x)/2 \in A$.

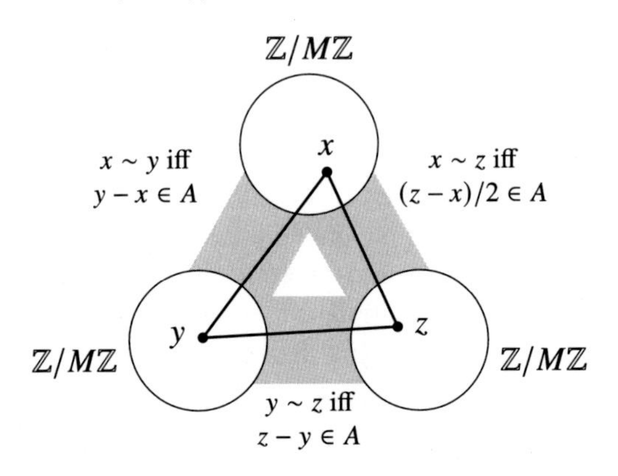

In this graph, $(x, y, z) \in X \times Y \times Z$ is a triangle if and only if

$$y - x, \frac{z - x}{2}, z - y \in A.$$

The graph was designed so that the preceding three numbers form an arithmetic progression in the listed order. Since A is 3-AP-free, these three numbers must all be equal. So, every edge of G lies in a unique triangle, formed by setting the three numbers to equal.

The graph G has exactly $3M = 6N + 3$ vertices and $3M\,|A|$ edges. Corollary 2.3.3 implies that G has $o(N^2)$ edges. So, $|A| = o(N)$. □

Now we prove a higher-dimensional generalization of Roth's theorem.

A ***corner*** in $\mathbb{Z}^2$ is a three-element set of the form $\{(x, y), (x + d, y), (x, y + d)\}$ with $d > 0$.

(Note that one could relax the assumption $d > 0$ to $d \neq 0$, allowing "negative" corners. As shown in the first step in the following proof, the assumption $d > 0$ is inconsequential.)

Theorem 2.4.2 (Corner-free)
Every corner-free subset of $[N]^2$ has size $o(N^2)$.

***Remark* 2.4.3** (History). The theorem is due to Ajtai and Szemerédi (1974), who originally proved it by invoking the full power of Szemerédi's theorem. Here we present a much simpler proof using the triangle removal lemma due to Solymosi (2003).

Proof. First, we show how to relax the assumption in the definition of a corner from $d > 0$ to $d \neq 0$.

Let $A \subseteq [N]^2$ be a corner-free set. For each $z \in \mathbb{Z}^2$, let $A_z = A \cap (z - A)$. Then $|A_z|$ is the number of ways that one can write $z = a + b$ for some $(a, b) \in A \times A$. So, $\sum_{z \in [2N]^2} |A_z| = |A|^2$, therefore there is some $z \in [2N]^2$ with $|A_z| \geq |A|^2 / (2N)^2$. To show that $|A| = o(N^2)$, it suffices to show that $|A_z| = o(N^2)$. Moreover, since $A_z = z - A_z$, its being corner-free implies that it does not contain three points $\{(x, y), (x + d, y), (x, y + d)\}$ with $d \neq 0$.

Write $A = A_z$ from now on. Build a tripartite graph G with parts $X = \{x_1, \ldots, x_N\}$, $Y = \{y_1, \ldots, y_N\}$, and $Z = \{z_1, \ldots, z_{2N}\}$, where each vertex x_i corresponds to a vertical line $\{x = i\} \subseteq \mathbb{Z}^2$, each vertex y_j corresponds to a horizontal line $\{y = j\}$, and each vertex z_k corresponds to a slanted line $\{y = -x + k\}$ with slope -1. Join two distinct vertices of G with an edge if and only if the corresponding lines intersect at a point belonging to A. Then, each triangle in the graph G corresponds to a set of three lines of slopes $0, \infty, -1$ pairwise intersecting at a point of A.

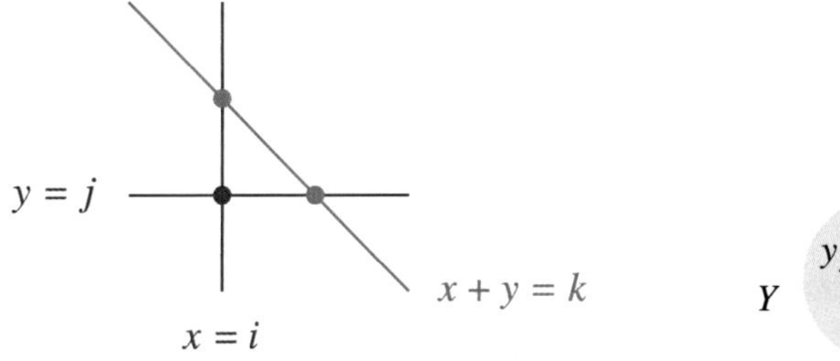

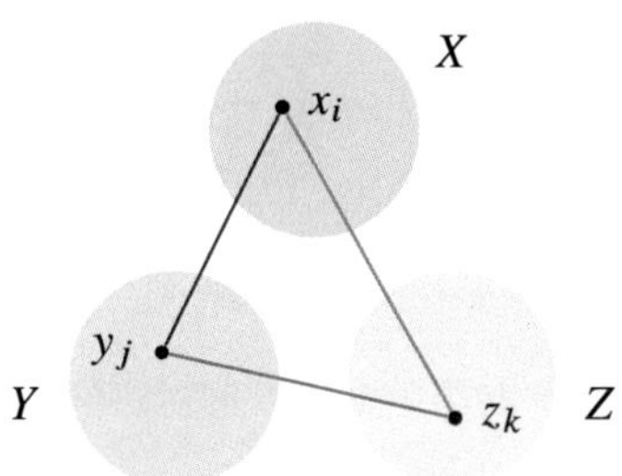

Since A is corner-free in the sense stated at the end of the previous paragraph, x_i, y_j, z_k form a triangle in G if and only if the three corresponding lines pass through the same point of A (i.e., forming a trivial corner with $d = 0$). Since there is exactly one line of each direction passing through every point of A, it follows that each edge of G belongs to exactly one triangle. Thus, by Corollary 2.3.3, $3|A| = e(G) = o(N^2)$. □

The upper bound on corner-free sets actually implies Roth's theorem, as shown in what follows. So, we now have a second proof of Roth's theorem (though this second proof is secretly the same as the first proof).

Proposition 2.4.4 (Corner-free sets vs. 3-AP-free sets)
Let $r_3(N)$ be the size of the largest subset of $[N]$ that contains no three-term arithmetic progression, and $r_\llcorner(N)$ be the size of the largest subset of $[N]^2$ that contains no corner. Then, $r_3(N)N \le r_\llcorner(2N)$.

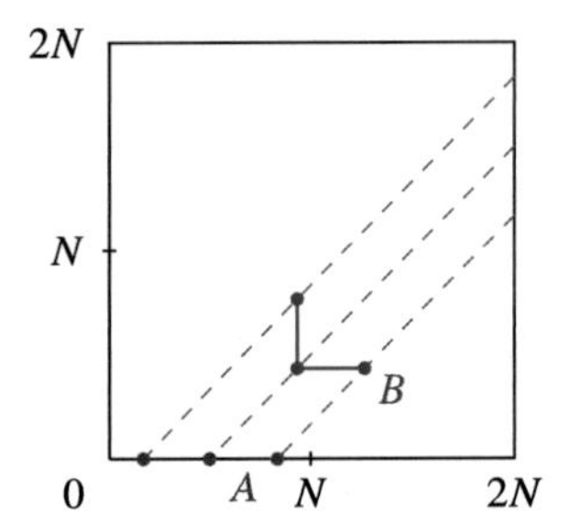

Proof. Given a 3-AP-free set $A \subseteq [N]$ of size $r_3(N)$, define a set

$$B := \left\{(x, y) \in [2N]^2 : x - y \in A\right\}.$$

Each element $a \in A$ gives rise to $\ge N$ different elements (x, y) of B with $x - y = a$. So $|B| \ge N|A|$. Furthermore, B is corner-free, since each corner $(x + d, y)$, (x, y), $(x, y + d)$ in B gives rise to a 3-AP $x - y - d$, $x - y$, $x - y + d$ with common difference d. So, $|B| \le r_\llcorner(2N)$. Thus, $r_3(N)N \le |A|N \le |B| \le r_\llcorner(2N)$. □

Remark **2.4.5** (Quantitative bounds). Both of the preceding proofs rely on the graph regularity lemma and hence give poor quantitative bounds. They tell us that a 3-AP-free $A \subseteq [N]$ has $|A| \le N/(\log^* N)^c$, where $\log^* N$ is the iterated logarithm (the number of times the logarithm function must be applied to bring N to less than or equal to 1). Later, in Chapter 6, we discuss Roth's original Fourier analytic proof, which uses different methods (though sharing the structure and randomness dichotomy theme) and gives much better quantitative bounds.

The current best upper bound on the size of a 3-AP-free subset of $[N]$ is $N/(\log N)^{1+c}$ for some constant $c > 0$ (Bloom and Sisask 2020). The current best upper bound on the size of corner-free subsets of $[N]^2$ is $N^2/(\log \log N)^c$ for some constant $c > 0$ (Shkredov 2006). Both use Fourier analysis.

For the next exercise, apply the triangle removal lemma to an appropriate graph.

Exercise 2.4.6* (Arithmetic triangle removal lemma). Show that for every $\varepsilon > 0$, there exists $\delta > 0$ such that if $A \subseteq [n]$ has fewer than δn^2 many triples $(x, y, z) \in A^3$ with $x + y = z$, then there is some $B \subseteq A$ with $|A \setminus B| \le \varepsilon n$ such that B is sum-free (i.e., no $x, y, z \in B$ with $x + y = z$).

2.5 Large 3-AP-Free Sets: Behrend's Construction

How can we construct a large 3-AP-free subset of $[N]$?

We can do it greedily. Starting with 0 (which produces a nicer pattern), we successively put in each positive integer if adding it does not create a 3-AP with the already chosen integers. This would produce the following sequence:

$$0 \quad 1 \quad 3 \quad 4 \quad 9 \quad 10 \quad 12 \quad 13 \quad 27 \quad 28 \quad 30 \quad 31 \quad \dots.$$

The above sequence is known as a ***Stanley sequence***. It consists of all nonnegative integers whose ternary representations have only the digits 0 and 1 (why?). Up to $N = 3^k$, the subset $A \subseteq [N]$ so constructed has size $|A| = 2^k = N^{\log_3 2}$.

For quite some time, people thought the preceding example was close to the optimal. Salem and Spencer (1942) then found a much larger 3-AP-free subset of $[N]$, with size $N^{1-o(1)}$. Their result was further improved by Behrend (1946), whose construction we present in what follows. This construction has not yet been substantially improved. (See Elkin [2011] and Green and Wolf [2010] for some lower-order improvements.)

Behrend's construction has surprising applications, such as in the design of fast matrix multiplication algorithms (Coppersmith and Winograd 1990).

Theorem 2.5.1 (Behrend's construction)
There exists a constant $C > 0$ such that for every positive integer N, there exists a 3-AP-free $A \subseteq [N]$ with $|A| \ge N e^{-C\sqrt{\log N}}$.

The rough idea is to first find a high-dimensional sphere with many lattice points via the pigeonhole principle. The sphere contains no 3-AP due to convexity. We then project these lattice points onto $\mathbb{Z}$ in a way that creates no additional 3-APs. This is done by treating the coordinates as the base-q expansion of an integer with some large q.

Proof. Let m and d be two positive integers depending on N to be specified later. Consider the lattice points of $X = \{0, 1, \dots, m-1\}^d$ that lie on a sphere of radius $\sqrt{L}$:

$$X_L := \left\{(x_1, \dots, x_d) \in X : x_1^2 + \cdots + x_d^2 = L\right\}.$$

Then, $X = \bigcup_{i=1}^{dm^2} X_i$. So by the pigeonhole principle, there exists an $L \in [dm^2]$ such that $|X_L| \ge m^d/(dm^2)$. Define the base $2m$ digital expansion

$$\phi(x_1, \dots, x_d) := \sum_{i=1}^{d} x_i (2m)^{i-1}.$$

Then, ϕ is injective on X. Furthermore, $x, y, z \in [m]^d$ satisfy $x + z = 2y$ if and only if $\phi(x) + \phi(z) = 2\phi(y)$. (There are no wraparounds in base $2m$ with all coordinates less than m.) Since X_L is a subset of a sphere, it is 3-AP-free.Thus, $\phi(X) \subseteq [(2m)^d]$ is a 3-AP-free set of size $\geq m^d/(dm^2)$. We can optimize the parameters and take $m = \lfloor e^{\sqrt{\log N}}/2 \rfloor$ and $d = \lfloor \sqrt{\log N} \rfloor$, thereby producing a 3-AP-free subset of $[N]$ with size $\geq Ne^{-C\sqrt{\log N}}$, where C is some absolute constant. □

The Behrend construction also implies lower bound constructions for the other problems we saw earlier. For example, since we used the diamond-free lemma (Corollary 2.3.3) to deduce an upper bound on the size of a 3-AP-free set, turning this implication around, we see that having a large 3-AP-free set implies the following quantitative limitation on the diamond-free lemma.

Corollary 2.5.2 (Lower bound for the diamond-free lemma)
For every $n \geq 3$, there is some n-vertex graph with at least $n^2 e^{-C\sqrt{\log n}}$ edges where every edge lies on a unique triangle. Here C is some absolute constant.

Proof. In the proof of Theorem 2.4.1, starting from a 3-AP-free set $A \subseteq [N]$, we constructed a graph with $6N+3$ vertices and $(6N+3)\,|A|$ edges such that every edge lies in a unique triangle. Choosing $N = \lfloor (n-3)/6 \rfloor$ and letting A be the Behrend construction of Theorem 2.5.1 with $|A| \geq Ne^{-C\sqrt{\log N}}$, we obtain the desired graph. □

***Remark* 2.5.3** (More lower bounds from Behrend's construction). The same graph construction also shows, after examining the proof of Corollary 2.3.3, that in the triangle removal lemma, Theorem 2.3.1, one cannot take $\delta = e^{-c(\log(1/\varepsilon))^2}$ if the constant $c > 0$ is too small.

In Proposition 2.4.4 we deduced an upper bound $r_3(N)N \leq r_{\llcorner}(2N)$ on corner-free sets using 3-AP-free sets. The Behrend construction then also gives a corner-free subset of $[N]^2$ of size $\geq N^2 e^{-C\sqrt{\log N}}$.

Exercise 2.5.4 (Modifying Behrend's construction). Prove that there is some constant $C > 0$ so that for all N, there exists $A \subseteq [N]$ with $|A| \geq N\exp(-C\sqrt{\log N})$ so that there do not exist $w, y, x, z \in A$ not all equal and satisfying $x + y + z = 3w$.

Exercise 2.5.5* (Avoiding 5-term quadratic configurations). Prove that there is some constant $C > 0$ so that for all N, there exists $A \subseteq [N]$ with $|A| \geq N\exp(-C\sqrt{\log N})$ so that there does not exist a nonconstant quadratic polynomial P so that $P(0)$, $P(1)$, $P(2)$, $P(3)$, $P(4) \in A$.

2.6 Graph Counting and Removal Lemmas

In this section, we generalize the triangle counting lemma from triangles to other graphs and discuss applications.

Graph Counting Lemma

Let us first illustrate the technique for K_4. Similar to the triangle counting lemma, we embed the vertices of K_4 one at a time. At each stage we ensure that many eligible vertices remain for the yet to be embedded vertices.

Proposition 2.6.1 (K_4 counting lemma)
Let $0 < \varepsilon < 1$. Let $X_1, \ldots, X_4$ be vertex subsets of a graph G such that (X_i, X_j) is ε-regular with edge-density $d_{ij} := d(X_i, X_j) \geq 3\sqrt{\varepsilon}$ for each pair $i < j$. Then the number of quadruples $(x_1, x_2, x_3, x_4) \in X_1 \times X_2 \times X_3 \times X_4$ such that $x_1x_2x_3x_4$ is a clique in G is

$$\geq (1 - 3\varepsilon)(d_{12} - 3\varepsilon)(d_{13} - \varepsilon)(d_{14} - \varepsilon)(d_{23} - \varepsilon)(d_{24} - \varepsilon)(d_{34} - \varepsilon)\, |X_1|\, |X_2|\, |X_3|\, |X_4|\, .$$

Proof. We repeatedly apply the following statement, which is a simple consequence of the definition of ε-regularity (and a small extension of Lemma 2.2.3):

Given an ε-regular pair (X, Y), and $B \subseteq Y$ with $|B| \geq \varepsilon\, |Y|$, the number of vertices in X with $< (d(X,Y) - \varepsilon)\, |B|$ neighbors in B is $< \varepsilon\, |X|$.

The number of vertices X_1 with $\geq (d_{1i} - \varepsilon)\, |X_i|$ neighbors in X_i for each $i = 2, 3, 4$ is $\geq (1 - 3\varepsilon)\, |X_1|$. Fix a choice of such an $x_1 \in X_1$. For each $i = 2, 3, 4$, let Y_i be the neighbors of x_1 in X_i, so that $|Y_i| \geq (d_{1i} - \varepsilon)\, |X_i|$.

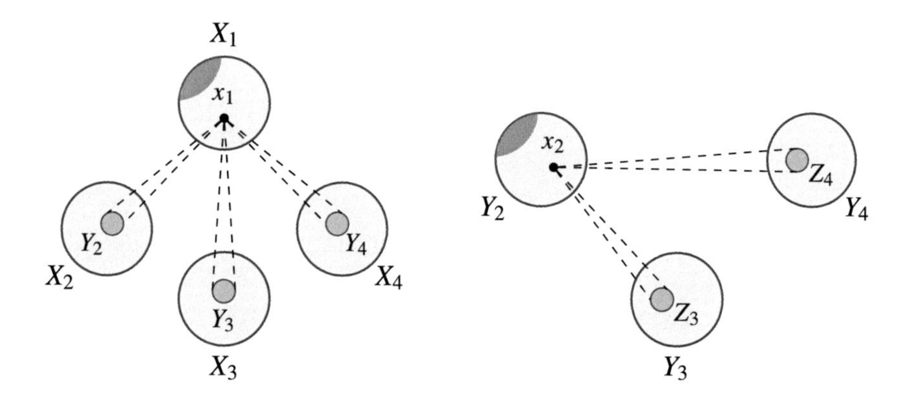

The number of vertices in Y_2 with $\geq (d_{2i} - \varepsilon)\, |Y_i|$ common neighbors in Y_i for each $i = 3, 4$ is $\geq |Y_2| - 2\varepsilon\, |X_2| \geq (d_{12} - 3\varepsilon)\, |X_2|$. Fix a choice of such an $x_2 \in Y_2$. For each $i = 3, 4$, let Z_i be the neighbors of x_2 in Y_i.

For each $i = 3, 4$, $|Z_i| \geq (d_{1i} - \varepsilon)(d_{2i} - \varepsilon)\, |X_i| \geq \varepsilon\, |X_i|$, and so

$$\begin{aligned} e(Z_3, Z_4) &\geq (d_{34} - \varepsilon)\, |Z_3|\, |Z_4| \\ &\geq (d_{34} - \varepsilon) \cdot (d_{13} - \varepsilon)(d_{23} - \varepsilon)\, |X_3| \cdot (d_{14} - \varepsilon)(d_{24} - \varepsilon)\, |X_4|\, . \end{aligned}$$

Any edge between Z_3 and Z_4 forms a K_4 together with x_1 and x_2. Multiplying the preceding quantity with the earlier lower bounds on the number of choices of x_1 and x_2 gives the result. □

The same strategy works more generally for counting any graph. To find copies of H, we embed vertices of H one at a time.

Theorem 2.6.2 (Graph counting lemma)

For every graph H and real $\delta > 0$, there exists an $\varepsilon > 0$ such that the following is true.

Let G be a graph, and $X_i \subseteq V(G)$ for each $i \in V(H)$ such that for each $ij \in E(H)$, (X_i, X_j) is an ε-regular pair with edge density $d_{ij} := d(X_i, X_j) \geq \delta$. Then the number of graph homomorphisms $H \to G$ where each $i \in V(H)$ is mapped to X_i is

$$\geq (1-\delta) \prod_{ij \in E(H)} (d_{ij} - \delta) \prod_{i \in V(H)} |X_i|.$$

***Remark* 2.6.3.** (a) For a fixed H, as $|X_i| \to \infty$ for each i, all but a negligible fraction of such homomorphisms from H are injective (i.e., yielding a copy of H as a subgraph).

(b) It is useful (and in fact equivalent) to think about the setting where G is a multipartite graph with parts X_i, as illustrated in the following diagram.

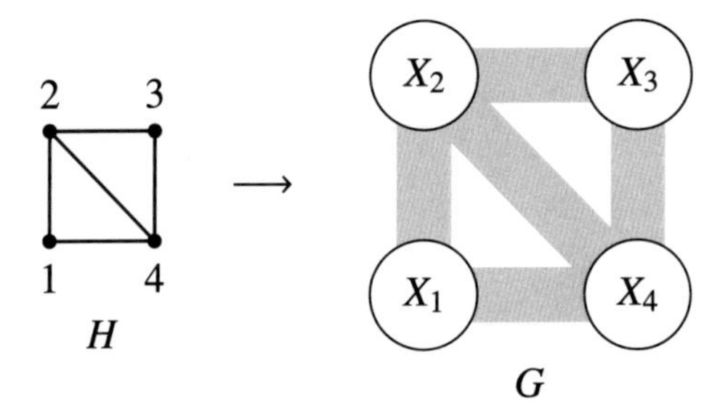

In the multipartite setting, we see that the graph counting lemma can be adapted to variants such as counting induced copies of H. Indeed, an induced copy of H is the same as a $v(H)$-clique in an auxiliary graph G' obtained by replacing the bipartite graph in G between X_i and X_j by its complementary bipartite graph between X_i and X_j for each $ij \notin E(H)$.

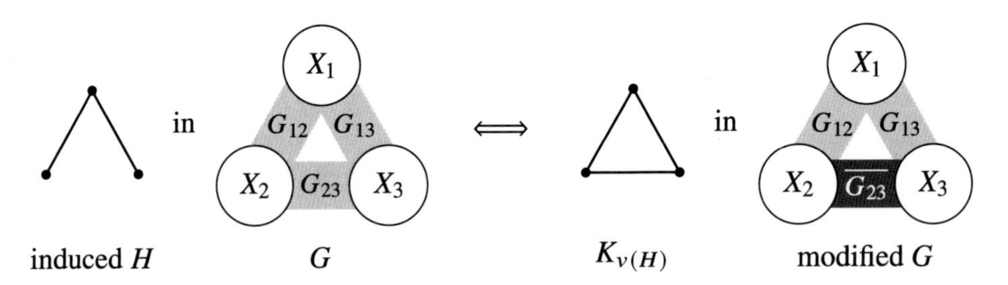

(c) We will see a different proof in Section 4.5 using the language of graphons. There, instead of embedding H one vertex at a time, we compare the density of H and $H \setminus \{e\}$.

We establish the following stronger statement, which has the additional advantage that one can choose the regularity parameter ε to depend on the maximum degree of H rather than H itself. You may wish to skip reading the proof, as it is notationally rather heavy. The main ideas were already illustrated in the K_4 counting lemma.

Theorem 2.6.4 (Graph counting lemma)
Let H be a graph with maximum degree $\Delta \geq 1$ and $c(H)$ connected components. Let $\varepsilon > 0$. Let G be a graph. Let $X_i \subseteq V(G)$ for each $i \in V(H)$. Suppose that for each $ij \in E(H)$, (X_i, X_j) is an ε-regular pair with edge density $d_{ij} := d(X_i, X_j) \geq (\Delta + 1)\varepsilon^{1/\Delta}$. Then the number of graph homomorphisms $H \to G$ where each $i \in V(H)$ is mapped to X_i is

$$\geq (1 - \Delta\varepsilon)^{c(H)} \prod_{ij \in E(H)} (d_{ij} - \Delta\varepsilon^{1/\Delta}) \cdot \prod_{i \in V(H)} |X_i| .$$

Furthermore, if $|X_i| \geq v(H)/\varepsilon$ for each i, then there exists such a homomorphism $H \to G$ that is injective (i.e., an embedding of H as a subgraph).

Proof. Let us order and label the vertices of H by $1, \ldots, v(H)$ arbitrarily. We will select vertices $x_1 \in X_1, x_2 \in X_2, \ldots$ in order. The idea is to always make sure that they have enough neighbors in G so that there are many ways to continue the embedding of H. We say that a partial embedding $x_1, \ldots, x_{s-1}$ (here *partial embedding* means that $x_i x_j \in E(G)$ whenever $ij \in E(H)$ for all the x_i's chosen so far) is *abundant* if for each $j \geq s$, the number of valid extensions $x_j \in X_j$ (meaning that $x_i x_j \in E(G)$ whenever $i < s$ and $ij \in E(H)$) is $\geq |X_j| \prod_{i<s: ij \in E(H)} (d_{ij} - \varepsilon)$.

For each $s = 1, 2, \ldots, v(H)$ in order, suppose we have already fixed an abundant partial embedding $x_1, \ldots, x_{s-1}$. For each $j \geq s$, let

$$Y_j = \{x_j \in X_j : x_i x_j \in E(G) \text{ whenever } i < s \text{ and } ij \in E(H)\}$$

be the set of valid extensions of the jth vertex in X_j given the partial embeddings of $x_1, \ldots, x_{s-1}$, so that the abundance hypothesis gives

$$|Y_j| \geq |X_j| \prod_{\substack{i<s \\ ij \in E(H)}} (d_{ij} - \varepsilon) \geq (\varepsilon^{1/\Delta})^{|\{i<s: ij \in E(H)\}|} |X_j| \geq \varepsilon |X_j|.$$

Thus, as in the proof of Proposition 2.6.1 for K_4, the number of choices $x_s \in X_s$ that would extend $x_1, \ldots, x_{s-1}$ to an abundant partial embedding is

$$\begin{aligned} &\geq |Y_s| - |\{i > s : si \in E(H)\}| \, \varepsilon \, |X_s| \\ &\geq |X_s| \prod_{\substack{i<s \\ is \in E(H)}} (d_{ij} - \varepsilon) - |\{i > s : si \in E(H)\}| \, \varepsilon \, |X_s| . \qquad (\dagger) \end{aligned}$$

If none of $1, \ldots, s-1$ is a neighbor of s in H, then the first term in $(\dagger)$ is $|X_s|$, and so

$$(\dagger) \geq (1 - \Delta\varepsilon) \, |X_s| .$$

Otherwise, we can absorb the second term into the product and obtain

$$(\dagger) \geq |X_s| \prod_{\substack{i<s \\ is \in E(H)}} (d_{ij} - \varepsilon) - (\Delta - 1)\varepsilon \, |X_s| \geq |X_s| \prod_{\substack{i<s \\ is \in E(H)}} (d_{ij} - \Delta\varepsilon^{1/\Delta}).$$

Fix such a choice of x_s. And now we move onto embedding the next vertex x_{s+1}.

Multiplying together these lower bounds for the number of choices of each x_s over all $s = 1, \ldots, v(H)$, we obtain the lower bound on the number of homomorphisms $H \to G$.

Finally, note that in both cases $(\dagger) \geq \varepsilon |X_s|$, and so if $|X_s| \geq v(H)/\varepsilon$, then $(\dagger) \geq v(H)$ and so we can choose each x_s to be distinct from the previously embedded vertices $x_1, \ldots, x_{s-1}$, thereby yielding an injective homomorphism. □

Graph Removal Lemma

As an application, we have the following graph removal lemma, generalizing the triangle removal lemma, Theorem 2.3.1. The proof is basically the same as Theorem 2.3.1 except with the previous graph counting lemma taking the role of the triangle counting lemma, so we will not repeat the proof here.

Theorem 2.6.5 (Graph removal lemma)
For every graph H and constant $\varepsilon > 0$, there exists a constant $\delta = \delta(H, \varepsilon) > 0$ such that every n-vertex graph G with fewer than $\delta n^{v(H)}$ copies of H can be made H-free by removing fewer than εn^2 edges.

The next exercise asks you to show that, if H is bipartite, then one can prove the H-removal lemma without using regularity, and thereby get a much better bound.

Exercise 2.6.6 (Removal lemma for bipartite graphs with polynomial bounds). Prove that for every bipartite graph H, there is a constant C such that for every $\varepsilon > 0$, every n-vertex graph with fewer than $\varepsilon^C n^{v(H)}$ copies of H can be made H-free by removing at most εn^2 edges.

Erdős–Stone–Simonovits Theorem

As another application, let us give a different proof of the Erdős–Stone–Simonovits theorem from Section 1.5, restated in what follows, which gives the asymptotics (up to a $+o(n^2)$ error term) for $\mathrm{ex}(n, H)$, the maximum number of edges in an n-vertex H-free graph. We saw a proof in Section 1.5 using supersaturation and the hypergraph KST theorem. The proof here follows the partition-clean-count strategy in Remark 2.3.2 combined with an application of Turán's theorem. A common feature of many regularity applications is that they "boost" an exact extremal graph theoretic result (e.g., Turán's theorem) to an asymptotic result involving more complex derived structures (e.g., from the existence of a copy of K_r to embedding a complete r-partite graph).

Theorem 2.6.7 (Erdős–Stone–Simonovits theorem)
Fix graph H with at least one edge. Then

$$\mathrm{ex}(n, H) = \left(1 - \frac{1}{\chi(H) - 1} + o(1)\right) \frac{n^2}{2}.$$

Proof. Fix $\varepsilon > 0$. Let G be any n-vertex graph with at least $\left(1 - \frac{1}{\chi(H)-1} + \varepsilon\right) \frac{n^2}{2}$ edges. The theorem is equivalent to the claim that for $n = n(\varepsilon, H)$ sufficiently large, G contains H as a subgraph.

Apply the graph regularity lemma to obtain an η-regular partition $V(G) = V_1 \cup \cdots \cup V_m$ for some sufficiently small $\eta > 0$ only depending on ε and H, to be decided later. Then the number m of parts is also bounded for fixed H and ε.

Remove an edge $(x, y) \in V_i \times V_j$ if

(a) (V_i, V_j) is not η-regular, or
(b) $d(V_i, V_j) < \varepsilon/8$, or
(c) $\min\{|V_i|, |V_j|\} < \varepsilon n/(8m)$.

Then, as in Theorem 2.3.1, the number of edges in (a) is $\leq \eta n^2 \leq \varepsilon n^2/8$, the number of edges in (b) is $< \varepsilon n^2/8$, and the number of edges in (c) is $< m\varepsilon n^2/(8m) \leq \varepsilon n^2/8$. Thus, the total number of edges removed is $\leq (3/8)\varepsilon n^2$. After removing all these edges, the resulting graph G' has still has $> \left(1 - \frac{1}{\chi(H)-1} + \frac{\varepsilon}{4}\right) \frac{n^2}{2}$ edges.

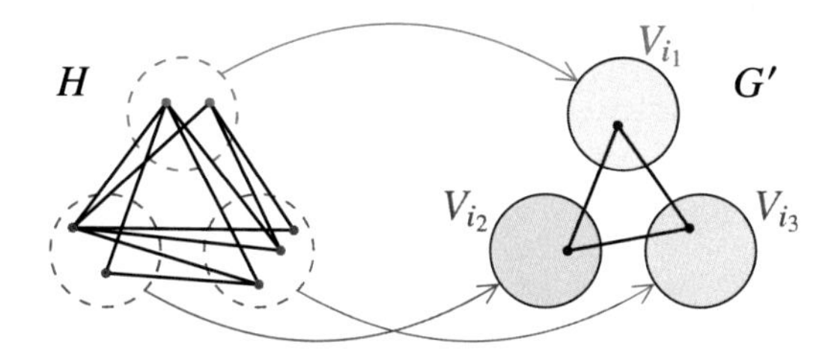

By Turán's theorem (Corollary 1.2.6), G' contains a copy of $K_{\chi(H)}$. Suppose that the $\chi(H)$ vertices of this $K_{\chi(H)}$ land in $V_{i_1}, \ldots, V_{i_{\chi(H)}}$ (allowing repeated indices). Since each pair of these sets is η-regular, has edge density $\geq \varepsilon/8$, and has size $\geq \varepsilon n/(8m)$, by applying the graph counting lemma, Theorem 2.6.2, we see that as long as η is sufficiently small in terms of ε and H, and n is sufficiently large, there exists an injective embedding of H into G' where the vertices of H in the rth color class are mapped into V_{i_r}. So, G contains H as a subgraph. □

2.7 Exercises on Applying Graph Regularity

The regularity method can be difficult at first to grasp conceptually. The following exercises are useful for gaining familiarity in applying the regularity lemma. For these exercises, you are welcome to use the equitable form of the graph regularity lemma (Theorem 2.1.20), which is more convenient to apply.

Exercise 2.7.1 (Ramsey–Turán).

(a) Show that for every $\varepsilon > 0$, there exists $\delta > 0$ such that every n-vertex K_4-free graph with at least $(\frac{1}{8} + \varepsilon)n^2$ edges contains an independent set of size at least δn.
(b) Show that for every $\varepsilon > 0$, there exists $\delta > 0$ such that every n-vertex K_4-free graph with at least $(\frac{1}{8} - \delta)n^2$ edges and independence number at most δn can be made bipartite by removing at most εn^2 edges.

Exercise 2.7.2 (Nearly homogeneous subset). Show that for every H and $\varepsilon > 0$ there exists $\delta > 0$ such that every graph on n vertices without an induced copy of H contains an induced subgraph on at least δn vertices whose edge density is at most ε or at least $1 - \varepsilon$.

Exercise 2.7.3 (Ramsey numbers of bounded degree graphs). Show that for every Δ there exists a constant C_Δ so that if H is a graph with maximum degree at most Δ, then every 2-edge-coloring of a complete graph on at least $C_\Delta v(H)$ vertices contains a monochromatic copy of H.

Exercise 2.7.4 (Counting H-free graphs).

(a) Show that the number of n-vertex triangle-free graphs is $2^{(1/4+o(1))n^2}$.
(b) More generally, show that for any fixed graph H, the number of n-vertex H-free graphs is $2^{\mathrm{ex}(n,H)+o(n^2)}$.

Exercise 2.7.5* (Induced Ramsey). Show that for every graph H there is some graph G such that if the edges of G are colored with two colors, then some induced subgraph of G is a monochromatic copy of H.

Exercise 2.7.6* (Finding a degree-regular subgraph). Show that for every $\alpha > 0$, there exists $\beta > 0$ such that every graph on n vertices with at least αn^2 edges contains a d-regular subgraph for some $d \geq \beta n$. (Here *d-regular* refers to every vertex having degree d.)

2.8 Induced Graph Removal and Strong Regularity

Recall that H is an ***induced subgraph*** of G if one can obtain H from G by deleting vertices from G (but you are not allowed to simply remove edges from G). We say that G is ***induced H-free*** if G contains no induced subgraph isomorphic to H. (See Notation and Conventions.)

The following removal lemma for induced subgraphs is due to Alon, Fischer, Krivelevich, and Szegedy (2000).

Theorem 2.8.1 (Induced graph removal lemma)
For any graph H and $\varepsilon > 0$, there exists $\delta > 0$ such that if an n-vertex graph has fewer than $\delta n^{v(H)}$ induced copies of H, then it can be made induced H-free by adding and/or deleting fewer than εn^2 edges.

Remark* 2.8.2.** Given two graphs on the same vertex set, the minimum number of edges that one needs to add/delete to obtain the second graph from the first graph is called the ***edit distance between the two graphs. The induced graph removal lemma can be rephrased as saying that every graph with few induced copies of H is close in edit distance to an induced H-free graph.

Unlike the previous graph removal lemma, for the induced version, it is important that we allow both adding and deleting edges. The statement would be false if we only allow edge deletion but not addition. For example, suppose $G = K_n \setminus K_3$ (i.e., a complete graph on n vertices with three edges of a single triangle removed). If H is an empty graph on three vertices, then G has exactly one induced copy of H, but G cannot be made induced H-free by only deleting edges.

To see why the earlier proof of the graph removal lemma (Theorem 2.6.5) does not apply in a straightforward way to prove the induced graph removal lemma, let us attempt to follow the earlier strategy and see where things go wrong.

First, we apply the graph regularity lemma. Then we need to *clean* up the graph. In the induced graph removal lemma, edges and nonedges play symmetric roles. We can handle low-density pairs (edge density less than ε) by removing edges between such pairs. Naturally, for the induced graph removal lemma, we also need to handle high-density pairs (density more than $1 - \varepsilon$), and we can add all the edges between such pairs. However, it is not clear what to do with irregular pairs. Earlier, we just removed all edges between irregular pairs. The problem is that this may create many induced copies of H that were not present previously (see the following illustration). Likewise, we cannot simply add all edges between irregular pairs.

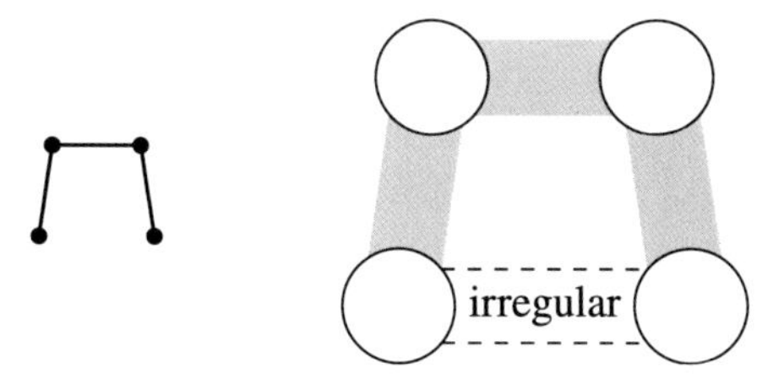

Perhaps we can always find a regularity partition without irregular pairs? Unfortunately, this is false, as shown in Exercise 2.1.24. One must allow for the possibility of irregular pairs.

Strong Regularity Lemma

We will iterate the regularity partitioning lemma to obtain a stronger form of the regularity lemma. Recall the energy $q(\mathcal{P})$ of a partition (Definition 2.1.10) as the mean-squared edge density between parts.

Theorem 2.8.3 (Strong regularity lemma)
For any sequence of constants $\varepsilon_0 \geq \varepsilon_1 \geq \varepsilon_2 \geq \ldots > 0$, there exists an integer M so that every graph has two vertex partitions $\mathcal{P}$ and $\mathcal{Q}$ so that

(a) $\mathcal{Q}$ refines $\mathcal{P}$,
(b) $\mathcal{P}$ is ε_0-regular and $\mathcal{Q}$ is $\varepsilon_{|\mathcal{P}|}$-regular,
(c) $q(\mathcal{Q}) \leq q(\mathcal{P}) + \varepsilon_0$, and
(d) $|\mathcal{Q}| \leq M$.

***Remark* 2.8.4.** One should think of the sequence $\varepsilon_1, \varepsilon_2, \ldots$ as rapidly decreasing. This strong regularity lemma outputs a refining pair of partitions $\mathcal{P}$ and $\mathcal{Q}$ such that $\mathcal{P}$ is regular, $\mathcal{Q}$ is *extremely* regular, and $\mathcal{P}$ and $\mathcal{Q}$ are close to each other (as captured by $q(\mathcal{P}) \leq q(\mathcal{Q}) \leq q(\mathcal{P}) + \varepsilon_0$; see Lemma 2.8.7 in what follows). A key point here is that we demand $\mathcal{Q}$ to be extremely regular relative to the number of parts of $\mathcal{P}$. The more parts $\mathcal{P}$ has, the more regular $\mathcal{Q}$ should be.

Proof. We repeatedly apply the following version of Szemerédi's regularity lemma:

Theorem 2.1.19 (restated): For all $\varepsilon > 0$ and k, there exists an integer $M_0(k, \varepsilon)$ so that for all partitions $\mathcal{P}$ of $V(G)$ with at most k parts, there exists a refinement $\mathcal{P}'$ of $\mathcal{P}$ with each part in $\mathcal{P}$ refined into $\leq M_0$ parts so that $\mathcal{P}'$ is ε-regular.

By iteratively applying the preceding regularity partition, we obtain a sequence of partitions $\mathcal{P}_0, \mathcal{P}_1, \ldots$ of $V(G)$ starting with $\mathcal{P}_0 = \{V(G)\}$ being the trivial partition. Each $\mathcal{P}_{i+1}$ is $\varepsilon_{|\mathcal{P}_i|}$-regular and refines $\mathcal{P}_i$. The regularity lemma guarantees that we can have $|\mathcal{P}_{i+1}| \le M_0(|\mathcal{P}_i|, \varepsilon_{|\mathcal{P}_i|})$.

Since $0 \le q(\cdot) \le 1$, there exists $i \le \varepsilon_0^{-1}$ so that $q(\mathcal{P}_{i+1}) \le q(\mathcal{P}_i) + \varepsilon_0$. Then, setting $\mathcal{P} = \mathcal{P}_i$ and $Q = \mathcal{P}_{i+1}$ satisfies the desired requirements. Indeed, the number of parts of Q is bounded by a function of the sequence $(\varepsilon_0, \varepsilon_1, \ldots)$ since there are a bounded number of iterations and each iteration produced a refining partition with a bounded number of parts. □

Remark* 2.8.5** (Bounds in the strong regularity lemma)**.** The bound on M produced by the proof depends on the sequence $(\varepsilon_0, \varepsilon_1, \ldots)$. In the following application, we use $\varepsilon_i = \varepsilon_0/\text{poly}(i)$. Then the size of M is comparable to applying M_0 to ε_0 in succession $1/\varepsilon_0$ times. Note that M_0 is a tower function, and this makes M a tower function iterated i times. This iterated tower function is called the ***wowzer function:

$$\textbf{wowzer}(\boldsymbol{k}) := \text{tower}(\text{tower}(\cdots(\text{tower}(2))\cdots))$$

(with 2 applications of tower). The wowzer function is one step up from the tower function in the Ackermann hierarchy. It grows extremely quickly.

***Remark* 2.8.6** (Equitability)**.** We can further ensure that the parts have nearly equal size. This can be done by adapting the ideas sketched in the proof sketch of Theorem 2.1.20.

The following lemma explains the significance of the inequality $q(Q) \le q(\mathcal{P}) + \varepsilon$ from earlier.

Lemma 2.8.7 (Energy and approximation)
Let $\mathcal{P}$ and Q both be vertex partitions of a graph G, with Q refining $\mathcal{P}$. For each $x \in V(G)$, write V_x for the part of $\mathcal{P}$ that x lies in and W_x for the part of Q that x lies in. If

$$q(Q) \le q(\mathcal{P}) + \varepsilon^3,$$

then

$$\left|d(V_x, V_y) - d(W_x, W_y)\right| \le \varepsilon$$

for all but εn^2 pairs $(x, y) \in V(G)^2$.

Proof. Let $x, y \in V(G)$ be chosen uniformly at random. As in the proof of Lemma 2.1.11, we have $q(\mathcal{P}) = \mathbb{E}[Z_{\mathcal{P}}^2]$, where $Z_{\mathcal{P}} = d(V_x, V_y)$. Likewise, $q(Q) = \mathbb{E}[Z_Q^2]$, where $Z_Q = d(W_x, W_y)$.

We have

$$q(Q) - q(\mathcal{P}) = \mathbb{E}[Z_Q^2] - \mathbb{E}[Z_{\mathcal{P}}^2] = \mathbb{E}[(Z_Q - Z_{\mathcal{P}})^2],$$

where the final step is a "Pythagorean identity."

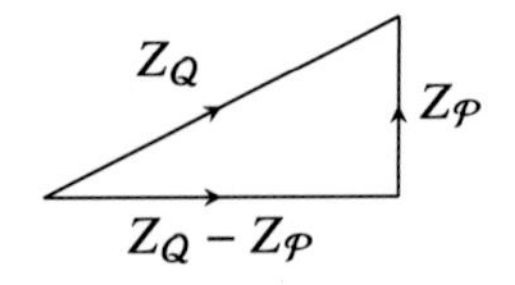

Indeed, the identity $\mathbb{E}[Z_Q^2] - \mathbb{E}[Z_{\mathcal{P}}^2] = \mathbb{E}[(Z_Q - Z_{\mathcal{P}})^2]$ is equivalent to $\mathbb{E}[Z_{\mathcal{P}}(Z_Q - Z_{\mathcal{P}})] = 0$, which is true since as x and y each vary over their own parts of $\mathcal{P}$, the expression $Z_Q - Z_{\mathcal{P}}$ averages to zero.

So, $q(Q) \le q(\mathcal{P}) + \varepsilon^3$ is equivalent to $\mathbb{E}[(Z_Q - Z_{\mathcal{P}})^2] \le \varepsilon^3$, which in turn implies, by Markov's inequality, that $\mathbb{P}(|Z_Q - Z_{\mathcal{P}}| > \varepsilon) \le \varepsilon$, which is the same as the desired conclusion. □

Exercise 2.8.8. Let $0 < \varepsilon < 1$. Using the notation of Lemma 2.8.7, show that if $|d(V_x, V_y) - d(W_x, W_y)| \le \varepsilon$ for all but εn^2 pairs $(x, y) \in V(G)^2$, then $q(Q) \le q(\mathcal{P}) + 2\varepsilon$.

We now deduce the following form of the strong regularity lemma, which considers only select subsets of vertex parts but does not require irregular pairs.

Theorem 2.8.9 (Strong regularity lemma)
For any sequences of constants $\varepsilon_0 \ge \varepsilon_1 \ge \varepsilon_2 \ge \cdots > 0$, there exists a constant $\delta > 0$ so that every n-vertex graph has an equitable vertex partition $V_1 \cup \cdots \cup V_k$ and a subset $W_i \subseteq V_i$ for each i satisfying

(a) $|W_i| \ge \delta n$,
(b) (W_i, W_j) is ε_k-regular for all $1 \le i \le j \le k$, and
(c) $\left|d(V_i, V_j) - d(W_i, W_j)\right| \le \varepsilon_0$ for all but $< \varepsilon_0 k^2$ pairs $(i, j) \in [k]^2$.

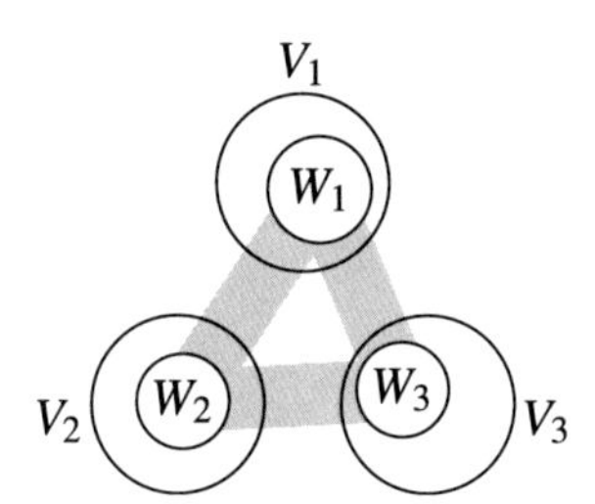

***Remark* 2.8.10.** It is significant that *all* (rather than nearly all) pairs (W_i, W_j) are regular. We will need this fact in the applications that follow.

Proof sketch. Here we show how to prove a slightly weaker result where $i \le j$ in (b) is replaced by $i < j$. In other words, this proof does not promise that each W_i is ε_k-regular. To obtain the stronger conclusion as stated (requiring each W_i to be regular with itself), we can adapt the ideas in Exercise 2.1.27. We omit the details.

By decreasing the ε_i's if needed (we can do this since a smaller sequence of ε_i's yields a stronger conclusion), we may assume that $\varepsilon_i \le 1/(10i^2)$ and $\varepsilon_i \le \varepsilon_0/4$ for every $i \ge 1$.

Let us apply the strong regularity lemma, Theorem 2.8.3, with equitable partitions (see Remark 2.8.6). That is, we have (with the simplifying assumption that all partitions are exactly equitable, to avoid unimportant technicalities):

- an equitable ε_0-regular partition $\mathcal{P} = \{V_1, \ldots, V_k\}$ of $V(G)$ and
- an equitable ε_k-regular partition Q refining $\mathcal{P}$

satisfying

- $q(Q) \le q(\mathcal{P}) + \varepsilon_0^3/8$, and
- $|Q| \le M = M(\varepsilon_0, \varepsilon_1, \ldots)$.

Inside each part V_i, let us choose a part W_i of Q uniformly at random. Since $|Q| \le M$, the equitability assumption implies that each part of Q has size $\ge \delta n$ for some constant $\delta = \delta(\varepsilon_0, \varepsilon_1, \dots)$. So, (a) is satisfied.

Since Q is ε_k-regular, all but at most an ε_k-fraction of pairs of parts of Q are ε_k-regular. Summing over all $i < j$, using linearity of expectations, the expected number of pairs (W_i, W_j) that are not ε_k-regular is $\le \varepsilon_k k^2 \le 1/10$. It follows that with probability $\ge 9/10$, (W_i, W_j) is ε_k-regular for all $i < j$, so (b) is satisfied. (This argument ignores $i = j$, as mentioned at the beginning of the proof.)

Let X denote the number of pairs $(i, j) \in [k]^2$ with $|d(V_i, V_j) - d(W_i, W_j)| > \varepsilon_0$. Since $q(Q) \le q(\mathcal{P}) + (\varepsilon_0/2)^3$, by Lemma 2.8.7 and linearity of expectations, $\mathbb{E}X \le (\varepsilon_0/2)k^2$. So, by Markov's inequality, $X \le \varepsilon_0 k^2$ with probability $\ge 1/2$, so that (c) is satisfied.

It follows that (a) and (b) are both satisfied with probability $\ge 1 - 1/10 - 1/2$. Therefore, there exist valid choices of W_i's. □

Induced Graph Removal Lemma

As with earlier regularity applications, we follow the partition-clean-count recipe from Remark 2.3.2.

Proof of the induced graph removal lemma (Theorem 2.8.1). Apply Theorem 2.8.9 to obtain a **partition** $V_1 \cup \cdots \cup V_k$ of the vertex set of the graph, along with $W_k \subseteq V_k$, so that:

(a) (W_i, W_j) is ε'-regular for every $i \le j$, with some sufficiently small constant $\varepsilon' > 0$ depending on ε and H,

(b) $|d(V_i, V_j) - d(W_i, W_j)| \le \varepsilon/8$ for all but $< \varepsilon k^2/8$ pairs $(i, j) \in [k]^2$, and

(c) $|W_i| \ge \delta_0 n$, for some constant δ_0 depending only on ε and H.

Now we **clean** the graph. For each pair $i \le j$ (including $i = j$),

- if $d(W_i, W_j) \le \varepsilon/8$, then remove all edges between (V_i, V_j), and
- if $d(W_i, W_j) \ge 1 - \varepsilon/8$, then add all edges between (V_i, V_j).

Note that we are not simply adding or removing edges within each pair (W_i, W_j), but rather all of (V_i, V_j). To bound the number of edges added or deleted, recall (b) from the previous paragraph. If $d(W_i, W_j) \le \varepsilon/8$ and $|d(V_i, V_j) - d(W_i, W_j)| \le \varepsilon/8$, then $d(V_i, V_j) \le \varepsilon/4$, and the number of edges in all such (V_i, V_j) is at most $\varepsilon n^2/4$. Likewise for $d(W_i, W_j) \ge 1 - \varepsilon/8$. For the remaining $< \varepsilon k^2/8$ pairs (i, j) not satisfying $|d(V_i, V_j) - d(W_i, W_j)| \le \varepsilon/8$, the total number of edges among all such pairs is at most $\varepsilon n^2/8$. Altogether, we added/deleted $< \varepsilon n^2$ edges from G. Call the resulting graph G'. There are no irregular pairs (W_i, W_j) for us to worry about.

It remains to show that G' is induced H-free. Suppose otherwise. Let us **count** induced copies of H in G as in the proof of the graph removal lemma, Theorem 2.6.5. We have some induced copy of H in G', with each vertex $v \in V(H)$ embedded in $V_{\phi(v)}$ for some $\phi \colon V(H) \to [k]$.

Consider a pair of distinct vertices u, v of H. If $uv \in E(H)$, there must be an edge in G' between $V_{\phi(u)}$ and $V_{\phi(v)}$. (Here $\phi(u)$ and $\phi(v)$ are not necessarily different.) So we must not have deleted all the edges in G between $V_{\phi(u)}$ and $V_{\phi(v)}$ in the cleaning step. By the cleaning algorithm presented earlier, this means that $d_G(W_i, W_j) > \varepsilon/8$. Likewise, if $uv \notin E(H)$ for any pair of distinct $u, v \in V(H)$, we have $d_G(W_i, W_j) < 1 - \varepsilon/8$.

Since (W_i, W_j) is ε'-regular in G for every $i \leq j$, provided that ε' is small enough (in terms of ε and H), the graph counting lemma (Theorem 2.6.2 with the induced variation as in Remark 2.6.3(b)) applied to G gives

$$\#\text{ induced copies of } H \text{ in } G \geq (1-\varepsilon)\left(\frac{\varepsilon}{10}\right)^{\binom{v(H)}{2}} (\delta_0 n)^{v(H)} =: \delta n^{v(H)}$$

(recall $|W_i| \geq \delta_0 n$). Setting δ as earlier, this contradicts the hypothesis that G has $< \delta n^{v(H)}$ copies of H. Thus, G' must be induced H-free. □

Infinite Graph Removal Lemma

Finally, let us prove a graph removal lemma with an infinite number of forbidden induced subgraphs (Alon and Shapira 2008). Given a (possibly infinite) set $\mathcal{H}$ of graphs, we say that G is ***induced $\mathcal{H}$-free*** if G is induced H-free for every $H \in \mathcal{H}$.

Theorem 2.8.11 (Infinite graph removal lemma)
For each (possibly infinite) set of graphs $\mathcal{H}$ and $\varepsilon > 0$, there exist h_0 and $\delta > 0$ so that if G is an n-vertex graph with fewer than $\delta n^{v(H)}$ induced copies of H for every $H \in \mathcal{H}$ with at most h_0 vertices, then G can be made induced $\mathcal{H}$-free by adding/removing fewer than εn^2 edges.

***Remark* 2.8.12.** The presence of h_0 may seem a bit strange at first. In the next section, we will see a reformulation of this theorem in the language of property testing, where h_0 comes up naturally.

Proof. The proof is mostly the same as the proof of the induced graph removal lemma that we just saw. The main tricky issue here is how to choose the regularity parameter ε' for every pair (W_i, W_j) in condition (a) of the earlier proof. Previously, we did not use the full strength of Theorem 2.8.9, which allowed ε' to depend on k, but now we are going to use it. Recall that we had to make sure that this ε' was chosen to be small enough for the H-counting lemma to work. Now that there are possibly infinitely many graphs in $\mathcal{H}$, we cannot naively choose ε' to be sufficiently small. The main point of the proof is to reduce the problem to a finite subset of $\mathcal{H}$ for each k.

Define a ***template*** T to be an edge-coloring of the looped k-clique (i.e., a complete graph on k vertices along with a loop at every vertex) where each edge is colored by one of {white, black, gray}. We say that a graph H is ***compatible*** with a template T if there exists a map $\phi\colon V(H) \to V(T)$ such that for every distinct pair u, v of vertices of H:

- if $uv \in E(H)$, then $\phi(u)\phi(v)$ is colored black or gray in T; and
- if $uv \notin E(H)$, then $\phi(u)\phi(v)$ is colored white or gray in T.

That is, a black edge in a template means an edge of H, a white edge means a nonedge of H, and a gray edge is a wildcard. An example is shown in the following graph:

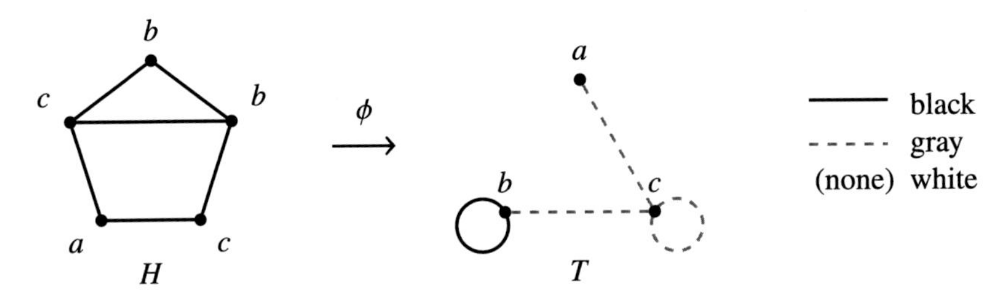

As another example, every graph is compatible with every completely gray template.

For every template T, pick some ***representative*** $H_T \in \mathcal{H}$ compatible with T, as long as such a representative exists (and ignore T otherwise). A graph in $\mathcal{H}$ is allowed to be the representative of more than one template. Let $\mathcal{H}_k$ be a set of all $H \in \mathcal{H}$ that arise as the representative of some k-vertex template. Note that $\mathcal{H}_k$ is finite since there are finitely many k-vertex templates. We can pick each $\varepsilon_k > 0$ to be small enough so that the conclusion of the counting step later can be guaranteed for all elements of $\mathcal{H}_k$.

Now we proceed nearly identically as in the proof of the induced removal lemma, Theorem 2.8.1, that we just saw. In applying Theorem 2.8.9 to obtain the partition $V_1 \cup \cdots \cup V_k$ and finding $W_i \subseteq V_i$, we ensure the following condition instead of the earlier (a):

(a) (W_i, W_j) is ε_k-regular for every $i \le j$.

We set h_0 to be the maximum number of vertices of a graph in $\mathcal{H}_k$.

Now we do the cleaning step. Along the way, we create a k-vertex template T with vertex set $[k]$ corresponding to the parts $\{V_1, \ldots, V_k\}$ of the partition. For each $1 \le i \le j \le n$,

- if $d(W_i, W_j) \le \varepsilon/4$, then remove all edges between (V_i, V_j) from G, and color the edge ij in template T white;
- if $d(W_i, W_j) \ge 1 - \varepsilon/4$, then add all edges between (V_i, V_j), and color the edge ij in template T black;
- otherwise, color the edge in ij in template T gray.

Finally, suppose some induced $H \in \mathcal{H}$ remains in G'. Due to our cleaning procedure, H must be compatible with the template T. Then the representative $H_T \in \mathcal{H}_k$ of T is a graph on at most h_0 vertices, and furthermore, the counting lemma guarantees that, provided $\varepsilon_k > 0$ is small enough (subject to a finite number of prechosen constraints, one for each element of $\mathcal{H}_k$), the number of copies of H_T in G is $\ge \delta n^{v(H_T)}$ for some constant $\delta > 0$ that only depends on ε and $\mathcal{H}$. This contradicts the hypothesis, and thus G' is induced $\mathcal{H}$-free. □

All the preceding techniques work nearly verbatim for a generalization to colored graphs.

Theorem 2.8.13 (Infinite edge-colored graph removal lemma)

For every $\varepsilon > 0$, positive integer r, and a (possibly infinite) set $\mathcal{H}$ of r-edge-colored graphs, there exists some h_0 and $\delta > 0$ such that if G is an r-edge-coloring of the complete graph on n vertices with $< \delta n^{v(H)}$ copies of H for every $\mathcal{H}$ with at most h_0 vertices, then G can be made $\mathcal{H}$-free by recoloring $< \varepsilon n^2$ edges (using the same palette of r colors throughout).

The induced graph removal lemma corresponds to the special case $r = 2$, with the two colors representing edges and nonedges respectively.

2.9 Graph Property Testing

We are given random query access to a very large graph. The graph may be too large for us to see every vertex or edge. What can we learn about the graph by sampling a constant number of vertices and the edges between them?

For example, we cannot distinguish two graphs if they only differ on a small number of vertices or edges. We also need some error tolerance.

A ***graph property*** $\mathcal{P}$ is simply a set of isomorphism classes of graphs. The graph properties that we usually encounter have some nice name and/or compact description, such as *triangle-free*, *planar*, and 3-*colorable*.

We say that an n-vertex graph G is ***ε-far*** from property $\mathcal{P}$ if one cannot change G into a graph in $\mathcal{P}$ by adding/deleting εn^2 edges.

The following theorem gives a straightforward algorithm, with a probabilistic guarantee, on testing triangle-freeness. It allows us to distinguish two types of graphs from each other:

triangle-free vs. **far from triangle-free**.

Theorem 2.9.1 (Triangle-freeness is testable)
For every $\varepsilon > 0$, there exists $K = K(\varepsilon)$ so that the following algorithm satisfies the probabilistic guarantees below.

Input: A graph G.

Algorithm: Sample K vertices from G uniformly at random without replacement. (If G has fewer than K vertices, then return the entire graph.) If G has no triangles among these K vertices, then output that G is triangle-free; else output that G is ε-far from triangle-free.

Probabilistic guarantees:

(a) If the input graph G is triangle-free, then the algorithm always correctly outputs that G is triangle-free;
(b) If the input graph G is ε-far from triangle-free, then with probability ≥ 0.99 the algorithm outputs that G is ε-far from triangle-free;
(c) We do not make any guarantees when the input graph is neither triangle-free nor ε-far from triangle-free.

Remark* 2.9.2.** This is an example of a ***one-sided tester, meaning that it always outputs a correct answer when G satisfies property $\mathcal{P}$ and only has a probabilistic guarantee when G does not satisfy property G. (In contrast, a two-sided tester would have probabilistic guarantees for both situations.)

For a one-sided tester, there is nothing special about the number 0.99 in (b). It can be any positive constant $\delta > 0$. If we run the algorithm m times, then the probability of success improves from $\geq \delta$ to $\geq 1 - (1 - \delta)^m$, which can be made arbitrarily close to 1 if we choose m large enough.

The probabilistic guarantee turns out to be essentially a rephrasing of the triangle removal lemma.

Proof. If the graph G is triangle-free, the algorithm clearly always outputs correctly. On the other hand, if G is ε-far from triangle-free, then by the triangle removal lemma (Theorem 2.3.1), G has $\geq \delta\binom{n}{3}$ triangles with some constant $\delta = \delta(\varepsilon) > 0$. If we sample three vertices from G uniformly at random, then they form a triangle with probability $\geq \delta$. And if run $K/3$ independent trials, then the probability that we see a triangle is $\geq 1 - (1-\delta)^{K/3}$, which is ≥ 0.99 as long as K is a sufficiently large constant (depending on δ, which in turn depends on ε).

In the algorithm as stated in the theorem, K vertices are sampled without replacement. Earlier we had K independent trials of picking a triple of vertices at random. But this difference hardly matters. We can couple the two processes by adding additional random vertices to the latter process until we see K distinct vertices. □

Just as the guarantee of the preceding algorithm is essentially a rephrasing of the triangle removal lemma, other graph removal lemmas can be rephrased as graph property testing theorems. For the infinite induced graph removal lemma, Theorem 2.8.11, we can rephrase the result in terms of graph property testing for hereditary properties.

A graph property $\mathcal{P}$ is ***hereditary*** if it is closed under vertex-deletion: if $G \in \mathcal{P}$, then every induced subgraph of G is in $\mathcal{P}$. Here are some examples of hereditary graph properties: *H-free*, *induced H-free*, *planar*, *3-colorable*, *perfect*. Every hereditary property $\mathcal{P}$ can be characterized as the set of induced $\mathcal{H}$-free graphs for some (possibly infinite) family of graphs $\mathcal{H}$; we can take $\mathcal{H} = \{H : H \notin \mathcal{P}\}$.

Theorem 2.9.3 (Every hereditary graph property is testable)

For every hereditary graph property $\mathcal{P}$, and constant $\varepsilon > 0$, there exists a constant $K = K(\mathcal{P}, \varepsilon)$ so that the following algorithm satisfies the probabilistic guarantees listed here.

Input: A graph G.

Algorithm: Sample K vertices from G uniformly at random without replacement and let H be the induced subgraph on these K vertices. If $H \in \mathcal{P}$, then output that G satisfies $\mathcal{P}$; else output that G is ε-far from $\mathcal{P}$.

Probabilistic guarantees:

(a) If the input graph G satisfies $\mathcal{P}$, then the algorithm always correctly outputs that G satisfies $\mathcal{P}$;

(b) If the input graph G is ε-far from $\mathcal{P}$, then with probability ≥ 0.99 the algorithm outputs that G is ε-far from $\mathcal{P}$;

(c) We do not make any guarantees when the input graph is neither in $\mathcal{P}$ nor ε-far from $\mathcal{P}$.

Proof. If $G \in \mathcal{P}$, then since $\mathcal{P}$ is hereditary, $H \in \mathcal{P}$, and so the algorithm always correctly outputs that $G \in \mathcal{P}$. So suppose G is ε-far from $\mathcal{P}$. Let $\mathcal{H}$ be such that $\mathcal{P}$ is the set of induced $\mathcal{H}$-free graphs. By the infinite induced graph removal lemma, there is some h_0 and $\delta > 0$ so that G has $\geq \delta\binom{n}{v(H)}$ copies of some $H \in \mathcal{H}$ with at most h_0 vertices. So with probability $\geq \delta$, a sample of h_0 vertices sees an induced subgraph not satisfying $\mathcal{P}$. Running K/h_0 independent trials, we see some induced subgraph not satisfying $\mathcal{P}$ with probability $\geq 1 - (1-\delta)^{K/h_0}$,which can be made arbitrarily close to 1 by choosing K to be sufficiently large. As with earlier, this implies the result about choosing K random points without replacement. □

2.10 Hypergraph Removal and Szemerédi's Theorem

We showed earlier how to deduce Roth's theorem from the triangle removal lemma. However, the graph removal lemma, or the graph regularity method more generally, is insufficient for understanding longer arithmetic progressions.

Szemerédi's theorem follows as a corollary of a hypergraph generalization of the triangle removal lemma. (Note that historically, Szemerédi's theorem was initially shown using other methods; see the discussion in Section 0.2.) The hypergraph removal lemma turns out to be substantially more difficult. The following theorem was proved by Rödl et al. (2005) and Gowers (2007). The special case of the tetrahedron removal lemma in 3-graphs was proved earlier by Frankl and Rödl (2002).

Theorem 2.10.1 (Hypergraph removal lemma)
For every r-graph H and $\varepsilon > 0$, there exists $\delta > 0$ so that every n-vertex r-graph with $< \delta n^{v(H)}$ copies of H can be made H-free by removing $< \varepsilon n^r$ edges.

Recall that Szemerédi's theorem says that for every fixed $k \geq 3$, every k-AP-free subset of $[N]$ has size $o(N)$. We will prove it as a corollary of the hypergraph removal lemma for $H = K_k^{(k-1)}$, the complete $(k-1)$-graph on k vertices (also known as a ***simplex***; when $k = 3$ it is called a ***tetrahedron***). For concreteness, we will show how the deduction works in the case $k = 4$ (it is straightforward to generalize).

Here is a corollary of the tetrahedron removal lemma. It is analogous to Corollary 2.3.3.

Corollary 2.10.2
If G is a 3-graph such that every edge is contained in a unique tetrahedron (i.e., a clique on four vertices), then G has $o(n^3)$ edges.

Proof of Szemerédi's theorem for 4-APs. Let $A \subseteq [N]$ be 4-AP-free. Let $M = 6N + 1$. Then A is also a 4-AP-free subset of $\mathbb{Z}/M\mathbb{Z}$. (There are no wraparounds.) Build a 4-partite 3-graph G with parts W, X, Y, Z, all of which are M-vertex sets indexed by the elements of $\mathbb{Z}/M\mathbb{Z}$. We define edges as follows, where w, x, y, z range over elements of W, X, Y, Z, respectively:

$$\begin{aligned}
wxy \in E(G) &\iff 3w + 2x + y \in A,\\
wxz \in E(G) &\iff 2w + x - z \in A,\\
wyz \in E(G) &\iff w - y - 2z \in A,\\
xyz \in E(G) &\iff -x - 2y - 3z \in A.
\end{aligned}$$

What is important here is that the ith expression does not contain the ith variable.

The vertices $xyzw$ form a tetrahedron if and only if

$$3w + 2x + y, 2w + x - z, w - y - 2z, -x - 2y - 3z \in A.$$

However, these values form a 4-AP with common difference $-x - y - z - w$. Since A is 4-AP-free, the only tetrahedra in A are trivial 4-APs (those with common difference zero).

For each triple $(w,x,y) \in W \times X \times Y$, there is exactly one $z \in \mathbb{Z}/M\mathbb{Z}$ such that $x+y+z+w = 0$. Thus, every edge of the hypergraph lies in exactly one tetrahedron.

By Corollary 2.10.2, the number of edges in the hypergraph is $o(M^3)$. On the other hand, the number of edges is exactly $4M^2 |A|$. (For example, for every $a \in A$, there are exactly M^2 triples $(w,x,y) \in (\mathbb{Z}/M\mathbb{Z})^3$ with $3w + 2x + y = a$.) Therefore $|A| = o(M) = o(N)$. □

The hypergraph removal lemma is proved using a substantial and difficult generalization of the graph regularity method to hypergraphs. We will not be able to prove it in this book. In the next section, we sketch some key ideas in hypergraph regularity.

It is instructive to work out the proof in the special cases below. For the next two exercises, you should assume Corollary 2.10.2.

Exercise 2.10.3 (3-dimensional corners). Suppose $A \subseteq [N]^3$ contains no four points of the form

$$(x,y,z),\ (x+d,y,z),\ (x,y+d,z),\ (x,y,z+d), \quad \text{with } d > 0.$$

Show that $|A| = o(N^3)$.

Exercise 2.10.4 (Multidimensional Szemerédi for axis-aligned squares). Suppose $A \subseteq [N]^2$ contains no four points of the form

$$(x,y),\ (x+d,y),\ (x,y+d),\ (x+d,y+d), \quad \text{with } d \neq 0.$$

Show that $|A| = o(N^2)$.

Exercise 2.10.5 (Multidimensional Szemerédi theorem from the hypergraph removal lemma). Generalizing the previous exercise, prove the multidimensional Szemerédi theorem (Theorem 0.2.6) using the hypergraph removal lemma.

2.11 Hypergraph Regularity

Hypergraph regularity is substantially more difficult to prove than graph regularity. We only sketch some key ideas here. For concreteness, we focus our discussion on 3-graphs. Throughout this section, G will be a 3-graph with vertex set V.

What should correspond to an "ε-regular pair" from the graph regularity lemma? Here is an initial attempt.

Definition 2.11.1 (Initial attempt at 3-graph regularity)
Given vertex subsets $V_1, V_2, V_3 \subseteq V$, we say that (V_1, V_2, V_3) is ***ε-regular*** if, for all $A_i \subseteq V_i$ such that $|A_i| \geq \varepsilon |V_i|$, we have

$$|d(V_1,V_2,V_3) - d(A_1,A_2,A_3)| \leq \varepsilon.$$

Here, the edge density $d(X,Y,Z)$ is the fraction of elements of $X \times Y \times Z$ that are edges of G.

By following the proof of the graph regularity lemma nearly verbatim, we can show the following.

Proposition 2.11.2 (Initial attempt at 3-graph regularity partition)
For all $\varepsilon > 0$, there exists $M = M(\varepsilon)$ such that every 3-graph has a partition into at most M parts so that all but at most an ε-fraction of triples of vertices lie in ε-regular triples of vertex parts.

Can this result be used to prove the hypergraph removal lemma? Unfortunately, no.

Recall that our graph regularity recipe (Remark 2.3.2) involves three steps: partition, clean, and count. It turns out that no counting lemma is possible for the preceding notion of 3-graph regularity.

The notion of ε-regularity is supposed to model pseudorandomness. So why don't we try truly random hypergraphs and see what happens? Let us consider two different random 3-graph constructions:

(a) First pick constants $p, q \in [0, 1]$. Build a random graph $G^{(2)} = \mathbf{G}(n, p)$, an ordinary Erdős–Rényi graph. Then construct $G^{(3)}$ by including each triangle of $G^{(2)}$ as an edge of $G^{(3)}$ with probability q. Call this 3-graph X.

(b) For each possible edge (i.e. triple of vertices), include the edge with probability $p^3 q$, independent of all other edges. Call this 3-graph Y.

The edge density in both X and Y are close to $p^3 q$, even when restricted to linearly sized triples of vertex subsets. So both graphs satisfy our preceding notion of ε-regularity with high probability. However, we can compute the tetrahedron densities in both of these graphs and see that they do not match.

The tetrahedron density in X is around q^4 times the K_4 density in the underlying random graph $G^{(2)}$. The K_4 density in $G^{(2)}$ is around p^6. So the tetrahedron density in X is around $p^6 q^4$.

On the other hand, the tetrahedron density in Y is around $(p^3 q)^4$, different from $p^6 q^4$ earlier. So we should not expect a counting lemma with this notion of ε-regularity (unless the 3-graph we are counting is linear, as in the exercise below).

Exercise 2.11.3. Under the notion of 3-graph regularity in Definition 2.11.1, formulate and prove an H-counting lemma for every linear 3-graph H. Here a hypergraph is said to be ***linear*** if every pair of its edges intersects in at most one vertex.

As hinted by the first random hypergraph earlier, a more useful notion of hypergraph regularity should involve both vertex subsets as well as subsets of vertex-pairs (i.e., an underlying 2-graph).

Given a 3-graph G, a regularity decomposition will consist of

(1) a partition of $\binom{V}{2}$ into 2-graphs $G_1^{(2)} \cup \cdots \cup G_l^{(2)}$ so that G sits in a random-like way on top of most triples of these 2-graphs (we won't try to make it precise), and

(2) a partition of V that gives an extremely regular partition for all 2-graphs $G_1^{(2)}, \ldots, G_l^{(2)}$. (This should be somewhat reminiscent of the strong graph regularity lemma from Section 2.8.)

For such a decomposition to be applicable, it should come with a corresponding *counting lemma*.

There are several ways to make the preceding notions precise. Certain formulations make the regularity partition easier to prove while the counting lemma is harder, and some vice versa. The interested readers should consult Rödl et al. (2005), Gowers (2007) (see Gowers [2006] for an exposition of the case of 3-uniform hypergraphs), and Tao (2006) for three different approaches to the hypergraph regularity lemma.

***Remark* 2.11.4** (Quantitative bounds). Whereas the proof of the graph regularity lemma gives tower-type bounds tower($\varepsilon^{-O(1)}$), the proof of the 3-graph regularity lemma has wowzer-type bounds. The 4-graph regularity lemma moves us one more step up in the Ackermann hierarchy (i.e., iterating wowzer), and so on. Just as with the tower-type lower bound (Theorem 2.1.17) for the graph regularity lemma, Ackermann-type bounds are necessary for hypergraph regularity as well (Moshkovitz and Shapira 2019).

Further Reading

For surveys on the graph regularity method and applications, see Komlós and Simonovits (1996) and Komlós, Shokoufandeh, Simonovits, and Szemerédi (2002).

The survey *Graph Removal Lemmas* by Conlon and Fox (2013) discusses many variants, extensions, and proof techniques of graph removal lemmas.

For a well-motivated introduction to the hypergraph regularity lemma, see the article *Quasirandomness, Counting and Regularity for 3-Uniform Hypergraphs* by Gowers (2006).

Chapter Summary

- **Szemerédi's graph regularity lemma.** For every $\varepsilon > 0$, there exists a constant M such that every graph has an ε-regular partition into at most M parts.
  - Proof method: **energy increment**.
- Regularity method recipe: **partition, clean, count**.
- **Graph counting lemma.** The number of copies of H among ε-regular parts is similar to random.
- **Graph removal lemma.** Fix H. Every n-vertex graph with $o(n^{v(H)})$ copies of H can be made H-free by removing $o(n^2)$ edges.
- **Roth's theorem** can be proved by applying the triangle removal lemma to a graph whose triangles correspond to 3-APs.
- **Szemerédi's theorem** follows from the **hypergraph removal lemma**, whose proof uses the **hypergraph regularity method** (not covered in this book).
- **Induced removal lemma.** Fix H. Every n-vertex graph with $o(n^{v(H)})$ induced copies of H can be made induced H-free by adding/removing $o(n^2)$ edges.
  - Proof uses a **strong regularity lemma**, which involves iterating the earlier graph regularity lemma.
- Every hereditary graph property is **testable**.
  - One can distinguish graphs that have property $\mathcal{P}$ from those that are ε-far from property $\mathcal{P}$ (far in the sense of edit distance $\geq \varepsilon n^2$) by sampling a subgraph induced by a constant number of random vertices.
  - The probabilistic guarantee is essentially equivalent to removal lemmas.

3

Pseudorandom Graphs

Chapter Highlights

- Equivalent notions of graph quasirandomness
- Role of eigenvalues in pseudorandomness
- Expander mixing lemma
- Eigenvalues of abelian Cayley graphs and the Fourier transform
- Quasirandom groups and representations theory
- Quasirandom Cayley graphs and Grothendieck's inequality
- Alon–Boppana bound on the second eigenvalue of a d-regular graph

In the previous chapter on the graph regularity method, we saw that every graph can be partitioned into a bounded number of vertex parts so that the graph looks "random-like" between most pairs of parts. In this chapter, we dive further into how a graph can be random-like.

Pseudorandomness is a concept prevalent in combinatorics, theoretical computer science, and in many other areas. It specifies how a nonrandom object can behave like a truly random object.

***Example* 3.0.1** (Pseudorandom generators). Suppose you want to generate a random number on a computer. In most systems and programming languages, you can do this easily with a single command (e.g., `rand()`). The output is not actually truly random. Instead, the output came from a *pseudorandom generator*, which is some function/algorithm that takes a *seed* as input and passes it through some sophisticated function, so that there is no practical way to distinguish the output from a truly random object. In other words, the output is not actually truly random, but for all practical purposes the output cannot be distinguished from a truly random output.

***Example* 3.0.2** (Primes). In number theory, the prime numbers behave like a random sequence in many ways. The celebrated *Riemann hypothesis* and its generalizations give quantitative predictions about how closely the primes behave in a certain specific way like a random sequence. There is also something called *Cramér's random model* for the primes that allows one to make predictions about the asymptotic density of certain patterns in the primes (e.g., how many twin primes up to N are there?). Empirical data support these predictions, and they have been proved in certain cases. Nevertheless, there are still notorious open problems such as the twin prime and Goldbach conjectures. Despite their pseudorandom behavior, the primes are not random!

***Example* 3.0.3** (Normal numbers). It is very much believed that the digits of π behave in a random-like way, where every digit or block of digits appears with frequency similar to that of a truly random number. Such numbers are called *normal*. It is widely believed that numbers such as $\sqrt{2}$, π, and e are normal, but proofs remain elusive. Again, the digits of π are deterministic, not random, but they are believed to behave pseudorandomly. On the other hand, nearly all real numbers are normal, with the exceptions occupying only a measure zero subset of the reals.

Coming back to graph theory. The ***Erdős–Rényi random graph*** $\boldsymbol{G(n,p)}$ is a random n-vertex graph where each edge appears with probability p independently. Now, given some specific graph (perhaps an instance of the random graph, or perhaps generated via some other means), we can ask whether this graph, for the purpose of some intended application, behaves similarly to that of a typical random graph. What are some useful ways to measure the pseudorandomness of a graph? This is the main theme that we explore in this chapter.

3.1 Quasirandom Graphs

Here are several natural notions of how a graph (or rather, a sequence of graphs) can look random. The main theorem of this section says that, surprisingly, these notions are all equivalent. This result is due to Chung, Graham, and Wilson (1989), who coined the term ***quasirandom graphs***. Similar ideas also appeared in the work of Thomason (1987). These results had an important impact in the field.

Theorem 3.1.1 (Quasirandom graphs)

Let $p \in [0,1]$ be fixed. Let (G_n) be a sequence of graphs with G_n having n vertices and $(p + o(1))\binom{n}{2}$ edges. (Here $n \to \infty$ along some subsequence of integers, and is allowed to skip some integers.) Denote G_n by G. The following properties are all equivalent:

DISC (discrepancy) $e(X,Y) = p\,|X|\,|Y| + o(n^2)$ for all $X,Y \subseteq V(G)$.

DISC' $e(X) = p\binom{|X|}{2} + o(n^2)$ for all $X \subseteq V(G)$.

COUNT For every graph H, the number of labeled copies of H in G is $(p^{e(H)} + o(1))n^{v(H)}$.

(Here a labeled copy of H is the same as an injective map $V(H) \to V(G)$ that sends every edge of H to an edge of G. The rate that the $o(1)$ goes to zero is allowed to depend on H.)

$\mathbf{C_4}$ (4-cycle) The number of labeled 4-cycles is at most $(p^4 + o(1))n^4$.

CODEG (codegree) Letting $\operatorname{codeg}(u,v)$ denote the number of common neighbors of u and v,

$$\sum_{u,v \in V(G)} \left|\operatorname{codeg}(u,v) - p^2 n\right| = o(n^3).$$

EIG (eigenvalue) If $\lambda_1 \geq \lambda_2 \geq \cdots \geq \lambda_n$ are the eigenvalues of the adjacency matrix of G, then $\lambda_1 = pn + o(n)$ and $\max_{i \neq 1} |\lambda_i| = o(n)$.

Definition 3.1.2 (Quasirandom graphs)

We say a sequence of graphs is ***quasirandom*** (at edge density p) if it satisfies the preceding conditions for some constant $p \in [0, 1]$.

***Remark* 3.1.3** (Single graph vs. a sequence of graphs). Strictly speaking, it does not make sense to say whether a *single* graph is quasirandom, but we will abuse the definition as such when it is clear that the graph we are referring to is part of a sequence.

***Remark* 3.1.4** ($\mathbf{C_4}$ condition). The $\mathbf{C_4}$ condition is surprising. It says that the 4-cycle density, a single statistic, is equivalent to all the other quasirandomness conditions.

We will soon see shortly in Proposition 3.1.14 that the $\mathbf{C_4}$ can be replaced by the equivalent condition that the number of labeled 4-cycles is $(p^4+o(1))n^4$ (rather than at most this quantity).

***Remark* 3.1.5** (Checking quasirandomness). The discrepancy conditions are hard to verify since they involve checking exponentially many sets. The other conditions can all be checked in time polynomial in the size of the graph. So the equivalence gives us an algorithmically efficient way to certify the discrepancy condition.

***Remark* 3.1.6** (Quantitative equivalences). Rather than stating these properties for a sequence of graphs using a decaying error term $o(1)$, we can state a quantitative quasirandomness hypothesis for a specific graph using an error tolerance parameter ε. For example, we can restate the discrepancy condition as follows.

DISC(ε): For all $X, Y \subseteq V(G)$, $|e(X,Y) - p\,|X|\,|Y|| < \varepsilon n^2$.

Similar statements can be made for other quasirandom graph notions. The proof that shortly follows will show that these notions are equivalent up to a polynomial change in ε; that is, for each pair of properties, **Prop1**(ε) implies **Prop2**($C\varepsilon^c$) for some constants $C, c > 0$.

Examples of Quasirandom Graphs

First let us check that random graphs are quasirandom (hence justifying the name). Recall the following basic tail bound for a sum of independent random variables.

Theorem 3.1.7 (Chernoff bound)

Let X be a sum of m independent Bernoulli random variables (not necessarily identically distributed). Then, for every $t > 0$,

$$\mathbb{P}(|X - \mathbb{E}X| \geq t) \leq 2e^{-t^2/(2m)}.$$

Proposition 3.1.8 (Edge densities in a random graph)

Let $p \in [0, 1]$ and $\varepsilon > 0$. With probability at least $1 - 2^{n+1}e^{-\varepsilon^2 n^2}$, the Erdős–Rényi random graph $\mathbf{G}(n, p)$ has the property that for every vertex subset X,

$$\left| e(X) - p\binom{|X|}{2} \right| \leq \varepsilon n^2.$$

Proof. Applying the Chernoff bound to $e(X)$, we see that

$$\mathbb{P}\left(\left|e(X) - p\binom{|X|}{2}\right| > \varepsilon n^2\right) \le 2\exp\left(\frac{-(\varepsilon n^2)^2}{2\binom{|X|}{2}}\right) \le 2\exp\left(-\varepsilon^2 n^2\right).$$

The result then follows by taking a union bound over all 2^n subsets X of the n-vertex graph. □

Applying the Borel–Cantelli lemma with the preceding bound, we obtain the following consequence.

Corollary 3.1.9 (Random graphs are quasirandom)
Fix $p \in [0, 1]$. With probability 1, a sequence of random graphs $G_n \sim \mathbf{G}(n, p)$ is quasirandom at edge density p.

It would be somewhat disappointing if the only interesting example of a quasirandom graph were actual random graphs. Fortunately, we have more explicit constructions. In the rest of the chapter, we will see several constructions using Cayley graphs on groups. A notable example, which we will prove in Section 3.3, is that the Paley graph is quasirandom.

***Example* 3.1.10** (Paley graph). Let $p \equiv 1 \pmod 4$ be a prime. Form a graph with vertex set $\mathbb{F}_p$, with two vertices x, y joined if $x - y$ is a quadratic residue. Then this graph is quasirandom at edge density $1/2$ as $p \to \infty$. (By a standard fact from elementary number theory, since $p \equiv 1 \pmod 4$, -1 is a quadratic residue, and hence $x - y$ is a quadratic residue if and only if $y - x$ is. So the graph is well defined.)

In Section 3.4, we will show that for certain sequences of groups, the Cayley graphs are always quasirandom provided that the edge densities converge. We will call such groups *quasirandom*. We will later prove the following important example.

***Example* 3.1.11** (PSL(2, p)). Let p be a prime. Let $S \subseteq \mathrm{PSL}(2, p)$ be a subset of nonzero elements with $S = S^{-1}$. Let G be the Cayley graph on $\mathrm{PSL}(2, p)$ with generator S, meaning that the vertices are elements of $\mathrm{PSL}(2, p)$, and two vertices x, y are adjacent if $x^{-1}y \in S$. Then G is quasirandom as $p \to \infty$ as long as $|S|/p^3$ converges.

Finally, here is an explicit construction using finite geometry. We leave it as an exercise to verify its quasirandomness using the conditions given earlier.

***Example* 3.1.12.** Let p be a prime. Let $S \subseteq \mathbb{F}_p \cup \{\infty\}$. Let G be a graph on vertex set $\mathbb{F}_p^2$ where two points are joined if the slope of the line connecting them lies in S. Then G is quasirandom as $p \to \infty$ as long as $|S|/p$ converges.

Exercise 3.1.13. Prove that the construction in Example 3.1.12 is quasirandom.

Proof of Equivalence of Graph Quasirandomness Conditions

We will now start to prove Theorem 3.1.1. Let us begin with a warm-up on how to apply the Cauchy–Schwarz inequality in graph theory since it will come up several times in the proof. (We will revisit this topic in Section 5.2.)

The following statement says that the 4-cycle density is always roughly at least as much as random. Later in Chapter 5, we will see Sidorenko's conjecture, which says that all bipartite graphs have this property.

As a consequence, the $\mathbf{C_4}$ condition is equivalent to saying that the number of labeled 4-cycles *is* $(p^4 + o(1))n^4$ (rather than *at most*).

Proposition 3.1.14 (Minimum 4-cycle density)
Every n-vertex graph with at least $pn^2/2$ edges has at least p^4n^4 labeled closed walks of length 4.

***Remark* 3.1.15.** Since all but $O(n^3)$ such closed walks use four distinct vertices, the preceding statement implies that the number of labeled 4-cycles is at least $(p^4 - o(1))n^4$.

Proof. The number of closed walks of length 4 is

$$\begin{aligned}
|\{(w,x,y,z) \text{ closed walk}\}| &= \sum_{w,y} |\{x : w \sim x \sim y\}|^2 \\
&\geq \frac{1}{n^2}\left(\sum_{w,y} |\{x : w \sim x \sim y\}|\right)^2 \\
&= \frac{1}{n^2}\left(\sum_{x} |\{(w,y) : w \sim x \sim y\}|\right)^2 \\
&= \frac{1}{n^2}\left(\sum_{x} (\deg x)^2\right)^2 \\
&\geq \frac{1}{n^4}\left(\sum_{x} \deg x\right)^4 \\
&= (2e(G))^4/n^4 \geq p^4n^4
\end{aligned}$$

Here both inequality steps are due to Cauchy–Schwarz. On the right column is a pictorial depiction of what is being counted by the inner sum on each line. These diagrams are a useful way to keep track of the graph inequalities, especially when dealing with much larger graphs, where the algebraic expressions get unwieldy. Note that each application of the Cauchy–Schwarz inequality corresponds to "folding" the graph along a line of reflection. □

We shall prove the equivalences of Theorem 3.1.1 in the following way:

$$\begin{array}{ccccc}
\mathbf{DISC'} \longleftrightarrow & \mathbf{DISC} & \longrightarrow & \mathbf{COUNT} & \\
& \uparrow & & \downarrow & \\
& \mathbf{CODEG} & \longleftarrow & \mathbf{C_4} & \longleftrightarrow \mathbf{EIG}
\end{array}$$

Proof That **DISC** *Implies* **DISC′**. Take $Y = X$ in **DISC**. (Note that $e(X, X) = 2e(X)$ and $\binom{|X|}{2} = |X|^2/2 - O(n)$.) □

Proof that **DISC′** *implies* **DISC**. We have the following "polarization identity" together with a proof by picture (recall $2e(X) = e(X,X)$):

$$e(X,Y) = e(X \cup Y) + e(X \cap Y) - e(X \setminus Y) - e(Y \setminus X).$$

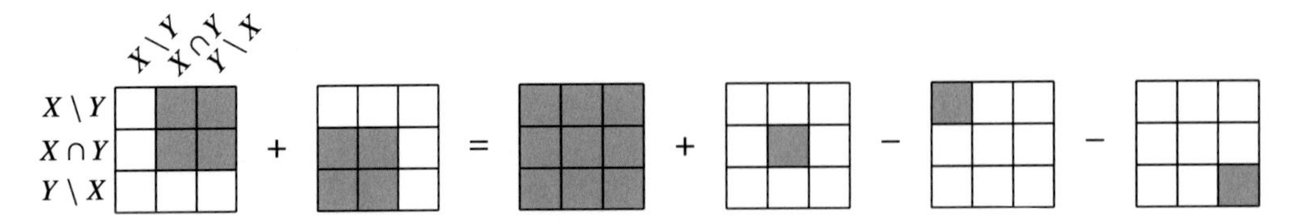

If **DISC′** holds, then the right-hand side in the preceding picture equals to

$$p\binom{|X \cup Y|}{2} + p\binom{|X \cap Y|}{2} - p\binom{|X \setminus Y|}{2} - p\binom{|Y \setminus X|}{2} + o(n^2) = p\,|X|\,|Y| + o(n^2),$$

where the final step applies the polarization identity again, this time on the complete graph. So, we have $e(X,Y) = p\,|X|\,|Y| + o(n^2)$, thereby confirming **DISC**. □

Proof (deferred) that **DISC** *implies* **COUNT**. This is essentially a counting lemma. In Section 2.6 we proved a version of the counting lemma but for lower bounds. The same proof can be modified to a two-sided bound. We will see another proof of a counting lemma (Theorem 4.5.1) in the next chapter on graph limits, which gives us a convenient language to set up a more streamlined proof. So, we will defer this proof until then. □

Proof that **COUNT** *implies* $\mathbf{C_4}$. $\mathbf{C_4}$ is a special case of **COUNT**. □

Proof that $\mathbf{C_4}$ *implies* **CODEG**. Assuming $\mathbf{C_4}$, we have

$$\sum_{u,v} \operatorname{codeg}(u,v) = \sum_{x \in G} \deg(x)^2 \ge \frac{1}{n}\left(\sum_{x \in G} \deg(x)\right)^2 = \frac{1}{n}\left(pn^2 + o(n^2)\right)^2 = p^2n^3 + o(n^3).$$

We also have (the $O(n^3)$ error term is due to walks of length 4 that use repeated vertices)

$$\begin{aligned}\sum_{u,v} \operatorname{codeg}(u,v)^2 &= \#\text{ labeled } C_4 + O(n^3)\\ &\le p^4n^4 + o(n^4).\end{aligned}$$

Thus, by the Cauchy–Schwarz inequality,

$$\begin{aligned}\frac{1}{n^2}\left(\sum_{u,v}\left|\operatorname{codeg}(u,v) - p^2n\right|\right)^2 &\le \sum_{u,v}\left(\operatorname{codeg}(u,v) - p^2n\right)^2\\ &= \sum_{u,v}\operatorname{codeg}(u,v)^2 - 2p^2n\sum_{u,v}\operatorname{codeg}(u,v) + p^4n^4\\ &\le p^4n^4 - 2p^2n\cdot p^2n^3 + p^4n^4 + o(n^4)\\ &= o(n^4).\end{aligned}$$

□

***Remark* 3.1.16.** These calculations share the spirit of the *second moment method* in probabilistic combinatorics. The condition $\mathbf{C_4}$ says that the variance of the codegree of two random vertices is small.

Exercise 3.1.17. Show that if we modify the **COEG** condition to

$$\sum_{u,v\in V(G)} \left(\operatorname{codeg}(u,v) - p^2 n\right) = o(n^3),$$

then it would not be enough to imply quasirandomness.

Proof that **CODEG** *implies* **DISC**. We first show that the codegree condition implies the concentration of degrees:

$$\begin{aligned}
\frac{1}{n}\left(\sum_u |\deg u - pn|\right)^2 &\le \sum_u (\deg u - pn)^2 \\
&= \sum_u (\deg u)^2 - 2pn \sum_u \deg u + p^2 n^3 \\
&= \sum_{x,y} \operatorname{codeg}(x,y) - 4pn\, e(G) + p^2 n^3 \\
&= p^2 n^3 - 2p^2 n^3 + p^2 n^3 + o(n^3) \\
&= o(n^3). \qquad (3.1)
\end{aligned}$$

Now we bound the expression in **DISC**. We have

$$\begin{aligned}
\frac{1}{n} |e(X,Y) - p\,|X|\,|Y||^2 &= \frac{1}{n}\left(\sum_{x\in X} (\deg(x,Y) - p\,|Y|)\right)^2 \\
&\le \sum_{x\in X} (\deg(x,Y) - p\,|Y|)^2 .
\end{aligned}$$

The preceding Cauchy–Schwarz step turned all the summands nonnegative, which allows us to expand the domain of summation from X to all of $V = V(G)$ in the next step. Continuing,

$$\begin{aligned}
&\le \sum_{x\in V} (\deg(x,Y) - p\,|Y|)^2 \\
&= \sum_{x\in V} \deg(x,Y)^2 - 2p\,|Y| \sum_{x\in V} \deg(x,Y) + p^2 n\,|Y|^2 \\
&= \sum_{y,y'\in Y} \operatorname{codeg}(y,y') - 2p\,|Y| \sum_{y\in Y} \deg y + p^2 n\,|Y|^2 \\
&= |Y|^2 p^2 n - 2p\,|Y| \cdot |Y|\, pn + p^2 n\,|Y|^2 + o(n^3) \qquad \text{[by } \textbf{CODEG} \text{ and (3.1)]} \\
&= o(n^3).
\end{aligned}$$

□

Finally, let us consider the ***graph spectrum***, which consists of the eigenvalues of the graph adjacency matrix, accounting for eigenvalue multiplicities. Eigenvalues are core to the study of pseudorandomness and they will play a central role in the rest of this chapter.

In this book, when we talk about the ***eigenvalues of a graph***, we always mean the eigenvalues of the adjacency matrix of the graph. In other contexts, it may be useful to consider other related matrices, such as the Laplacian matrix or a normalized adjacency matrix.

We will generally consider only real symmetric matrices, whose eigenvalues are always all real. (Hermitian matrices also have this property.) Our usual convention is to list all the eigenvalues in order (including multiplicities): $\lambda_1 \geq \lambda_2 \geq \cdots \geq \lambda_n$. We refer to λ_1 as the ***top eigenvalue*** (or ***largest eigenvalue***), and λ_i as the ***ith eigenvalue*** (or the ***ith largest eigenvalue***). The second eigenvalue plays an important role. We write $\lambda_i(A)$ for the ith eigenvalue of the matrix A and $\lambda_i(G) = \lambda_i(A_G)$, where A_G is the adjacency matrix of G.

***Remark* 3.1.18** (Linear algebra review). For every $n \times n$ real symmetric matrix A with eigenvalues $\lambda_1 \geq \cdots \geq \lambda_n$, we can choose an eigenvector $v_i \in \mathbb{R}^n$ for each eigenvalue λ_i (so that $Av_i = \lambda_i v_i$) and such that $\{v_1, \ldots, v_n\}$ is an orthogonal basis of $\mathbb{R}^n$. (This is false for general nonsymmetric matrices.)

The ***Courant–Fischer min-max theorem*** is an important characterization of eigenvalues in terms of a variational problem. Here we only state some consequences most useful for us. We have

$$\lambda_1 = \max_{v \in \mathbb{R}^n \setminus \{0\}} \frac{\langle v, Av \rangle}{\langle v, v \rangle}.$$

Once we have fixed a choice of an eigenvector v_1 for the top eigenvalue λ_1, we have

$$\lambda_2 = \max_{\substack{v \perp v_1 \\ v \in \mathbb{R}^n \setminus \{0\}}} \frac{\langle v, Av \rangle}{\langle v, v \rangle}.$$

In particular, if G is a d-regular graph, then the all-1 vector, denoted $\mathbf{1} \in \mathbb{R}^{v(G)}$, is an eigenvector for the top eigenvalue d.

The ***Perron–Frobenius theorem*** tells us some important information about the top eigenvector and eigenvalue of a nonnegative matrix. For every connected graph G, the top eigenvector is simple (i.e., multiplicity one), so that $\lambda_i < \lambda_1$ for all $i > 1$. We also have $|\lambda_i| \leq \lambda_1$ for all i (one has $\lambda_n = -\lambda_1$ if and only if G is bipartite; see Remark 3.1.23). Also, the top eigenvector v_1 (which is unique up to scalar multiplication) has all coordinates positive.

If G has multiple connected components $G_1, \ldots, G_k$, then the eigenvalues of G (with multiplicities) are obtained by taking a multiset union of the eigenvalues of its connected components. An orthogonal system of eigenvectors can also be derived as such, by extending each eigenvector of G_i to an eigenvector of G via padding the eigenvector by zeros outside the vertices of G_i.

Here is a useful formula:

$$\operatorname{tr} A^k = \lambda_1^k + \cdots + \lambda_n^k.$$

When A is the adjacency matrix of a graph G, $\operatorname{tr} A^k$ counts the number of closed walks of length k. In particular, $\operatorname{tr} A^2 = 2e(G)$.

Proof that **EIG** *implies* $\mathbf{C_4}$. Let A denote the adjacency matrix of G. The number of labeled 4-cycles is within $O(n^3)$ of the number of closed walks of length 4, and the latter equals

$$\operatorname{tr} A^4 = \lambda_1^4 + \cdots + \lambda_n^4 = p^4 n^4 + o(n^4) + \sum_{i=2}^{n} \lambda_i^4.$$

Since $\operatorname{tr} A^2 = 2e(G) \leq n^2$, we have

$$\sum_{i=2}^{n} \lambda_i^4 \leq \max_{i \neq 1} \lambda_i^2 \cdot \sum_{i=1}^{n} \lambda_i^2. = o(n^2) \cdot \operatorname{tr} A^2 = o(n^4).$$

So $\operatorname{tr} A^4 \leq p^4 n^4 + o(n^4)$. □

***Remark* 3.1.19.** A rookie error would be to bound $\sum_{i\geq 2}\lambda_i^4$ by $n\max_{i\geq 2}\lambda_i^4 = o(n^5)$, but this would not be enough. (Where do we save in the above proof?) We will see a similar situation later in Chapter 6 when we discuss the Fourier analytic proof of Roth's theorem.

> **Lemma 3.1.20** (Top eigenvalue and average degree)
> The top eigenvalue of the adjacency matrix of a graph is always at least its average degree.

Proof. Let $\mathbf{1} \in \mathbb{R}^n$ be the all-1 vector. By the Courant–Fischer min-max theorem, the adjacency matrix A of the graph G has top eigenvalue

$$\lambda_1 = \sup_{\substack{x\in\mathbb{R}^n\\ x\neq 0}} \frac{\langle x, Ax\rangle}{\langle x, x\rangle} \geq \frac{\langle \mathbf{1}, A\mathbf{1}\rangle}{\langle \mathbf{1},\mathbf{1}\rangle} = \frac{2e(G)}{v(G)} = \operatorname{avgdeg}(G). \qquad \square$$

Proof that $\mathbf{C_4}$ *implies* **EIG**. Again writing A for the adjacency matrix,

$$\sum_{i=1}^{n} \lambda_i^4 = \operatorname{tr} A^4 = \#\{\text{closed walks of length } 4\} \leq p^4n^4 + o(n^4).$$

On the other hand, by Lemma 3.1.20, we have $\lambda_1 \geq pn + o(n)$. So we must have $\lambda_1 = pn + o(n)$ and $\max_{i\geq 2}|\lambda_i| = o(n)$. $\square$

This completes all the implications in the proof of Theorem 3.1.1.

Additional Remarks

***Remark* 3.1.21** (Forcing graphs). The $\mathbf{C_4}$ hypothesis says that having 4-cycle density asymptotically the same as random implies quasirandomness. Which other graphs besides C_4 have this property?

Chung, Graham, and Wilson (1989) called a graph F ***forcing*** if every graph with edge density $p + o(1)$ and F-density $p^{e(F)} + o(1)$ (i.e., asymptotically the same as random) is automatically quasirandom. Theorem 3.1.1 implies that C_4 is forcing. Here is a conjectural characterization of forcing graphs (Skokan and Thoma 2004; Conlon, Fox, and Sudakov 2010).

> **Conjecture 3.1.22** (Forcing conjecture)
> A graph is forcing if and only if it is bipartite and not a tree.

We will revisit this conjecture in Chapter 5 where we will reformulate it using the language of graphons.

More generally, one says that a family of graphs $\mathcal{F}$ is forcing if having F-density being $p^{e(F)} + o(1)$ for each $F \in \mathcal{F}$ implies quasirandomness. So $\{K_2, C_4\}$ is forcing. It seems to be a difficult problem to classify forcing families.

Even though many other graphs can potentially play the role of the 4-cycle, the 4-cycle nevertheless occupies an important role in the study of quasirandomness. The 4-cycle comes up naturally in the proofs, as we will see below. It also is closely tied to other important pseudorandomness measurements such as the Gowers U^2 uniformity norm in additive combinatorics.

Let us formulate a **bipartite analogue** of Theorem 3.1.1 since we will need it later. It is easy to adapt the preceding proofs to the bipartite version – we encourage the reader to think about the differences between the two settings.

***Remark* 3.1.23** (Eigenvalues of bipartite graphs). Given a bipartite graph G with vertex bipartition $V \cup W$, we can write its adjacency matrix as

$$A = \begin{pmatrix} \mathbf{0} & B \\ B^\intercal & \mathbf{0} \end{pmatrix}, \tag{3.2}$$

where B is an $|V| \times |W|$ matrix with rows indexed by V and columns indexed by W. The eigenvalues $\lambda_1 \geq \cdots \geq \lambda_n$ of A always satisfy

$$\lambda_i = \lambda_{n+1-i} \qquad \text{for every } 1 \leq i \leq n.$$

In other words, the eigenvalues are symmetric around zero. One way to see this is that if $x = (v, w)$ is an eigenvector of A, where $v \in \mathbb{R}^V$ is the restriction of x to the first $|V|$ coordinates, and w is the restriction of x to the last $|W|$ coordinates, then

$$\begin{pmatrix} \lambda v \\ \lambda w \end{pmatrix} = \lambda x = Ax = \begin{pmatrix} \mathbf{0} & B \\ B^\intercal & \mathbf{0} \end{pmatrix} \begin{pmatrix} v \\ w \end{pmatrix} = \begin{pmatrix} Bw \\ B^\intercal v \end{pmatrix},$$

so that

$$Bw = \lambda v \qquad \text{and} \qquad B^\intercal v = \lambda w.$$

Then the vector $x' = (v, -w)$ satisfies

$$Ax' = \begin{pmatrix} \mathbf{0} & B \\ B^\intercal & \mathbf{0} \end{pmatrix} \begin{pmatrix} v \\ -w \end{pmatrix} = \begin{pmatrix} -Bw \\ B^\intercal v \end{pmatrix} = \begin{pmatrix} -\lambda v \\ \lambda w \end{pmatrix} = -\lambda x'.$$

So, we can pair each eigenvalue of A with its negation.

Exercise 3.1.24. Using the notation from (3.2), show that the positive eigenvalues of the adjacency matrix A coincide with the positive singular values of B (the singular values of B are also the positive square roots of the eigenvalues of $B^\intercal B$).

Theorem 3.1.25 (Bipartite quasirandom graphs)

Fix $p \in [0, 1]$. Let $(G_n)_{n \geq 1}$ be a sequence of bipartite graphs G_n. Write G_n as G, with vertex bipartition $V \cup W$. Suppose $|V|, |W| \to \infty$ and $|E| = (p + o(1)) |V| |W|$ as $n \to \infty$. The following properties are all equivalent:

DISC $e(X, Y) = p |X| |Y| + o(n^2)$ for all $X \subseteq V$ and $Y \subseteq W$.

COUNT For every bipartite graph H with vertex bipartition (S, T), the number of labeled copies of H in G with S embedded in V and T embedded in W is $(p^{e(H)} + o(1)) |V|^{|S|} |W|^{|T|}$.

$\mathbf{C_4}$ The number of closed walks of length 4 in G starting in V is at most $(p^4 + o(1)) |V|^2 |W|^2$.

Left-CODEG $\sum_{x,y \in V} \left|\operatorname{codeg}(x, y) - p^2 |W|\right| = o(|V|^2 |W|)$.

Right-CODEG $\sum_{x,y \in W} \left|\operatorname{codeg}(x, y) - p^2 |V|\right| = o(|V| |W|^2)$.

EIG The adjacency matrix of G has top eigenvalue $(p + o(1))\sqrt{|V|\,|W|}$ and second largest eigenvalue $o(\sqrt{|V|\,|W|})$.

The bipartite discrepancy condition **DISC** is equivalent to being an $o(1)$-regular pair (Definition 2.1.2, Exercise 2.1.5).

***Remark* 3.1.26** (Bipartite double cover). Theorem 3.1.25 implies the nonbipartite version Theorem 3.1.1, since every graph G can be transformed into a bipartite graph $G \times K_2$ (a graph tensor power) whose two vertex parts are both copies of $V(G)$. Each edge $u \sim v$ of G lifts to two edges $(u,0) \sim (v,1)$ and $(u,1) \sim (v,0)$ in $G \times K_2$. An example is shown below.

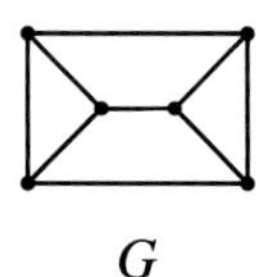

G $\qquad$ $G \times K_2$

Exercise 3.1.27. Show that a graph G satisfies each property in Theorem 3.1.1 if and only if $G \times K_2$ satisfies the corresponding bipartite property in Theorem 3.1.25.

Like earlier, random bipartite graphs are bipartite quasirandom. The proof (omitted) is essentially the same as Proposition 3.1.8 and Corollary 3.1.9.

Proposition 3.1.28 (Random bipartite graphs are typically quasirandom.)
Fix $p \in [0,1]$. With probability 1, a sequence of bipartite random graphs $G_n \sim \mathbf{G}(n,n,p)$ (obtained by keeping every edge of $K_{n,n}$ with probability p independently) is quasirandom in the sense of Theorem 3.1.25.

***Remark* 3.1.29** (Sparse graphs). We stated quasirandom properties so far only for graphs of constant order density (i.e., p is a constant). Let us think about what happens if we allow $p = p_n$ to depend on n and decay to zero as $n \to \infty$. Such graphs are sometimes called *sparse* (although some other authors reserve the word "sparse" for bounded degree graphs). Theorems 3.1.1 and 3.1.25 as stated do hold for a constant $p = 0$, but the results are not as informative as we would like. For example, the error tolerance on the **DISC** is $o(n^2)$, which does not tell us much since the graph already has much fewer edges due to its sparseness anyway.

To remedy the situation, the natural thing to do is to adjust the error tolerance relative to the edge density $p = p_n \to 0$. Here are some representative examples (all of these properties should also depend on p):

SparseDISC $|e(X,Y) - p\,|X|\,|Y|| = o(pn^2)$ for all $X, Y \subseteq V(G)$.

SparseCOUNT$_H$ The number of labeled copies of H is $(1 + o(1))p^{e(H)}n^{v(H)}$.

SparseC$_4$ The number of labeled 4-cycles is at most $(1 + o(1))p^4n^4$.

SparseEIG $\lambda_1 = (1 + o(1))pn$ and $\max_{i \neq 1} |\lambda_i| = o(pn)$.

Warning: these sparse pseudorandomness conditions are *not* all equivalent to each other. Some of the implications still hold. (The reader is encouraged to think about which ones.) However, some crucial implications such as the counting lemma fail quite miserably. For example:

SparseDISC does not imply **SparseCOUNT**.

Indeed, suppose $p = n^{-c}$ for some constant $1/2 < c < 1$. In a typical random graph $\mathbf{G}(n,p)$, the number of triangles is close to $\binom{n}{3}p^3$, while the number of edges is close to $\binom{n}{2}p$. We have $p^3n^3 = o(pn^2)$ as long as $p = o(n^{-1/2})$, so there are significantly fewer triangles than there are edges. Now remove an edge from every triangle in this random graph. We will have removed $o(pn^2)$ edges, a negligible fraction of the $(p + o(1))\binom{n}{2}$ edges, and this edge removal should not significantly affect **SparseDISC**. However, we have changed the triangle count significantly as a result.

Fortunately, this is not the end of the story. With additional hypotheses on the sparse graph, we can sometimes salvage a counting lemma. **Sparse counting lemmas** play an important role in the proof of the Green–Tao theorem on arithmetic progressions in the primes, as we will explain in Chapter 9.

Exercise 3.1.30 (Nearly optimal C_4-free graphs are sparse quasirandom). Let G_n be a sequence of n-vertex C_4-free graphs with $(1/2 - o(1))n^{3/2}$ edges. Prove that $e_{G_n}(A, B) = n^{-1/2}|A||B| + o(n^{3/2})$ for every $A, B \subseteq V(G_n)$.

Hint: Revisit the CODEG $\Longleftarrow$ DISC proof and the proof of the KST theorem (Theorem 1.4.2).

Exercise 3.1.31* (Quasirandomness through fixed sized subsets). Fix $p \in [0,1]$. Let (G_n) be a sequence of graphs with $v(G_n) = n$ (here $n \to \infty$ along a subsequence of integers).

(a) Fix a single $\alpha \in (0,1)$. Suppose

$$e(S) = \frac{p\alpha^2 n^2}{2} + o(n^2) \qquad \text{for all } S \subseteq V(G) \text{ with } |S| = \lfloor \alpha n \rfloor .$$

Prove that G is quasirandom.

(b) Fix a single $\alpha \in (0,1/2)$. Suppose

$$e(S, V(G) \setminus S) = p\alpha(1-\alpha)n^2 + o(n^2) \qquad \text{for all } S \subseteq V(G) \text{ with } |S| = \lfloor \alpha n \rfloor .$$

Prove that G is quasirandom. Furthermore, show that the conclusion is false for $\alpha = 1/2$.

Exercise 3.1.32 (Quasirandomness and regularity partitions). Fix $p \in [0,1]$. Let (G_n) be a sequence of graphs with $v(G_n) \to \infty$. Suppose that for every $\varepsilon > 0$, there exists $M = M(\varepsilon)$ so that each G_n has an ε-regular partition where all but ε-fraction of vertex pairs lie between pairs of parts with edge density $p + o(1)$ (as $n \to \infty$). Prove that G_n is quasirandom.

Exercise 3.1.33* (Triangle counts on induced subgraphs). Fix $p \in (0,1]$. Let (G_n) be a sequence of graphs with $v(G_n) = n$. Let $G = G_n$. Suppose that for every $S \subseteq V(G)$, the number of triangles in the induced subgraph $G[S]$ is $p^3\binom{|S|}{3} + o(n^3)$. Prove that G is quasirandom.

Exercise 3.1.34* (Perfect matchings). Prove that there are constant $\beta, \varepsilon > 0$ such that for every positive even integer n and real $p \geq n^{-\beta}$, if G is an n-vertex graph where every vertex has degree $(1 \pm \varepsilon)pn$ (meaning within εpn of pn) and every pair of vertices has codegree $(1 \pm \varepsilon)p^2 n$, then G has a perfect matching.

3.2 Expander Mixing Lemma

We dive further into the relationship between graph eigenvalues and its pseudorandomness properties. We focus on d-regular graphs since they occur often in practice (e.g., from Cayley graphs), and they are also cleaner to work with. Unlike the previous section, the results here are effective for any value of d (not just when d is on the same order as n).

As we saw earlier, the magnitudes of eigenvalues are related to the pseudorandomness of a graph. In a d-regular graph, the top eigenvalue is always exactly d. The following condition says that all other eigenvalues are bounded by λ in absolute value.

Definition 3.2.1 ((n, d, λ)-graph)
An ***(n, d, λ)-graph*** is an n-vertex, d-regular graph whose adjacency matrix eigenvalues $d = \lambda_1 \geq \cdots \geq \lambda_n$ satisfy

$$\max_{i \neq 1} |\lambda_i| \leq \lambda.$$

***Remark* 3.2.2** (Notation). Rather than saying "an $(n, 7, 6)$-graph" we prefer to say "an (n, d, λ)-graph with $d = 7$ and $\lambda = 6$" for clarity as the name "(n, d, λ)" is quite standard and recognizable.

Remark* 3.2.3** (Linear algebra review). The ***operator norm of a matrix $A \in \mathbb{R}^{m \times n}$ is defined by

$$\|A\| = \sup_{x \in \mathbb{R}^n \setminus \{0\}} \frac{|Ax|}{|x|} = \sup_{\substack{x \in \mathbb{R}^n \setminus \{0\} \\ y \in \mathbb{R}^m \setminus \{0\}}} \frac{\langle y, Ax \rangle}{|x|\,|y|}.$$

Here $|x| = \sqrt{\langle x, x \rangle}$ denotes the length of vector x. The operator norm of A is the maximum ratio that A can amplify the length of a vector by. If A is a real symmetric matrix, then

$$\|A\| = \max_i |\lambda_i(A)|.$$

For general matrices, the operator norm of A equals the largest singular value of A.

Here is the main result of this section.

Theorem 3.2.4 (Expander mixing lemma)
If G is an (n, d, λ)-graph, then

$$\left| e(X, Y) - \frac{d}{n} |X|\,|Y| \right| \leq \lambda \sqrt{|X|\,|Y|} \qquad \text{for all } X, Y \subseteq V(G).$$

On the left-hand side, $(d/n)\,|X|\,|Y|$ is the number of edges that one should expect between X and Y purely based on the edge density d/n of the graph and the sizes of X and Y. Note that unlike the discrepancy condition (**DISC**) from quasirandom graphs (Theorem 3.1.1), the error bound on the right-side hand depends on the sizes of X and Y. We can apply the expander mixing lemma to small subsets X and Y and still obtain useful estimates on $e(X, Y)$, unlike the dense quasirandom graph conditions.

Proof. Let J be the $n \times n$ all-1 matrix. Since the all-1 vector $\mathbf{1} \in \mathbb{R}^n$ is an eigenvector of A_G with eigenvalue d, we see that $\mathbf{1}$ is an eigenvector of $A_G - \frac{d}{n} J$ with eigenvalue 0. Any other eigenvector v of A_G, with $v \perp \mathbf{1}$, satisfies $Jv = 0$, and thus v is also an eigenvector of $A_G - \frac{d}{n}J$ with the same eigenvalue as in A_G. Therefore, the eigenvalues of $A_G - \frac{d}{n}J$ are obtained by taking the eigenvalues of A_G, then replacing one top eigenvalue d by zero. All the other eigenvalues of $A_G - \frac{d}{n}J$ are therefore at most λ in absolute value, so $\left\|A_G - \frac{d}{n}J\right\| \le \lambda$. Therefore,

$$\begin{aligned}\left|e(X,Y) - \frac{d}{n}|X|\,|Y|\right| &= \left|\left\langle \mathbf{1}_X, \left(A_G - \frac{d}{n}J\right)\mathbf{1}_Y\right\rangle\right| \\ &\le \left\|A_G - \frac{d}{n}J\right\| |\mathbf{1}_X|\,|\mathbf{1}_Y| \\ &\le \lambda\sqrt{|X|\,|Y|}.\end{aligned}$$

□

Exercise 3.2.5. Prove the following strengthening the expander mixing lemma.

Theorem 3.2.6 (Expander mixing lemma – slightly strengthened)
If G is an (n,d,λ)-graph, then

$$\left|e(X,Y) - \frac{d}{n}|X|\,|Y|\right| \le \frac{\lambda}{n}\sqrt{|X|\,(n-|X|)\,|Y|\,(n-|Y|)} \qquad \text{for all } X, Y \subseteq V(G).$$

We also have a bipartite analogue. (The nomenclature used here is less standard.) Recall from Remark 3.1.23 that the eigenvalues of a bipartite graph are symmetric around zero.

Definition 3.2.7 (Bipartite-(n,d,λ)-graph)
An ***bipartite-(n,d,λ)-graph*** is a d-regular bipartite graph with n vertices in each part, such that its second largest eigenvalue is at most λ.

Exercise 3.2.8. Show that G is an (n,d,λ)-graph if and only if $G \times K_2$ is a bipartite-(n,d,λ)-graph.

Theorem 3.2.9 (Bipartite expander mixing lemma)
Let G be a bipartite-(n,d,λ)-graph with vertex bipartition $V \cup W$. Then

$$\left|e(X,Y) - \frac{d}{n}|X|\,|Y|\right| \le \lambda\sqrt{|X|\,|Y|} \qquad \text{for all } X \subseteq V \text{ and } Y \subseteq W.$$

Exercise 3.2.10. Prove Theorem 3.2.9.

***Remark* 3.2.11.** The following partial converse to the expander mixing lemma was shown by Bilu and Linial (2006). The extra log factor turns out to be necessary.

Theorem 3.2.12 (Converse to expander mixing lemma)
There exists an absolute constant C such that if G is a d-regular graph, and β satisfies

$$\left|e(X,Y) - \frac{d}{n}|X|\,|Y|\right| \le \beta\sqrt{|X|\,|Y|} \qquad \text{for all } X, Y \subseteq V(G),$$

then G is an (n,d,λ)-graph with $\lambda \le C\beta\log(2d/\beta)$.

Cheeger's Inequality: Edge Expansion versus Spectral Gap

Given a graph and its adjacency matrix, the ***spectral gap*** is defined to be the difference between the two most significant eigenvalues; that is, $\lambda_1 - \lambda_2$. This quantity turns out to be closely related to expansion in graphs. We define the ***edge-expansion ratio*** of a graph $G = (V, E)$ to be the quantity

$$\boldsymbol{h(G)} := \min_{\substack{S \subseteq V \\ 0 < |S| \leq |V|/2}} \frac{e_G(S, V \setminus S)}{|S|}.$$

In other words, a graph with edge-expansion ratio at least h has the property that for every nonempty subset of vertices S with $|S| \leq |V|/2$, there are at least $h|S|$ edges leaving S.

Cheeger's inequality, stated in Theorem 3.2.13, tells us that among d-regular graphs for a fixed d, having spectral gap bounded away from zero is equivalent to having edge-expansion ratio bounded away from zero. Cheeger (1970) originally developed this inequality for Riemannian manifolds. The graph theoretic analogue was proved by Dodziuk (1984), and independently by Alon and Milman (1985) and Alon (1986).

Theorem 3.2.13 (Cheeger's inequality)
Let G be an n-vertex d-regular graph with adjacency matrix spectral gap $\kappa = d - \lambda_2$. Then its edge-expansion ratio $h = h(G)$ satisfies

$$\kappa/2 \leq h \leq \sqrt{2d\kappa}.$$

The two bounds of Cheeger's inequality are tight up to constant factors. For the lower bound, taking G to be the skeleton of the d-dimensional cube with vertex set $\{0,1\}^d$ gives $h = 1$ (achieved by the $d-1$ dimensional subcube) and $\kappa = 2$. For the upper bound, taking G to be an n-cycle gives $h = 2/(n/2) = \Theta(1/n)$ while $d = 2$ and $\kappa = 2 - 2\cos(2\pi/n)) = \Theta(1/n^2)$.

We call a family of d-regular graphs ***expanders*** if there is some constant $\kappa_0 > 0$ so that each graph in the family has spectral gap $\geq \kappa_0$; by Cheeger's inequality, this is equivalent to the existence of some $h_0 > 0$ so that each graph in the family has edge expansion ratio $\geq h_0$. Expander graphs are important objects in mathematics and computer science. For example, expander graphs have rapid mixing properties, which are useful for designing efficient Monte Carlo algorithms for sampling and estimation.

The following direction of Cheeger's inequality is easier to prove. It is similar to the expander mixing lemma.

Exercise 3.2.14 (Spectral gap implies expansion). Prove the $\kappa/2 \leq h$ part of Cheeger's inequality.

The other direction, $h \leq \sqrt{2d\kappa}$, is more difficult and interesting. The proof is outlined in the following exercise.

Exercise 3.2.15 (Expansion implies spectral gap). Let $G = (V, E)$ be a connected d-regular graph with spectral gap κ. Let $x = (x_v)_{v \in V} \in \mathbb{R}^V$ be an eigenvector associated to the second largest eigenvalue $\lambda_2 = d - \kappa$ of the adjacency matrix of G. Assume that $x_v > 0$ on at most half of the vertex set (or else we replace x by $-x$). Let $y = (y_v)_{v \in V} \in \mathbb{R}^V$ be obtained from x by replacing all its negative coordinates by zero.

(a) Prove that

$$d - \frac{\langle y, Ay \rangle}{\langle y, y \rangle} \le \kappa.$$

Hint: Recall that $\lambda_2 x_v = \sum_{u \sim v} x_u$.

(b) Let

$$\Theta = \sum_{uv \in E} \left| y_u^2 - y_v^2 \right|.$$

Prove that

$$\Theta^2 \le 2d(d \langle y, y \rangle - \langle y, Ay \rangle) \langle y, y \rangle .$$

Hint: $y_u^2 - y_v^2 = (y_u - y_v)(y_u + y_v)$. Apply Cauchy–Schwarz.

(c) Relabel the vertex set V by $[n]$ so that $y_1 \ge y_2 \cdots \ge y_t > 0 = y_{t+1} = \cdots = y_n$. Prove

$$\Theta = \sum_{k=1}^{t} (y_k^2 - y_{k+1}^2)\, e([k], [n] \setminus [k]).$$

(d) Prove that for some $1 \le k \le t$,

$$\frac{e([k], [n] \setminus [k])}{k} \le \frac{\Theta}{\langle y, y \rangle}.$$

(e) Prove the $h \le \sqrt{2d\kappa}$ claim of Cheeger's inequality.

Exercises

Exercise 3.2.16 (Independence numbers)**.** Prove that every independent set in a (n, d, λ)-graph has size at most $n\lambda/(d + \lambda)$.

Exercise 3.2.17 (Diameter)**.** Prove that the diameter of an (n, d, λ)-graph is at most $\lceil \log n / \log(d/\lambda) \rceil$. (The ***diameter*** of a graph is the maximum distance between a pair of vertices.)

Exercise 3.2.18 (Counting cliques)**.** For each part, prove that for every $\varepsilon > 0$, there exists $\delta > 0$ such that the conclusion holds for every (n, d, λ)-graph G with $d = pn$.

(a) If $\lambda \le \delta p^2 n$, then the number of triangles of G is within a $1 \pm \varepsilon$ factor of $p^3 \binom{n}{3}$.

(b*) If $\lambda \le \delta p^3 n$, then the number of K_4's in G is within a $1 \pm \varepsilon$ factor of $p^6 \binom{n}{4}$.

3.3 Abelian Cayley Graphs and Eigenvalues

Many important constructions of pseudorandom graphs come from groups.

Definition 3.3.1 (Cayley graph)

Let Γ be a finite group, and let $S \subseteq \Gamma$ be a subset with $S = S^{-1}$ (i.e., $s^{-1} \in S$ for all $s \in S$) and not containing the identity element. We write $\mathbf{Cay}(\Gamma, S)$ to denote the ***Cayley graph*** on Γ generated by S, which has elements of Γ as vertices, and

$$g \sim gs \quad \text{for all } g \in \Gamma \text{ and } s \in S$$

as edges.

In this section, we consider only abelian groups, specifically $\mathbb{Z}/p\mathbb{Z}$ for concreteness (though everything here generalizes easily to all finite abelian groups). For abelian groups, we write the group operation additively as $g + s$. So, edges join elements whose difference lies in S.

***Remark* 3.3.2.** In later sections when we consider a nonabelian group Γ, one needs to make a choice whether to define edges by left- or right-multiplication (i.e., gs or sg; we chose gs here). It does not matter which choice one makes (as long as one is consistent) since the resulting Cayley graphs are isomorphic. (Why?) However, some careful bookkeeping is sometimes required to make sure that later computations are consistent with the initial choice.

***Example* 3.3.3.** $\mathrm{Cay}(\mathbb{Z}/n\mathbb{Z}, \{-1, 1\})$ is a cycle of length n. The graph for $n = 8$ is shown.

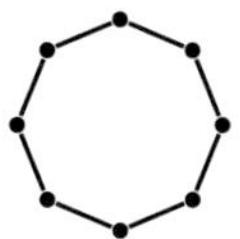

***Example* 3.3.4.** $\mathrm{Cay}(\mathbb{F}_2^n, \{e_1, \ldots, e_n\})$ is the skeleton of an n-dimensional cube. Here e_i is the ith standard basis vector. The graphs for $n = 1, 2, 3, 4$ are shown.

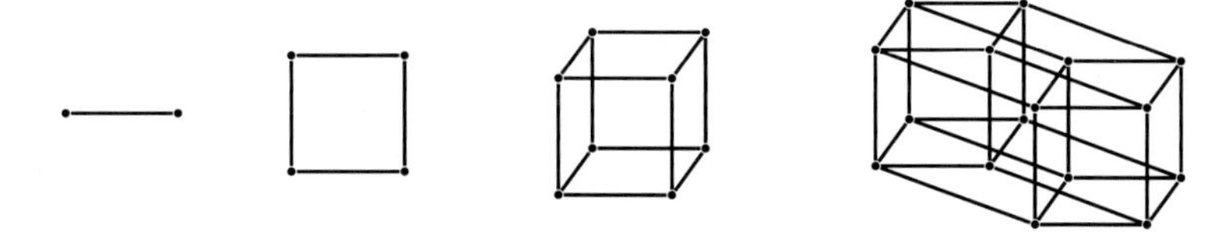

Here is an explicitly constructed family of quasirandom graphs with edge density $1/2+o(1)$.

> **Definition 3.3.5** (Paley graph)
> Let $p \equiv 1 \pmod 4$ be a prime. The ***Paley graph*** of order p is $\mathrm{Cay}(\mathbb{Z}/p\mathbb{Z}, S)$, where S is the set of nonzero quadratic residues in $\mathbb{Z}/p\mathbb{Z}$. (Here $\mathbb{Z}/p\mathbb{Z}$ is viewed as an additive group.)

***Example* 3.3.6.** The Paley graphs for $p = 5$ and $p = 13$ are shown.

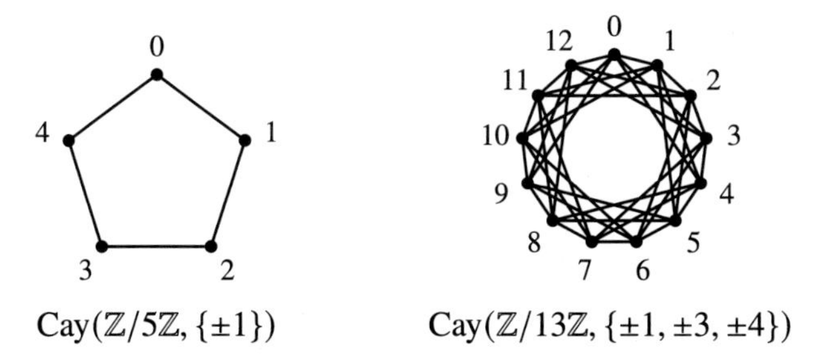

$\mathrm{Cay}(\mathbb{Z}/5\mathbb{Z}, \{\pm 1\})$ $\mathrm{Cay}(\mathbb{Z}/13\mathbb{Z}, \{\pm 1, \pm 3, \pm 4\})$

***Remark* 3.3.7** (Quadratic residues). Here we recall some facts from elementary number theory. For every odd prime p, the set $S = \{a^2 : a \in \mathbb{F}_p^\times\}$ of quadratic residues is a multiplicative subgroup of $\mathbb{F}_p^\times$ with index two. In particular, $|S| = (p-1)/2$. We have $-1 \in S$ if and only if $p \equiv 1 \pmod 4$ (which is required to define a Cayley graph, as the generating set needs to be symmetric in the sense that $S = -S$).

We will show that Paley graphs are quasirandom by verifying the **EIG** condition, which says that all eigenvalues, except the top one, are small. Here is a general formula for computing the eigenvalues of any Cayley graph on $\mathbb{Z}/p\mathbb{Z}$.

Theorem 3.3.8 (Eigenvalues of abelian Cayley graphs on $\mathbb{Z}/n\mathbb{Z}$)
Let n be a positive integer. Let $S \subseteq \mathbb{Z}/n\mathbb{Z}$ with $0 \notin S$ and $S = -S$. Let

$$\omega = \exp(2\pi i/n).$$

Then we have an orthonormal basis $v_0, \ldots, v_{n-1} \in \mathbb{C}^n$ of eigenvectors of $\mathrm{Cay}(\mathbb{Z}/n\mathbb{Z}, S)$ where

$$v_j \in \mathbb{C}^n \text{ has } x\text{-coordinate } \omega^{jx}/\sqrt{n}, \text{ for each } x \in \mathbb{Z}/n\mathbb{Z}.$$

The eigenvalue (not sorted by size) associated to the eigenvector v_j equals to

$$\lambda_j = \sum_{s \in S} \omega^{js}.$$

In particular, $\lambda_0 = |S|$ and v_0 has all coordinates $1/\sqrt{n}$.

Remark **3.3.9** (Eigenvalues and the Fourier transform)**.** The coordinates of the eigenvectors are shown below.

	$\mathbb{Z}/n\mathbb{Z}$				
	0	1	2	$\cdots$	$n-1$
$\sqrt{n}\ v_0$	1	1	1	$\cdots$	1
$\sqrt{n}\ v_1$	1	ω	ω^2	$\cdots$	ω^{n-1}
$\sqrt{n}\ v_2$	1	ω^2	ω^4	$\cdots$	$\omega^{2(n-1)}$
$\vdots$	$\vdots$	$\vdots$	$\vdots$	$\ddots$	$\vdots$
$\sqrt{n}\ v_{n-1}$	1	ω^{n-1}	$\omega^{2(n-1)}$	$\cdots$	$\omega^{(n-1)^2}$

Viewed as a matrix, this is sometimes known as the ***discrete Fourier transform matrix***. We will study the Fourier transform in Chapter 6. These two topics are closely tied. The eigenvalues of an abelian Cayley graph $\mathrm{Cay}(\Gamma, S)$ are precisely the Fourier transform in Γ of the generating set S, up to normalizing factors:

$$\text{eigenvalues of } \mathrm{Cay}(\Gamma, S) \quad \longleftrightarrow \quad \text{Fourier transform } \widehat{1_S} \text{ in } \Gamma.$$

We will say more about this in Remark 3.3.11 shortly.

Proof. Let A be the adjacency matrix of $\mathrm{Cay}(\mathbb{Z}/n\mathbb{Z}, S)$. First we check that each v_j is an eigenvector of A with eigenvalue λ_j. The coordinate of $\sqrt{n}Av_j$ at $x \in \mathbb{Z}/n\mathbb{Z}$ equals to

$$\sum_{s \in S} \omega^{j(x+s)} = \left(\sum_{s \in S} \omega^{js}\right)\omega^{jx} = \lambda_j \omega^{jx}.$$

So, $Av_j = \lambda_j v_j$.

Next we check that $\{v_0, \ldots, v_{n-1}\}$ is an orthonormal basis. We have the inner product

$$\langle v_j, v_k \rangle = \frac{1}{n}\left(1 \cdot 1 + \overline{\omega^j}\omega^k + \overline{\omega^{2j}}\omega^{2k} + \cdots + \overline{\omega^{(n-1)j}}\omega^{(n-1)k}\right)$$
$$= \frac{1}{n}\left(1 + \omega^{k-j} + \omega^{2(k-j)} + \cdots + \omega^{(n-1)(k-j)}\right) = \begin{cases} 1 & \text{if } j = k, \\ 0 & \text{if } j \neq k. \end{cases}$$

For the $i \neq j$ case, we use that for any mth root of unity $\zeta \neq 1$, $\sum_{j=0}^{m-1} \zeta^j = 0$. So, $\{v_0, \ldots, v_{n-1}\}$ is an orthonormal basis. □

***Remark* 3.3.10** (Real vs. complex eigenbases). The adjacency matrix of a graph is a real symmetric matrix, so all its eigenvalues are real, and it always has a real orthogonal eigenbasis. The eigenbasis given in Theorem 3.3.8 is complex, but it can always be made real. Looking at the formulas in Theorem 3.3.8, we have $\lambda_j = \lambda_{n-j}$, and v_j is the complex conjugate of v_{n-j}. So we can form a real orthogonal eigenbasis by replacing, for each $j \notin \{0, n/2\}$, the pair (v_j, v_{n-j}) by $((v_j + v_{n-j})/\sqrt{2}, i(v_j - v_{n-j})/\sqrt{2})$. Equivalently, we can separate the real and imaginary parts of each v_j, which are both eigenvectors with eigenvalue λ_j. All the real eigenvalues and eigenvectors can be expressed in terms of sines and cosines.

Remark* 3.3.11** (Every abelian Cayley graph has an eigenbasis independent of the generators.). The preceding theorem and its proof generalizes to all finite abelian groups, not just $\mathbb{Z}/n\mathbb{Z}$. For every finite abelian group Γ, we have a set $\widehat{\Gamma}$ of characters, where each ***character is a homomorphism $\chi : \Gamma \to \mathbb{C}^\times$. Then $\widehat{\Gamma}$ turns out to be a group isomorphic to Γ. (One can check this by first writing Γ as a direct product of cyclic groups.) For each $\chi \in \widehat{\Gamma}$, define the vector $v_\chi \in \mathbb{C}^\Gamma$ by setting the coordinate at $g \in \Gamma$ to be $\chi(g)/\sqrt{|\Gamma|}$. Then $\{v_\chi : \chi \in \widehat{\Gamma}\}$ is an orthonormal basis for the adjacency matrix of every Cayley graph on Γ. The eigenvalue corresponding to v_χ is $\lambda_\chi(S) = \sum_{s \in S} \chi(s)$. Up to normalization, $\lambda_\chi(S)$ is the Fourier transform of the indicator function of S on the abelian group Γ. (Theorem 3.3.8 is a special case of this construction.) In particular, this eigenbasis $\{v_\chi : \chi \in \widehat{\Gamma}\}$ depends only on the finite abelian group and not on the generating set S. In other words, we have a *simultaneous diagonalization* for all adjacency matrices of Cayley graphs on a fixed finite abelian group.

If Γ is a nonabelian group, then there does not exist a simultaneous eigenbasis for all Cayley graphs on Γ. There is a corresponding theory of nonabelian Fourier analysis, which uses group representation theory. We will discuss more about nonabelian Cayley graphs in Section 3.4.

Now we apply the preceding formula to compute eigenvalues of Paley graphs. In particular, the following tells us that Paley graphs satisfy the quasirandomness condition **EIG** from Theorem 3.1.1.

Theorem 3.3.12 (Eigenvalues of Paley graphs)
Let $p \equiv 1 \pmod 4$ be a prime. The adjacency of matrix of the Paley graph of order p has top eigenvalue $(p-1)/2$, and all other eigenvalues are either $(\sqrt{p}-1)/2$ or $(-\sqrt{p}-1)/2$.

Proof. Applying Theorem 3.3.8, we see that the eigenvalues are given by, for $j = 0, 1, \ldots, p-1$,

$$\lambda_j = \sum_{s \in S} \omega^{js} = \frac{1}{2}\left(-1 + \sum_{x \in \mathbb{F}_p} \omega^{jx^2}\right),$$

since each quadratic residue s appears as x^2 for exactly two nonzero x. Clearly $\lambda_0 = (p-1)/2$. For $j \neq 0$, the next result shows that the inner sum on the right-hand side is $\pm\sqrt{p}$. (Note that the preceding sum is real when $p \equiv 1 \pmod 4$ since $S = S^{-1}$ and so the sum equals its own complex conjugate; alternatively, the sum must be real since all eigenvalues of a symmetric matrix are real.) □

***Remark* 3.3.13.** Since the trace of the adjacency matrix is zero and equals the sum of eigenvalues, we see that the eigenvalues other than the top one are equally split between $(\sqrt{p} - 1)/2$ and $(-\sqrt{p} - 1)/2$.

Theorem 3.3.14 (Gauss sum)
Let p be an odd prime, $\omega = \exp(2\pi i/p)$, and $j \in \mathbb{F}_p \setminus \{0\}$. Then

$$\left| \sum_{x \in \mathbb{F}_p} \omega^{jx^2} \right| = \sqrt{p}.$$

Proof. We have

$$\left| \sum_{x \in \mathbb{F}_p} \omega^{jx^2} \right|^2 = \sum_{x,y \in \mathbb{Z}/p\mathbb{Z}} \omega^{j((x+y)^2 - x^2)} = \sum_{x,y \in \mathbb{Z}/p\mathbb{Z}} \omega^{j(2xy+y^2)}.$$

For each fixed y, we have

$$\sum_{x \in \mathbb{Z}/p\mathbb{Z}} \omega^{j(2xy+y^2)} = \begin{cases} p & \text{if } y = 0, \\ 0 & \text{if } y \neq 0. \end{cases}$$

Summing over y yields the claim. □

***Remark* 3.3.15** (Sign of the Gauss sum). The determination of this sign is a more difficult problem. Gauss conjectured the sign in 1801, and it took him four years to prove it. When j is a nonzero quadratic residue mod p, the inner sum in the preceding equation turns out to equal $\sqrt{p}$ if $p \equiv 1 \pmod 4$ and $i\sqrt{p}$ if $p \equiv 3 \pmod 4$. When j is a quadratic nonresidue, it is $-\sqrt{p}$ and $-i\sqrt{p}$ in the two cases respectively. See Ireland and Rosen (1990, Section 6.4) for a proof.

Exercise 3.3.16. Let p be an odd prime and $A, B \subseteq \mathbb{Z}/p\mathbb{Z}$. Show that

$$\left| \sum_{a \in A} \sum_{b \in B} \left(\frac{a+b}{p} \right) \right| \leq \sqrt{p\,|A|\,|B|}$$

where (a/p) is the Legendre symbol defined by

$$\left(\frac{a}{p}\right) = \begin{cases} 0 & \text{if } a \equiv 0 \pmod{p} \\ 1 & \text{if } a \text{ is a nonzero quadratic residue mod } p \\ -1 & \text{if } a \text{ is a quadratic nonresidue mod } p \end{cases}$$

Exercise 3.3.17. Prove that in a Paley graph of order p, every clique has size at most $\sqrt{p}$.

Exercise 3.3.18 (No spectral gap if too few generators). Prove that for every $\varepsilon > 0$ there is some $c > 0$ such that for every $S \subseteq \mathbb{Z}/n\mathbb{Z}$ with $0 \notin S = -S$ and $|S| \leq c \log n$, the second largest eigenvalue of the adjacency matrix of $\mathrm{Cay}(\mathbb{Z}/n\mathbb{Z}, S)$ is at least $(1 - \varepsilon)|S|$.

Exercise 3.3.19*. Let p be a prime and let S be a multiplicative subgroup of $\mathbb{F}_p^\times$. Suppose $-1 \in S$. Prove that all eigenvalues of the adjacency matrix of $\mathrm{Cay}(\mathbb{Z}/p\mathbb{Z}, S)$, other than the top one, are at most $\sqrt{p}$ in absolute value.

3.4 Quasirandom Groups

In the previous section, we saw that certain Cayley graphs on cyclic groups are quasirandom. Note that not all Cayley graphs on cyclic groups are quasirandom. For example, the Cayley graph with $\Gamma = \mathbb{Z}/n\mathbb{Z}$ and $S = \{x : |x| \leq n/4\} \subseteq \mathbb{Z}/n\mathbb{Z}$ is not quasirandom.

In this section, we will see that for certain families of nonabelian groups, *every* Cayley graph on the group is quasirandom, regardless of the Cayley graph generators. Gowers (2008) called such groups ***quasirandom groups*** and showed that they are precisely groups with no small nontrivial representations. He came up with this notion while solving the following problem about product-free sets in groups.

Question 3.4.1 (Product-free subset of groups)
Given a group of order n, what is the size of its largest product-free subset? Is it always $\geq cn$ for some constant $c > 0$?

***Remark* 3.4.2** (Representations of finite groups). We need some basic concepts from group representation theory in this section – mostly just some definitions. Feel free to skip this remark if you have already seen group representations.

Given a finite group Γ, it is often useful to study its actions as linear transformations on some vector space. For example, if Γ is a cyclic or dihedral group, it is natural to think of elements of Γ as rotations and reflection of a plane, which are linear transformations on $\mathbb{R}^2$. The theory turns out to be much nicer over $\mathbb{C}$ than $\mathbb{R}$ since $\mathbb{C}$ is algebraically closed. We are interested in ways that Γ can be represented as a group of linear transformations acting on some $\mathbb{C}^d$.

A ***representation*** of a finite group Γ is a group homomorphism $\rho\colon \Gamma \to \mathrm{GL}(V)$, where V is a complex vector space (everything will take place over $\mathbb{C}$) and $\mathrm{GL}(V)$ is the group of invertible linear transformations of V. We sometimes omit ρ from the notation and just say that V is a representation of Γ, and also that Γ ***acts*** on V (via ρ). For each $g \in \Gamma$ and $v \in V$, we write $gv = \rho(g)v$ for the image of the g-action on v. We write $\dim \rho = \dim V$ for the ***dimension*** of the representation.

The fact that $\rho\colon \Gamma \to \mathrm{GL}(V)$ is a group homomorphism means that the action of Γ on V is compatible with group operations in Γ in the following sense: if $g, h \in \Gamma$, then the expression ghx does not depend on whether we first apply h to x and then g to hx, or if we first multiply g and h in Γ and then apply their product gh to x.

For example, suppose Γ is a subgroup of permutations of $[n]$, with each element $g \in \Gamma$ viewed as a permutation $g\colon [n] \to [n]$. We can define a representation of Γ on $\mathbb{C}^n$ by letting Γ permute the coordinates: for any $x = (x_1, \dots, x_n) \in \mathbb{C}^n$, set $gx = (x_{g(1)}, \dots, x_{g(n)})$. As an element of $\mathrm{GL}(n, \mathbb{C})$, $\rho(g)$ is the $n \times n$ permutation matrix of the permutation g, and $gx = \rho(g)x$ for each $x \in \mathbb{C}^n$.

We say that the representation V of Γ is ***trivial*** if $gv = v$ for all $g \in \Gamma$ and $v \in V$, and ***nontrivial*** otherwise.

We say that a subspace W of V is ***Γ-invariant*** if $gw \in W$ for all $w \in W$. In other words, the image of W under Γ is contained in W (and actually must equal W due to the invertibility of group elements). Then W is a representation of Γ, and we call it a ***subrepresentation*** of V.

For an introduction to group representation theory, see any standard textbook such as the classic *Linear Representations of Finite Groups* by Serre (1977). Also, the lectures notes titled *Representation Theory of Finite Groups, and Applications* by Wigderson (2012) is a friendly introduction with applications to combinatorics and theoretical computer science.

Recall from Definition 3.2.1 that an ***(n, d, λ)-graph*** is an n-vertex d-regular graph all of whose eigenvalues, except the top one, are at most λ in absolute value.

The main theorem of this section, below, says that a group with no small nontrivial representations always produces quasirandom Cayley graphs (Gowers 2008).

Theorem 3.4.3 (Cayley graphs on quasirandom groups)
Let Γ be a group of order n with no nontrivial representations of dimension less than K. Then every d-regular Cayley graph on Γ is an (n, d, λ)-graph for some $\lambda < \sqrt{dn/K}$.

***Remark* 3.4.4** (Abelian groups and one-dimensional representations)**.** If Γ is abelian, then it has many one-dimensional nontrivial representations, namely its multiplicative characters. For example, if $\Gamma = \mathbb{Z}/n\mathbb{Z}$, then the map $\rho\colon \Gamma \to \mathbb{C}^\times$ sending $g \in \mathbb{Z}/n\mathbb{Z}$ to ω^g, where ω is some nontrivial root of unity, is a nontrivial one-dimensional representation. In fact, one can vary ω over all roots of unity to obtain all nonisomorphic one-dimensional representations of Γ.

So the hypothesis of having no low-dimensional nontrivial representations can be viewed as a statement that the group is highly nonabelian in some sense.

A representation is ***irreducible*** if it contains no subrepresentations other than itself and the zero-dimensional subrepresentation. Irreducible representations are the basic building blocks of group representations, and so understanding all irreducible representations of a group is a fundamental objective. Among finite groups, a group is abelian if and only if all its irreducible representations are one-dimensional.

More generally, we will prove the result for vertex-transitive groups, of which Cayley graphs are a special case.

Definition 3.4.5 (Vertex-transitive graphs)
Let G be a graph. An ***automorphism*** of G is a permutation of $V(G)$ that induces an isomorphism of G to itself (i.e., sending edges to edges). Let Γ be a group of automorphisms of G (not necessarily the whole automorphism group). We say that ***Γ acts vertex-transitively on G*** if, for every pair $v, w \in V(G)$, there is some $g \in \Gamma$ such that $gv = w$. We that G is a ***vertex-transitive graph*** if the automorphism group of G acts vertex-transitively on G.

In particular, every group Γ acts vertex-transitively on its Cayley graph $\mathrm{Cay}(\Gamma, S)$ by left-multiplication: the action of $g \in \Gamma$ sends each vertex $x \in \Gamma$ to $gx \in \Gamma$, which sends each edge (x, xs) to (gx, gxs), for all $x \in \Gamma$ and $s \in S$.

Theorem 3.4.6 (Vertex-transitive graphs and quasirandom groups)
Let Γ be a finite group with no nontrivial representations of dimension less than K. Then every n-vertex d-regular graph that admits a vertex-transitive Γ action is an (n, d, λ)-graph with $\lambda < \sqrt{dn/K}$.

Note that $\sqrt{dn/K} \le n/\sqrt{K}$, so that a sequence of such Cayley graphs is quasirandom (Definition 3.1.2) as long as $K \to \infty$ as $n \to \infty$.

Proof. Let A denote the adjacency matrix of the graph, whose vertices are indexed by $\{1, \ldots, n\}$. Each $g \in \Gamma$ gives a permutation $(g(1), \ldots, g(n))$ of the vertex set, which induces a representation of Γ on $\mathbb{C}^n$ given by permuting coordinates, sending $v = (v_1, \ldots, v_n) \in \mathbb{C}^n$ to $gv = (v_{g(1)}, \ldots, v_{g(n)})$.

We know that the all-1 vector $\mathbf{1}$ is an eigenvector of A with eigenvalue d. Let $v \in \mathbb{R}^n$ be an eigenvector of A with eigenvalue μ such that $v \perp \mathbf{1}$. Since each $g \in \Gamma$ induces a graph automorphism, $Av = \mu v$ implies $A(gv) = \mu gv$. (Check this claim! Basically it is because g relabels vertices in an isomorphically indistinguishable way.)

Since $\Gamma v = \{gv : g \in \Gamma\}$ is Γ-invariant, its $\mathbb{C}$-span W is a Γ-invariant subspace (i.e., $gW \subseteq W$ for all $g \in \Gamma$), and hence a subrepresentation of Γ. Since v is not a constant vector, the Γ-action on v is nontrivial. So, W is a nontrivial representation of Γ. Hence $\dim W \ge K$ by hypothesis. Every nonzero vector in W is an eigenvector of A with eigenvalue μ. It follows that μ appears as an eigenvalue of A with multiplicity at least K. Recall that we also have an eigenvalue d from the eigenvector $\mathbf{1}$. Thus

$$d^2 + K\mu^2 \le \sum_{j=1}^{n} \lambda_j(A)^2 = \operatorname{tr} A^2 = nd.$$

Therefore

$$|\mu| \le \sqrt{\frac{d(n-d)}{K}} < \sqrt{\frac{dn}{K}}. \qquad \square$$

The preceding proof can be modified to prove a bipartite version, which will be useful for certain applications.

Given a finite group Γ and a subset $S \subseteq \Gamma$ (not necessarily symmetric), we define the ***bipartite Cayley graph*** $\mathbf{BiCay}(\boldsymbol{\Gamma}, \boldsymbol{S})$ as the bipartite graph with vertex set Γ on both parts, with an edge joining g on the left with gs on the right for every $g \in \Gamma$ and $s \in S$.

Theorem 3.4.7 (Bipartite Cayley graphs on quasirandom groups)
Let Γ be a group of order n with no nontrivial representations of dimension less than K. Let $S \subseteq \Gamma$ with $|S| = d$. Then the bipartite Cayley graph $\mathrm{BiCay}(\Gamma, S)$ is a bipartite-(n, d, λ)-graph for some $\lambda < \sqrt{nd/K}$.

In other words, the second largest eigenvalue of the adjacency matrix of this bipartite Cayley graph is less than $\sqrt{nd/K}$.

Exercise 3.4.8. Prove Theorem 3.4.7.

As an application of the expander mixing lemma, we show that in a quasirandom group, the number of solutions to $xy = z$ with x, y, z lying in three given sets $X, Y, Z \subseteq \Gamma$ is close to what one should predict from density alone.

Theorem 3.4.9 (Mixing in quasirandom groups)
Let Γ be a finite group with no nontrivial representations of dimension less than K. Let $X, Y, Z \subseteq \Gamma$. Then

$$\left| |\{(x, y, z) \in X \times Y \times Z : xy = z\}| - \frac{|X|\,|Y|\,|Z|}{|\Gamma|} \right| < \sqrt{\frac{|X|\,|Y|\,|Z|\,|\Gamma|}{K}}.$$

Proof. Every solution to $xy = z$, with $(x, y, z) \in X \times Y \times Z$ corresponds to an edge (x, z) in $\mathrm{BiCay}(\Gamma, Y)$ between vertex subset X on the left and vertex subset Z on the right.

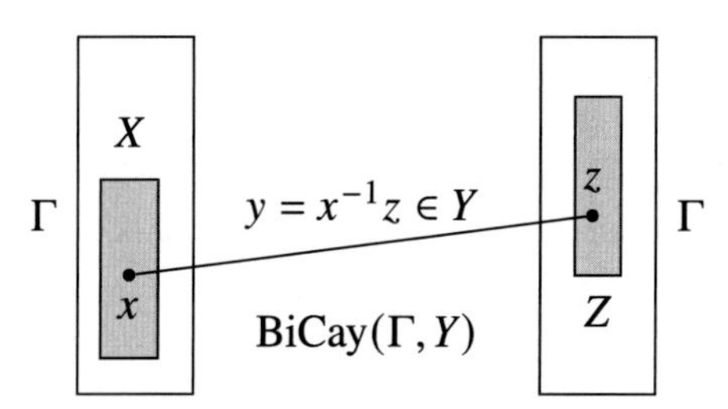

By Theorem 3.4.7, $\mathrm{BiCay}(\Gamma, Y)$ is a bipartite-(n, d, λ)-graph with $n = |\Gamma|$, $d = |Y|$, and some $\lambda < \sqrt{|\Gamma|\,|Y|\,/|K|}$. The claimed inequality then follows from applying the bipartite expander mixing lemma, Theorem 3.2.9, to $\mathrm{BiCay}(\Gamma, Y)$. □

Corollary 3.4.10 (Product-free sets)
Let Γ be a finite group with no nontrivial representations of dimension less than K. Let $X, Y, Z \subseteq \Gamma$. If there is no solution to $xy = z$ with $(x, y, z) \in X \times Y \times Z$, then

$$|X|\,|Y|\,|Z| < \frac{|\Gamma|^3}{K}.$$

In particular, every product-free $X \subseteq \Gamma$ has size less than $|\Gamma|\,/K^{1/3}$. (Here ***product-free*** means that there is no solution to $xy = z$ with $x, y, z \in X$.)

Proof. If there is no solution to $xy = z$, then the left-hand side of the inequality in Theorem 3.4.9 is $|X|\,|Y|\,|Z|\,/|\Gamma|$. Rearranging gives the result. □

The preceding result already shows that all product-free subsets of a quasirandom group must be small. This sharply contrasts the abelian setting. For example, in $\mathbb{Z}/n\mathbb{Z}$ (written additively), there is a sum-free subset of size around $n/3$ consisting of all group elements strictly between $n/3$ and $2n/3$.

Exercise 3.4.11 (Growth and expansion in quasirandom groups). Let Γ be a finite group with no nontrivial representations of dimension less than K. Let $X, Y, Z \subseteq \Gamma$. Suppose $|X|\,|Y|\,|Z| \geq |\Gamma|^3\,/K$. Then $XYZ = \Gamma$ (i.e., every element of Γ can be expressed as xyz for some $(x, y, z) \in X \times Y \times Z$).

Examples of Quasirandom Groups

***Example* 3.4.12** (Quasirandom groups). Here are some examples of groups with no small nontrivial representations.

(a) A classic result of Frobenius from around 1900 shows that every nontrivial representation of $\mathrm{PSL}(2, p)$ has dimension at least $(p-1)/2$ for all prime p. We will see a proof shortly. Jordan (1907) and Schur (1907) computed the character tables for $\mathrm{PSL}(2, q)$ for all prime power q. In particular, we know that every nontrivial representation of $\mathrm{PSL}(2, q)$ has dimension $\geq (q-1)/2$ for all prime power q.
(b) The alternating group A_m for $m \geq 2$ has order $m!/2$, and its smallest nontrivial representation has dimension $m - 1 = \Theta(\log n/\log\log n)$. The representations of symmetric and alternating groups have a nice combinatorial description using Young diagrams. See Sagan (2001) and Fulton and Harris (1991) for expository accounts of this theory.
(c) Gowers (2008, Theorem 4.7) gives an elementary proof that in every noncyclic simple group of order n, the smallest nontrivial representation has dimension at least $\sqrt{\log n}/2$.

Recall that the special linear group $\mathrm{SL}(2, p)$ is the group of 2×2 matrices (under multiplication) with determinant 1:

$$\mathrm{SL}(2, p) = \left\{ \begin{pmatrix} a & b \\ c & d \end{pmatrix} : a, b, c, d \in \mathbb{F}_p, ad - bc = 1 \right\}.$$

The projective special linear group $\mathrm{PSL}(2, p)$ is a quotient of $\mathrm{SL}(2, p)$ by all scalars; that is,

$$\mathrm{PSL}(2, p) = \mathrm{SL}(2, p)/\{\pm I\}\,.$$

The following result is due to Frobenius.

Theorem 3.4.13 ($\mathrm{PSL}(2, p)$ is quasirandom)
Let p be a prime. Then all nontrivial representations of $\mathrm{SL}(2, p)$ and $\mathrm{PSL}(2, p)$ have dimension at least $(p-1)/2$.

Proof. The claim is trivial for $p = 2$, so we can assume that p is odd. It suffices to prove the claim for $\mathrm{SL}(2,p)$. Indeed, any nontrivial representation of $\mathrm{PSL}(2,p)$ can be made into a representation of $\mathrm{SL}(2,p)$ by first passing through the quotient $\mathrm{SL}(2,p) \to \mathrm{SL}(2,p)/\{\pm I\} = \mathrm{PSL}(2,p)$.

Now suppose ρ is a nontrivial representation of $\mathrm{SL}(2,p)$. The group $\mathrm{SL}(2,p)$ is generated by the elements

$$g = \begin{pmatrix} 1 & 1 \\ 0 & 1 \end{pmatrix} \quad \text{and} \quad h = \begin{pmatrix} 1 & 0 \\ -1 & 1 \end{pmatrix}.$$

(Exercise: check!) These two elements are conjugate in $\mathrm{SL}(2,p)$ via $z = \left(\begin{smallmatrix} 1 & -1 \\ 1 & 0 \end{smallmatrix}\right)$ as $gz = zh$. If $\rho(g) = I$, then $\rho(h) = I$ by conjugation, and ρ would be trivial since g and h generate the group. So, $\rho(g) \neq I$. Since $g^p = I$, we have $\rho(g)^p = I$. So $\rho(g)$ is diagonalizable. (Here we use the fact that a matrix is diagonalizable if and only if its minimal polynomial has distinct roots, and that the minimal polynomial of $\rho(g)$ divides $X^p - 1$.) Since $\rho(g) \neq I$, $\rho(g)$ has an eigenvalue $\lambda \neq 1$. Since $\rho(g)^p = I$, λ is a primitive pth root of unity.

For every $a \in \mathbb{F}_p^\times$, g is conjugate to

$$\begin{pmatrix} a & 0 \\ 0 & a^{-1} \end{pmatrix}\begin{pmatrix} 1 & 1 \\ 0 & 1 \end{pmatrix}\begin{pmatrix} a^{-1} & 0 \\ 0 & a \end{pmatrix} = \begin{pmatrix} 1 & a^2 \\ 0 & 1 \end{pmatrix} = g^{a^2}.$$

Thus, $\rho(g)$ is conjugate to $\rho(g)^{a^2}$. Hence these two matrices have the same set of eigenvalues. So λ^{a^2} is an eigenvalue of $\rho(g)$ for every $a \in \mathbb{F}_p^\times$, and by ranging over all $a \in \mathbb{F}_p^\times$, this gives $(p-1)/2$ distinct eigenvalues of $\rho(g)$. (Recall that λ is a primitive pth root of unity.) It follows that $\dim \rho \geq (p-1)/2$. □

Applying Corollary 3.4.10 with Theorem 3.4.13 yields the following corollary (Gowers 2008). Note that the order of $\mathrm{PSL}(2,p)$ is $(p^3 - p)/2$.

Corollary 3.4.14 (Product-free subset of $\mathrm{PSL}(2,p)$)
The largest product-free subset of $\mathrm{PSL}(2,p)$ has size $O(p^{3-1/3})$.

In particular, there exist infinitely many groups of order n whose largest product-free subset has size $O(n^{8/9})$.

Before Gowers' work, it was not known whether every order n group has a product-free subset of size $\geq cn$ for some absolute constant $c > 0$. (This was Question 3.4.1, asked by Babai and Sós.) Gowers' result shows that the answer is no.

In the other direction, Kedlaya (1997; 1998) showed that every finite group of order n has a product-free subset of size $\gtrsim n^{11/14}$. In fact, he showed that if the group has a proper subgroup H of index m, then there is a product-free subset that is a union of $\gtrsim m^{1/2}$ cosets of H.

Equivalence of Quasirandomness Conditions

We saw that having no small nontrivial representations is a useful property of groups. Gowers further showed that this group representation theoretic property is equivalent to several other characterizations of the group.

Theorem 3.4.15 (Quasirandom groups)
Let Γ_n be a sequence of finite groups of increasing order. The following are equivalent:
REP The dimension of the smallest nontrivial representation of Γ_n tends to infinity.
GRAPH Every sequence of bipartite Cayley graphs on Γ_n, as $n \to \infty$, is quasirandom in the sense of Theorem 3.1.25.
PRODFREE The largest product-free subset of Γ_n has size $o(|\Gamma_n|)$.
QUOTIENT For every proper normal subgroup H of Γ_n, the quotient Γ_n/H is nonabelian and has order tending to infinity as $n \to \infty$.

Let us comment on the various implications.

By Theorem 3.4.7, **REP** implies **GRAPH**. For the converse, we need to construct a nonquasirandom Cayley graph on each group with a nontrivial representation of bounded dimension. One can first construct a weighted analogue of a bipartite Cayley graph with large eigenvalues by appealing to formulas from nonabelian Fourier transform (see Remark 3.4.17). And then one can sample a genuine bipartite Cayley graph from the weighted version.

By Corollary 3.4.10, **REP** implies **PRODFREE**. The converse is proved in Gowers (2008) using elementary methods. It was later proved with better polynomial quantitative dependence in Nikolov and Pyber (2011), who proved the following result.

Theorem 3.4.16 (PRODFREE implies REP)
Let Γ be a group with a nontrivial representation of dimension K. Then Γ has a product-free subset of size at least $c\,|\Gamma|\,/K$, where $c > 0$ is some absolute constant.

To see that **REP** implies **QUOTIENT**, note that any nontrivial representation of Γ/H is automatically a representation of Γ after passing through the quotient. Furthermore, every nontrivial abelian group has a nontrivial 1-dimensional representation, and every group of order $m > 1$ has a nontrivial representation of dimension $< \sqrt{m}$. For the proof of the converse, see Gowers (2008, Theorem 4.8). (This implication has an exponential dependence of parameters.)

***Remark* 3.4.17** (Nonabelian Fourier analysis). (This is an advanced remark and can be skipped over.) Section 3.3 discussed the Fourier transform on finite abelian groups. The topic of this section can be alternatively viewed through the lenses of the nonabelian Fourier transform. We refer to Wigderson (2012) for a tutorial on the nonabelian Fourier transform from a combinatorial perspective.

Let us give here the recipe for computing the eigenvalues and an orthonormal basis of eigenvectors of $\mathrm{Cay}(\Gamma, S)$.

For each irreducible representation ρ of Γ (always working over $\mathbb{C}$), let

$$M_\rho := \sum_{s \in S} \rho(s),$$

viewed as a $\dim \rho \times \dim \rho$ matrix over $\mathbb{C}$. Then M_ρ has $\dim \rho$ eigenvalues $\lambda_{\rho,1}, \dots, \lambda_{\rho,\dim\rho}$.

Here is how to list all the eigenvalues of the adjacency matrix of $\mathrm{Cay}(\Gamma, S)$: repeating each $\lambda_{\rho,i}$ with multiplicity $\dim \rho$, ranging over all irreducible representations ρ and all $1 \le i \le \dim \rho$.

To emphasize, the eigenvalues always come in bundles with multiplicities determined by the dimensions of the irreducible representations of Γ. (Although it is possible for there to be additional coalescence of eigenvalues.)

One can additionally recover a system of eigenvectors of $\mathrm{Cay}(\Gamma, S)$. For each eigenvector v with eigenvalue λ of M_ρ, and every $w \in \mathbb{C}^{\dim \rho}$, set $x^{\rho,v,w} \in \mathbb{C}^\Gamma$ with coordinates

$$x_g^{\rho,v,w} = \langle \rho(g)v, w \rangle$$

for all $g \in \Gamma$. Then x is an eigenvector of $\mathrm{Cay}(\Gamma, S)$ with eigenvalue λ. Now if we ρ range over all irreducible representations of Γ, and let v range over an orthonormal basis of eigenvectors of M_ρ (letting λ be the corresponding eigenvalue), and let w range over an orthonormal basis of eigenvectors of $\mathbb{C}^{\dim \rho}$, then $x^{\rho,v,w}$ ranges over an orthogonal system of eigenvectors of $\mathrm{Cay}(\Gamma, S)$. The eigenvalue associated to $x^{\rho,v,w}$ is λ.

A basic theorem in representation theory tells us that the regular representation decomposes into a direct sum of $\dim \rho$ copies of ρ ranging over every irreducible representation ρ of Γ. This decomposition then corresponds to a block diagonalization (simultaneously for all S) of the adjacency matrix of $\mathrm{Cay}(\Gamma, S)$ into blocks M_ρ, repeated $\dim \rho$ times, for each ρ. The preceding statement comes from interpreting this block diagonalization.

The matrix M_ρ, appropriately normalized, is the ***nonabelian Fourier transform*** of the indicator vector of S at ρ. Many basic and important formulas for Fourier analysis over abelian groups, for example inversion and Parseval (which we will see in Chapter 6), have nonabelian analogs.

3.5 Quasirandom Cayley Graphs and Grothendieck's Inequality

Let us examine the following two sparse quasirandom graph conditions (see Remark 3.1.29).

Definition 3.5.1 (Sparse quasirandom graphs)
Let G be an n-vertex d-regular graph. We say that G satisfies property
SparseDISC(ε) If $\left|e(X,Y) - \frac{d}{n}|X|\,|Y|\right| \le \varepsilon dn$ for all $X, Y \subseteq V(G)$;
SparseEIG(ε) If G is an (n, d, λ)-graph for some $\lambda \le \varepsilon d$.

In Section 3.1, we saw that when d grows linearly in n, then these two conditions are equivalent up to a polynomial change in the constant ε. As discussed in Remark 3.1.29, many quasirandomness equivalences break down for sparse graphs, meaning $d = o(n)$ here. Some still hold, for example:

Proposition 3.5.2 (SparseEIG implies SPARSEDISC)
Among regular graphs,

$$\textbf{SparseEIG}(\varepsilon) \quad \text{implies} \quad \textbf{SparseDISC}(\varepsilon).$$

Proof. In an (n, d, λ) graph with $\lambda \le \varepsilon d$, by the expander mixing lemma (Theorem 3.2.4), for every vertex subset X and Y,

$$\left| e(X,Y) - \frac{d}{n}|X|\,|Y| \right| \le \lambda\sqrt{|X|\,|Y|} \le \varepsilon d\sqrt{|X|\,|Y|} \le \varepsilon dn.$$

So, the graph satisfies **SparseDISC**(ε). □

The converse fails badly. Consider the disjoint union of a large random d-regular graph and a K_{d+1} (here $d = o(n)$).

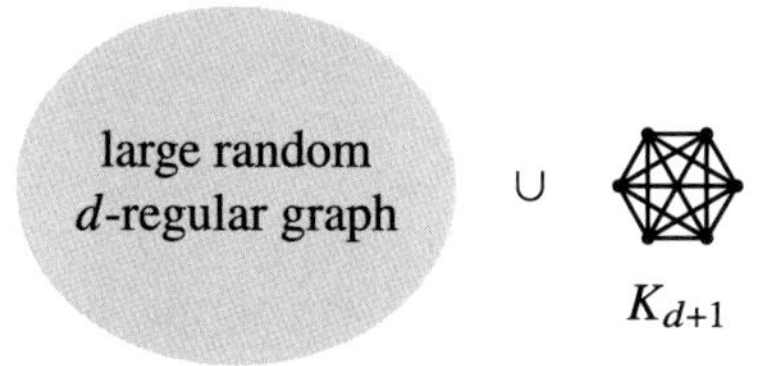

This graph satisfies **SparseDISC**($o(1)$) since it is satisfied by the large component, and the small component K_{d+1} contributes negligibly to discrepancy due to its size. On the other hand, each connected component contributes an eigenvalue of d (by taking the all-1 vector supported on each component), and so **SparseEIG**(ε) fails for any $\varepsilon < 1$.

The main result of this section is that despite the preceding example, if we restrict ourselves to Cayley graphs (abelian or nonabelian), **SparseDISC**(ε) and **SparseEIG**(ε) are always equivalent up to a linear change in ε. This result is due to Conlon and Zhao (2017).

Theorem 3.5.3 (SparseDISC implies SparseEIG for Cayley graphs)
Among Cayley graphs,

$$\textbf{SparseDISC}(\varepsilon) \quad \text{implies} \quad \textbf{SparseEIG}(8\varepsilon).$$

As in Section 3.4, we prove the preceding result more generally for vertex-transitive graphs. (See Definition 3.4.5).

Theorem 3.5.4 (SparseDISC implies SparseEIG for vertex-transitive graphs)
Among vertex-transitive graphs,

$$\textbf{SparseDISC}(\varepsilon) \quad \text{implies} \quad \textbf{SparseEIG}(8\varepsilon).$$

Grothendieck's Inequality

The proof of the preceding theorem uses the following important inequality from functional analysis due to Grothendieck (1953).

Given a matrix $A = (a_{i,j}) \in \mathbb{R}^{m\times n}$, we can consider its $\ell^\infty \to \ell^1$ norm

$$\sup_{\|y\|_\infty \le 1} \|Ay\|_{\ell^1},$$

which can also be written as (exercise: check! Also see Lemma 4.5.3 for a related fact about the cut norm of graphons)

$$\sup_{\substack{x\in\{-1,1\}^m \\ y\in\{-1,1\}^n}} \langle x, Ay\rangle = \sup_{\substack{x_1,\cdots,x_m\in\{-1,1\} \\ y_1,\ldots,y_n\in\{-1,1\}}} \sum_{i=1}^{n}\sum_{j=1}^{n} a_{i,j}x_iy_j. \tag{3.3}$$

This quantity is closely related to discrepancy.

One can consider a ***semidefinite relaxation*** of the preceding quantity:

$$\sup_{\substack{\|x_1\|,\dots,\|x_m\|\le 1\\ \|y_1\|,\dots,\|y_n\|\le 1}} \sum_{i=1}^{m}\sum_{j=1}^{n} a_{i,j}\left\langle x_i, y_j\right\rangle, \tag{3.4}$$

where the surpremum is taken over vectors $x_1, \dots, x_m, y_1, \dots, y_n$ in the unit ball of some real Hilbert space, whose norm is denoted by $\| \ \|$. Without loss of generality, we can assume that these vectors lie in $\mathbb{R}^{m+n}$ with the usual Euclidean norm. (Here $m+n$ dimensions are enough, since $x_1, \dots, x_m, y_1, \dots, y_n$ span a real subspace of dimension at most $m+n$.)

We always have

$$(3.3) \le (3.4)$$

by restricting the vectors in (3.4) to $\mathbb{R}$. There are efficient algorithms (both in theory and in practice) using **semidefinite programming** to solve (3.4), whereas no efficient algorithm is believed to exist for computing (3.3) (Alon and Naor 2006).

Grothendieck's inequality says that this semidefinite relaxation never loses more than a constant factor.

Theorem 3.5.5 (Grothendieck's inequality)
There exists a constant $K > 0$ ($K = 1.8$ works) such that for all matrices $A = (a_{i,j}) \in \mathbb{R}^{m\times n}$,

$$\sup_{\|x_i\|,\|y_j\|\le 1} \sum_{i=1}^{m}\sum_{j=1}^{n} a_{i,j}\left\langle x_i, y_j\right\rangle \le K \sup_{x_i,y_j\in\{\pm 1\}} \sum_{i=1}^{m}\sum_{j=1}^{n} a_{i,j}x_iy_j,$$

where the left-hand-side supremum is taken over vectors $x_1, \dots, x_n, y_1, \dots, y_m$ in the unit ball of some real Hilbert space.

Remark* 3.5.6.** The optimal constant K is known as the ***real Grothendieck's constant. Its exact value is unknown. It is known to lie within $[1.676, 1.783]$. There is also a complex version of Grothendieck's inequality, where the left-hand side uses a complex Hilbert space (and places an absolute value around the final sum). The corresponding ***complex Grothendieck's constant*** is known to lie within $[1.338, 1.405]$.

We will not prove Grothendieck's inequality here. See Alon and Naor (2006) for three proofs of the inequality, along with algorithmic discussions.

Proof That SparseDISC Implies SparseEIG for Vertex-Transitive Graphs

Proof of Theorem 3.5.4. Let G be an n-vertex d-regular graph with a vertex-transitive group Γ of automorphisms. Suppose G satisfies **SparseDISC**(ε). Let A be the adjacency matrix of G. Write

$$B = A - \frac{d}{n}J,$$

where J is the $n \times n$ all-1 matrix. To show that G is an (n, d, λ)-graph with $\lambda \le \varepsilon d$, it suffices to show that B has operator norm $\|B\| \le \varepsilon d$. (Here we are using that G is d-regular, so the all-1 eigenvector of A with eigenvalue d becomes an eigenvector of B with eigenvalue zero 0.)

For any $X, Y \subseteq V(G)$, the corresponding indicator vectors $x = \mathbf{1}_X \in \mathbb{R}^n$ and $y = \mathbf{1}_Y \in \mathbb{R}^n$ satisfy, by **SparseDISC**(ε),

$$|\langle x, By\rangle| = \left|e(X,Y) - \frac{d}{n}|X|\,|Y|\right| \le \varepsilon dn.$$

Then, for any $x, y \in \{-1,1\}^n$, we can write $x = x^+ - x^-$ and $y = y^+ - y^-$ with $x^+, x^-, y^+, y^- \in \{0,1\}^n$. Since

$$\langle x, By\rangle = \langle x^+, By^+\rangle - \langle x^+, By^-\rangle - \langle x^-, By^+\rangle + \langle x^-, By^-\rangle$$

and each term on the right-hand side is at most εdn in absolute value, we have

$$|\langle x, By\rangle| \le 4\varepsilon dn \quad \text{for all } x, y \in \{-1,1\}^n. \tag{3.5}$$

For any graph automorphism $g \in \Gamma$ and any $x = (x_1, \ldots, x_n) \in \mathbb{R}^n$ and $j \in [n]$, write

$$x^j = \left(\sqrt{\frac{n}{|\Gamma|}}\, x_{g(j)} : g \in \Gamma\right) \in \mathbb{R}^\Gamma.$$

For every unit vector $x \in \mathbb{R}^n$, the vector $x^j \in \mathbb{R}^\Gamma$ is a unit vector, since $x_1^2 + \cdots + x_n^2 = 1$ and the map $g \mapsto g(j)$ is $n/|\Gamma|$-to-1 for each j. Similarly define y^j for any $y \in \mathbb{R}^n$ and $j \in [n]$. Furthermore, $B_{i,j} = B_{g(i),g(j)}$ for any $g \in \Gamma$ and $j \in [n]$ due to g being a graph automorphism.

To prove the operator norm bound $\|B\| \le 8\varepsilon d$, it suffices to show that $\langle x, By\rangle \le 8\varepsilon d$ for every pair of unit vectors $x, y \in \mathbb{R}^n$. We have

$$\begin{aligned}\langle x, By\rangle = \sum_{i,j=1}^n B_{i,j} x_i y_j &= \frac{1}{|\Gamma|}\sum_{g\in\Gamma}\sum_{i,j=1}^n B_{g(i),g(j)} x_{g(i)} y_{g(j)} \\ &= \frac{1}{|\Gamma|}\sum_{g\in\Gamma}\sum_{i,j=1}^n B_{i,j} x_{g(i)} y_{g(j)} = \frac{1}{n}\sum_{i,j=1}^n B_{i,j}\langle x^i, y^j\rangle \le 8\varepsilon d.\end{aligned}$$

The final step follows from Grothendieck's inequality (applied with $K \le 2$) along with (3.5). This completes the proof of **SparseEIG**(8ε). □

3.6 Second Eigenvalue: Alon–Boppana Bound

The expander mixing lemma tells us that in an (n, d, λ)-graph, a smaller value of λ guarantees stronger pseudorandomness properties. In this chapter, we explore the following natural extremal question.

Question 3.6.1 (Minimum second eigenvalue)
Fix a positive integer d. What is the smallest possible λ (as a function of d alone) such that there exist infinitely many $(n, d, \lambda + o(1))$-graphs, where the $o(1)$ is some quantity that goes to zero as $n \to \infty$?

The answer turns out to be

$$\lambda = 2\sqrt{d-1}.$$

A significant fact about this quantity is that it is the spectral radius of the infinite d-regular tree.

The following result gives the lower bound on λ (Alon 1986).

Theorem 3.6.2 (Alon–Boppana second eigenvalue bound)
Fix a positive integer d. Let G be an n-vertex d-regular graph. If $\lambda_1 \geq \cdots \geq \lambda_n$ are the eigenvalues of its adjacency matrix, then

$$\lambda_2 \geq 2\sqrt{d-1} - o(1),$$

where $o(1) \to 0$ as $n \to \infty$.

In particular, the Alon–Boppana bound implies that $\max\{|\lambda_2|, |\lambda_n|\} \geq 2\sqrt{d-1} - o(1)$, which can be restated as below.

Corollary 3.6.3 (Alon–Boppana second eigenvalue bound)
For every fixed d and $\lambda < 2\sqrt{d-1}$, there are only finitely many (n, d, λ)-graphs.

We will see two different proofs. The first proof (Nilli 1991) constructs an eigenvector explicitly. The second proof (only for Corollary 3.6.3) uses the trace method to bound moments of the eigenvalues via counting closed walks.

Lemma 3.6.4 (Test vector)
Let $G = (V, E)$ be a d-regular graph. Let A be the adjacency matrix of G. Let r be a positive integer. Let st be an edge of G. For each $i \geq 0$, let V_i denote the set of all vertices at distance exactly i from $\{s,t\}$ (so that, in particular, $V_0 = \{s,t\}$). Let $x = (x_v)_{v \in V} \in \mathbb{R}^V$ be a vector with coordinates

$$x_v = \begin{cases} (d-1)^{-i/2} & \text{if } v \in V_i \text{ and } i \leq r, \\ 0 & \text{otherwise, i.e., } \operatorname{dist}(v, \{s,t\}) > r. \end{cases}$$

Then

$$\frac{\langle x, Ax \rangle}{\langle x, x \rangle} \geq 2\sqrt{d-1}\left(1 - \frac{1}{r+1}\right).$$

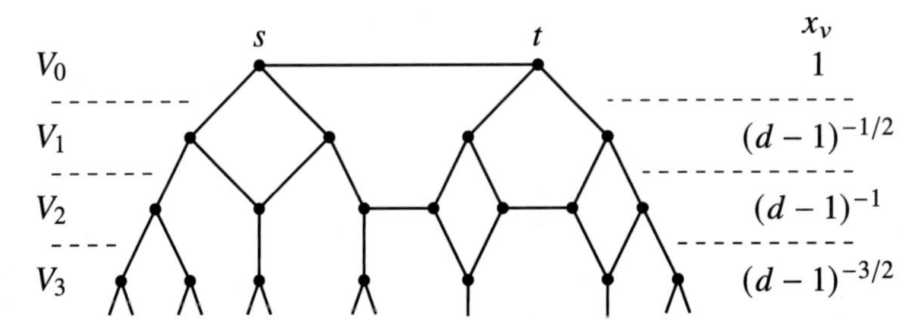

Proof. Let $L = dI - A$. (This is called the ***Laplacian matrix*** of G.) The claim can be rephrased as an upper bound on $\langle x, Lx \rangle / \langle x, x \rangle$. Here is an important and convenient formula (which can be easily proved by expanding):

$$\langle x, Lx \rangle = \sum_{uv \in E} (x_u - x_v)^2.$$

Since x_v is constant for all v in the same V_i, we only need to consider edges spanning consecutive V_i's. Using the formula for x, we obtain

$$\langle x, Lx \rangle = \sum_{i=0}^{r-1} e(V_i, V_{i+1}) \left(\frac{1}{(d-1)^{i/2}} - \frac{1}{(d-1)^{(i+1)/2}} \right)^2 + \frac{e(V_r, V_{r+1})}{(d-1)^r}.$$

For each $i \geq 0$, each vertex in V_i has at most $d-1$ neighbors in V_{i+1}, so $e(V_i, V_{i+1}) \leq (d-1)\,|V_i|$. Thus, continuing from the preceding equation,

$$\begin{aligned}
&\leq \sum_{i=0}^{r-1} |V_i|\,(d-1) \left(\frac{1}{(d-1)^{i/2}} - \frac{1}{(d-1)^{(i+1)/2}} \right)^2 + \frac{|V_r|\,(d-1)}{(d-1)^r} \\
&= \left(\sqrt{d-1} - 1 \right)^2 \sum_{i=0}^{r-1} \frac{|V_i|}{(d-1)^i} + \frac{|V_r|\,(d-1)}{(d-1)^r} \\
&= \left(d - 2\sqrt{d-1} \right) \sum_{i=0}^{r} \frac{|V_i|}{(d-1)^i} + \left(2\sqrt{d-1} - 1 \right) \frac{|V_r|}{(d-1)^r}.
\end{aligned}$$

We have $|V_{i+1}| \leq (d-1)\,|V_i|$ for every $i \geq 0$, so that $|V_r|\,(d-1)^{-r} \leq |V_i|\,(d-1)^{-i}$ for each $i \leq r$. So, continuing,

$$\begin{aligned}
&\leq \left(d - 2\sqrt{d-1} + \frac{2\sqrt{d-1} - 1}{r+1} \right) \sum_{i=0}^{r} \frac{|V_i|}{(d-1)^i} \\
&= \left(d - 2\sqrt{d-1} + \frac{2\sqrt{d-1} - 1}{r+1} \right) \langle x, x \rangle .
\end{aligned}$$

It follows that

$$\begin{aligned}
\frac{\langle x, Ax \rangle}{\langle x, x \rangle} = d - \frac{\langle x, Lx \rangle}{\langle x, x \rangle} &\geq \left(2\sqrt{d-1} - \frac{2\sqrt{d-1} - 1}{r+1} \right) \\
&\geq \left(1 - \frac{1}{r+1} \right) 2\sqrt{d-1}. \qquad \square
\end{aligned}$$

Proof of the Alon–Boppana bound (Theorem 3.6.2). Let $V = V(G)$. Let $\mathbf{1}$ be the all-1's vector, which is an eigenvector with eigenvalue d. To prove the theorem, it suffices to exhibit a nonzero vector $z \perp \mathbf{1}$ such that

$$\frac{\langle z, Az \rangle}{\langle z, z \rangle} \geq 2\sqrt{d-1} - o(1).$$

Let r be an arbitrary positive integer. When n is sufficiently large, there exist two edges st and $s't'$ in the graph with distance at least $2r+2$ apart (indeed, since the number of vertices within distance k of an edge is $\leq 2(1 + (d-1) + (d-1)^2 + \cdots + (d-1)^k)$). Let $x \in \mathbb{R}^V$ be the vector constructed as in Lemma 3.6.4 for st, and let $y \in \mathbb{R}^V$ be the corresponding vector constructed for $s't'$. Recall that x is supported on vertices within distance r from st, and likewise with y and $s't'$. Since st and $s't'$ are at distance at least $2r+2$ apart, the support of x is at distance at least 2 from the support of y. Thus

$$\langle x, y \rangle = 0 \quad \text{and} \quad \langle x, Ay \rangle = 0.$$

Choose a constant $c \in \mathbb{R}$ such that $z = x - cy$ has sum of its entries equal to zero. (This is possible since $\langle y, \mathbf{1} \rangle > 0$.) Then

$$\langle z, z \rangle = \langle x, x \rangle + c^2 \langle y, y \rangle$$

and so by Lemma 3.6.4

$$\begin{aligned}\langle z, Az \rangle &= \langle x, Ax \rangle + c^2 \langle y, Ay \rangle \\ &\geq \left(1 - \frac{1}{r+1}\right) 2\sqrt{d-1} \left(\langle x, x \rangle + c^2 \langle y, y \rangle\right) \\ &= \left(1 - \frac{1}{r+1}\right) 2\sqrt{d-1} \langle z, z \rangle .\end{aligned}$$

Taking $r \to \infty$ as $n \to \infty$ gives the theorem. □

***Remark* 3.6.5.** The preceding proof cleverly considers distance from an *edge* rather than from a single vertex. Why would the proof fail if we had instead considered distance from a vertex?

Now let us give another proof – actually we will prove only the slightly weaker statement of Corollary 3.6.3, which is equivalent to

$$\max \{|\lambda_2|, |\lambda_n|\} \geq 2\sqrt{d-1} - o(1). \tag{3.6}$$

As a warm-up, let us first prove (3.6) with $\sqrt{d} - o(1)$ on the right-hand side. We have

$$dn = 2e(G) = \operatorname{tr} A^2 = \sum_{i=1}^{n} \lambda_i^2 \leq d^2 + (n-1) \max \{|\lambda_2|, |\lambda_n|\}^2 .$$

So,

$$\max \{|\lambda_2|, |\lambda_n|\} \geq \sqrt{\frac{d(n-d)}{n-1}} = \sqrt{d} - o(1)$$

as $n \to \infty$ for fixed d.

To prove (3.6), we consider higher moments $\operatorname{tr} A^k$. This is a useful technique, sometimes called the **trace method** or the **moment method**.

Alternative Proof of (3.6). The quantity

$$\operatorname{tr} A^{2k} = \sum_{i=1}^{n} \lambda_i^{2k}$$

counts the number of closed walks of length $2k$ on G. Let $\mathbb{T}_d$ denote the infinite d-regular tree. Observe that

$$\begin{aligned}&\text{\# closed length-}2k\text{ walks in } G \text{ starting from a fixed vertex} \\ &\qquad \geq \text{\# closed length-}2k\text{ walks in } \mathbb{T}_d \text{ starting from a fixed vertex.}\end{aligned}$$

Indeed, at each vertex, for both G and $\mathbb{T}_d$, we can label its d incident edges arbitrarily from 1 to d. (The labels assigned from the two endpoints of the same edge do not have to match.) Then every closed length-$2k$ walk in $\mathbb{T}_d$ corresponds to a distinct closed length-$2k$ walk in G by tracing the same outgoing edges at each step. (Why?) Note that not all closed walks in G arise this way (e.g., walks that go around cycles in G).

The number of closed walks of length $2k$ on an infinite d-regular graph starting at a fixed root is at least $(d-1)^k C_k$, where $C_k = \frac{1}{k+1}\binom{2k}{k}$ is the kth Catalan number. To see this, note that each step in the walk is either "away from the root" or "toward the root." We record a sequence by denoting steps of the former type by + and of the latter type by −.

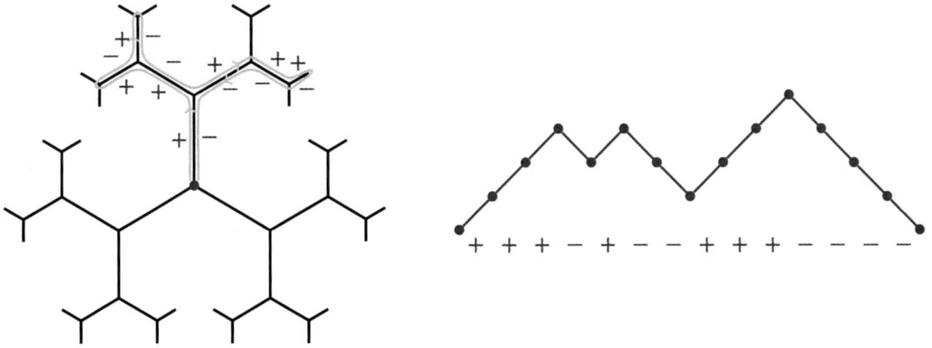

Then the number of valid sequences permuting k +'s and k −'s is exactly counted by the Catalan number C_k, as the only constraint is that there can never be more −'s than +'s up to any point in the sequence. Finally, there are at least $d-1$ choices for where to step in the walk at any + (there are d choices at the root), and exactly one choice for each −.

Thus, the number of closed walks of length $2k$ in G is at least

$$\operatorname{tr} A^{2k} \geq n(d-1)^k C_k \geq \frac{n}{k+1}\binom{2k}{k}(d-1)^k.$$

On the other hand, we have

$$\operatorname{tr} A^{2k} = \sum_{i=1}^{n} \lambda_i^{2k} \leq d^{2k} + (n-1)\max\left\{|\lambda_2|, |\lambda_n|\right\}^{2k}.$$

Thus,

$$\max\left\{|\lambda_2|, |\lambda_n|\right\}^{2k} \geq \frac{1}{k+1}\binom{2k}{k}(d-1)^k - \frac{d^{2k}}{n-1}.$$

The term $\frac{1}{k+1}\binom{2k}{k}$ is $(2-o(1))^{2k}$ as $k \to \infty$. Letting $k \to \infty$ slowly (e.g., $k = o(\log n)$) as $n \to \infty$ gives us $\max\{|\lambda_2|, |\lambda_n|\} \geq 2\sqrt{d-1} - o(1)$. □

***Remark* 3.6.6.** The infinite d-regular graph $\mathbb{T}_d$ is the universal cover of all d-regular graphs. (This fact is used in the first step of the argument.) The spectral radius of $\mathbb{T}_d$ is $2\sqrt{d-1}$, which is the fundamental reason why this number arises in the Alon–Boppana bound.

Graphs with $\lambda_2 \approx 2\sqrt{d-1}$

Let us return to Question 3.6.1: what is the smallest possible λ_2 for n-vertex d-regular graphs, with d fixed and n large? Is the Alon–Boppana bound tight? (The answer is yes.)

Alon's second eigenvalue conjecture says that random d-regular graphs match the Alon–Boppana bound. This was proved by Friedman (2008). We will not present the proof, as it is quite a difficult result.

Theorem 3.6.7 (Friedman's second eigenvalue theorem)
Fix positive integer d and $\lambda > 2\sqrt{d-1}$. With probability $1 - o(1)$ as $n \to \infty$ (with n even if d is odd), a uniformly chosen random n-vertex d-regular graph is an (n, d, λ)-graph.

In other words, the preceding theorem says that random d-random graphs on n vertices satisfy, with probability $1 - o(1)$ (for fixed $d \geq 3$ and $n \to \infty$),

$$\max\{|\lambda_2|, |\lambda_n|\} \leq 2\sqrt{d-1} + o(1).$$

Can we get $\leq 2\sqrt{d-1}$ exactly without an error term? This leads us to one of the biggest open problems of the field.

Definition 3.6.8 (Ramanujan graph)
A ***Ramanujan graph*** is an (n, d, λ)-graph with $\lambda = 2\sqrt{d-1}$. In other words, it is a d-regular graph whose adjacency matrix has all eigenvalues, except the top one, at most $2\sqrt{d-1}$ in absolute value.

A major open problem is to show the existence of infinite families of d-regular Ramanujan graphs.

Conjecture 3.6.9 (Existence of Ramanujan graphs)
For every positive integer $d \geq 3$, there exist infinitely many d-regular Ramanujan graphs.

While it is not too hard to construct small Ramanujan graphs (e.g., K_{d+1} has eigenvalues $\lambda_1 = d$ and $\lambda_2 = \cdots = \lambda_n = -1$), it is a major open problem to construct infinitely many d-regular Ramanujan graphs for each d.

The term "Ramanujan graph" was coined by Lubotzky, Phillips, and Sarnak (1988), who constructed infinite families of d-regular Ramanujan graphs when $d - 1$ is an odd prime. The same result was independently proved by Margulis (1988). The proof of the eigenvalue bounds uses deep results from number theory, namely solutions to the Ramanujan conjecture (hence the name). These constructions were later extended by Morgenstern (1994) whenever $d - 1$ is a prime power. The current state of Conjecture 3.6.9 is given below, and it remains open for all other d, with the smallest open case being $d = 7$.

Theorem 3.6.10 (Existence of Ramanujan graphs)
If $d - 1$ is a prime power, then there exist infinitely many d-regular Ramanujan graphs.

All known results are based on explicit constructions using Cayley graphs on $\mathrm{PSL}(2, q)$ or related groups. We refer the reader to Davidoff, Sarnak, and Valette (2003) for a gentle exposition of the construction.

Theorem 3.6.7 says that random d-regular graphs are "nearly-Ramanujan." Empirical evidence suggests that for each fixed d, a uniform random n-vertex d-regular graph is Ramanujan with probability bounded away from 0 and 1, for large n.

Conjecture 3.6.11 (A random d-regular graph is likely Ramanujan)
For every $d \geq 3$, there is some $c_d > 0$ so that for all sufficiently large n (with n even if d is odd), a uniformly chosen random n-vertex d-regular graph is Ramanujan with probability at least c_d.

If this were true, it would prove Conjecture 3.6.9 on the existence of Ramanujan graphs. However, no rigorous results are known in this vein.

One can formulate a bipartite analog.

Definition 3.6.12 (Bipartite Ramanujan graph)
A ***bipartite Ramanujan graph*** is some bipartite-(n, d, λ)-graph with $\lambda = 2\sqrt{d-1}$.

Given a Ramanujan graph G, we can turn it into a bipartite Ramanujan graph $G \times K_2$. So the existence of bipartite Ramanujan graphs is weaker than that of Ramanujan graphs. Nevertheless, for a long time, it was not known how to construct infinite families of bipartite Ramanujan graphs other than using Ramanujan graphs. A breakthrough by Marcus, Spielman, and Srivastava (2015) completely settled the bipartite version of the problem. Unlike earlier construction of Ramanujan graphs, their proof is existential (i.e., nonconstructive) and introduces an important technique of *interlacing families of polynomials*.

Theorem 3.6.13 (Bipartite Ramanujan graphs of every degree)
For every $d \geq 3$, there exist infinitely many d-regular bipartite Ramanujan graphs.

Exercise 3.6.14 (Alon–Boppana bound with multiplicity). Prove that for every positive integer d and real $\varepsilon > 0$, there is some constant $c > 0$ so that every n-vertex d-regular graph has at least cn eigenvalues greater than $2\sqrt{d-1} - \varepsilon$.

Exercise 3.6.15* (Net removal decreases top eigenvalue). Show that for every d and r, there is some $\varepsilon > 0$ such that if G is a d-regular graph, and $S \subseteq V(G)$ is such that every vertex of G is within distance r of S, then the top eigenvalue of the adjacency matrix of $G - S$ (i.e., remove S and its incident edges from G) is at most $d - \varepsilon$.

Further Reading

The survey *Pseudo-random Graphs* by Krivelevich and Sudakov (2006) discusses many combinatorial aspects of this topic.

Expander graphs are a large and intensely studied topic, partly due to many important applications in computer science. Here are two important survey articles:

- *Expander Graphs and Their Applications* by Hoory, Linial, and Wigderson (2006);
- *Expander Graphs in Pure and Applied Mathematics* by Lubotzky (2012).

For spectral graph theory, see the book *Spectral Graph Theory* by Chung (1997), or the book draft *Spectral and Algebraic Graph Theory* by Spielman.

The book *Elementary Number Theory, Group Theory and Ramanujan Graphs* by Davidoff, Sarnak, and Valette (2003) gives a gentle introduction to the construction of Ramanujan graphs.

The breakthrough by Marcus, Spielman, and Srivastava (2015) in constructing bipartite Ramanujan graphs via interlacing polynomials is an instant classic.

Chapter Summary

- We are interested in quantifying how a given graph can be similar to a random graph.
- The **Chung–Graham–Wilson quasirandom graphs theorem** says that several notions are equivalent, notably:
  - **DISC**: edge discrepancy,
  - **C_4**: 4-cycle count close to random, and
  - **EIG**: all eigenvalues (except the largest) small.

  These equivalences only apply to graphs at constant order edge density. Some of the implications break down for sparser graphs.
- An **(n, d, λ)-graph** is an n-vertex d-regular graph all of whose adjacency matrix eigenvalues are $\leq \lambda$ in absolute value except the top one (which must be d). The second eigenvalue plays an important role in pseudorandomness.
- **Expander mixing lemma.** An (n, d, λ)-graph satisfies

$$\left| e(X,Y) - \frac{d}{n} |X|\,|Y| \right| \leq \lambda \sqrt{|X|\,|Y|} \qquad \text{for all } X, Y \subseteq V(G).$$

- The **eigenvalues of an abelian Cayley graph** $\mathrm{Cay}(\Gamma, S)$ can be computed via the Fourier transform of $1_S S$. For example, using a Gauss sum, one can deduce that the Paley graph (generated by quadratic residues in $\mathbb{Z}/p\mathbb{Z}$) is quasirandom.
- A nonabelian group with no small nontrivial representations is called a **quasirandom group**.
  - Every Cayley graph on a quasirandom group is a quasirandom graph.
  - There are no large **product-free sets** in a quasirandom group.
  - Example of quasirandom group: $\mathrm{PSL}(2, p)$, which has order $(p^3 - p)/2$, and all nontrivial representations have dimension $\geq (p-1)/2$.
- Among vertex-transitive graphs (which includes all Cayley graphs), the sparse analogues of the discrepancy property (**SparseDISC**) and small second eigenvalue property (**SparseEIG**) are equivalent up to a linear change of the error tolerance parameter. This equivalence is false for general graphs.
  - The proof applies **Grothendieck's inequality**, which says that the semidefinite relaxation of the $\ell^\infty \to \ell^1$ norm (equivalent to the cut norm) gives a constant factor approximation.
- **Alon–Boppana second eigenvalue bound.** Every d-regular graph has second largest eigenvalue $\geq 2\sqrt{d-1} - o(1)$ for the adjacency matrix, with d fixed as the number of vertices goes to infinity.
  - Two spectral proof methods: (1) constructing a test vector and (2) trace/moment method.
  - The constant $2\sqrt{d-1}$ is best possible, as a random d-regular graph is typically an (n, d, λ)-graph with $\lambda = 2\sqrt{d-1} + o(1)$ (Friedman's theorem).
  - A **Ramanujan graph** is an (n, d, λ)-graph with $\lambda = 2\sqrt{d-1}$. It is conjectured that for every $d \geq 3$, there exist infinitely many d-regular Ramanujan graphs. (This is known to hold when $d - 1$ is a prime power.) A bipartite version of this conjecture is true.

4

Graph Limits

Chapter Highlights

- An analytic language for studying dense graphs
- Convergence and limit for a sequence of graphs
- Compactness of the graphon space with respect to the cut metric
- Applications of compactness
- Equivalence of cut metric convergence and left-convergence

The theory of graph limits was developed by Lovász and his collaborators in a series of works starting around 2003. The researchers were motivated by questions about very large graphs from several different angles, including from combinatorics, statistical physics, computer science, and applied math. Graph limits give an analytic framework for analyzing large graphs. The theory offers both a convenient mathematical language as well as powerful theorems.

Motivation

Suppose we live in a hypothetical world where we only had access to rational numbers and had no language for irrational numbers. We are given the following optimization problem:

$$\text{minimize } x^3 - x \text{ subject to } 0 \le x \le 1.$$

The minimum occurs at $x = 1/\sqrt{3}$, but this answer does not make sense over the rationals. With only access to rationals, we can state a progressively improving sequence of answers that converge to the optimum. This is rather cumbersome. It is much easier to write down a single real number expressing the answer.

Now consider an analogous question for graphs. Fix some real $p \in [0, 1]$. We want to

$$\begin{aligned} \text{minimize} \quad & (\text{\# closed walks of length 4})/n^4 \\ \text{among} \quad & n\text{-vertex graphs with} \ge pn^2/2 \text{ edges.} \end{aligned}$$

We know from Proposition 3.1.14 that every n-vertex graph with edge density $\ge p$ has at least n^4p^4 closed walks of length 4. On the other hand, every sequence of quasirandom graphs with edge density $p + o(1)$ has $p^4n^4 + o(n^4)$ closed walks of length 4. It follows that the minimum (or rather, infimum) is p^4 and is attained not by any single graph, but rather by a sequence of quasirandom graphs.

One of the purposes of graph limits is to provide an easy-to-use mathematical object that captures the limit of such graph sequences. The central object in the theory of dense graph limits is called a **graphon** (the word comes from combining *graph* and *function*), to be defined shortly. Graphons can be viewed as an analytic generalization of graphs.

Here are some questions that we will consider:

(1) What does it mean for a sequence of graphs (or graphons) to converge?
(2) Are different notions of convergence equivalent?
(3) Does every convergent sequence of graphs (or graphons) have a limit?

Note that it is possible to talk about convergence without a limit. In a first real analysis course, one learns about a ***Cauchy sequence*** in a metric space $(\mathcal{X}, d)$, which is some sequence $x_1, x_2, \dots \in \mathcal{X}$ such that for every $\varepsilon > 0$, there is some N so that $d(x_m, x_n) < \varepsilon$ for all $m, n \geq N$. For instance, one can have a Cauchy sequence without a limit in $\mathbb{Q}$. A metric space is ***complete*** if every Cauchy sequence has a limit. The ***completion*** of $\mathcal{X}$ is some complete metric space $\widetilde{\mathcal{X}}$ such that $\mathcal{X}$ is isometrically embedded in $\widetilde{\mathcal{X}}$ as a dense subset. The completion of $\mathcal{X}$ is in some sense the smallest complete space containing $\mathcal{X}$. For example, $\mathbb{R}$ is the completion of $\mathbb{Q}$. Intuitively, the completion of a space fills in all of its gaps. A basic result in analysis says that every space has a unique completion.

Here is a key result about graph limits that we will prove:

> The space of graphons is compact and is the completion of the set of graphs.

To make this statement precise, we also need to define a notion of similarity (i.e., distance) between graphs and also between graphons. We will see two different notions, one based on the *cut metric*, and another based on *subgraph densities*. Another important result in the theory of graph limits is that these two notions are equivalent. We will prove it at the end of the chapter once we have developed some tools.

4.1 Graphons

Here is the central object in the theory of dense graph limits.

Definition 4.1.1 (Graphon)
A ***graphon*** is a symmetric measurable function $W : [0,1]^2 \to [0,1]$. Here ***symmetric*** means $W(x,y) = W(y,x)$ for all x, y.

***Remark* 4.1.2.** More generally, we can consider an arbitrary probability space Ω and study symmetric measurable functions $\Omega \times \Omega \to [0,1]$. In practice, we do not lose much by restricting to $[0,1]$.

We will also sometimes consider symmetric measurable functions $[0,1]^2 \to \mathbb{R}$ (e.g., arising as the difference between two graphons). Such an object is sometimes called a ***kernel*** in the literature.

***Remark* 4.1.3** (Measure theoretic technicalities)**.** We try to sweep measure theoretic technicalities under the rug in order to focus on key ideas. If you have not seen measure theory before, do not worry. Just view "measure" as lengths of intervals or areas of boxes (or countable unions thereof) in the most natural sense. We always ignore measure zero differences. For example, we shall treat two graphons as the same if they only differ on a measure zero subset of the domain.

Turning a Graph into a Graphon

Here is a procedure to turn any graph G into a graphon W_G:

(1) Write down the adjacency matrix A_G of the graph;

(2) Replace the matrix by a black-and-white pixelated picture on $[0,1]^2$ by turning every 1-entry into a black square and every 0-entry into a white square.

(3) View the resulting picture as a graphon $W_G\colon [0,1]^2 \to [0,1]$ (with the axes labeled like a matrix with $x \in [0,1]$ running from top to bottom and $y \in [0,1]$ running from left to right), where we write $W_G(x,y) = 1$ if (x,y) is black and $W_G(x,y) = 0$ if (x,y) is white.

As with everything in this chapter, we ignore measure zero differences, and so it does not matter what we do with boundaries of the pixels.

Definition 4.1.4 (Associated graphon of a graph)
Given a graph G with n vertices labeled $1,\ldots,n$, we define its ***associated graphon*** $W_G\colon [0,1]^2 \to [0,1]$ by first partitioning $[0,1]$ into n equal-length intervals $I_1,\ldots,I_n$ and setting W_G to be 1 on all $I_i \times I_j$ where ij is an edge of G, and 0 on all other $I_i \times I_j$'s.

More generally, we can encode nonnegative vertex and edge weights in a graphon.

Definition 4.1.5 (Step graphon)
A ***step graphon*** W with k steps consists of first partitioning $[0,1]$ into k intervals $I_1,\ldots,I_k$, and then setting W to be a constant on each $I_i \times I_j$.

***Example* 4.1.6** (Half-graph)**.** Consider the bipartite graph on $2n$ vertices, with one vertex part $\{v_1,\ldots,v_n\}$ and the other vertex part $\{w_1,\ldots,w_n\}$, and edges v_iw_j whenever $i \le j$. Following are its adjacency matrix and associated graphon.

As $n \to \infty$, the associated graphons converge pointwise almost everywhere to the graphon

$$W(x,y) = \begin{cases} 1 & \text{if } x+y \le 1/2 \text{ or } x+y \ge 3/2, \\ 0 & \text{otherwise.} \end{cases}$$

In general, pointwise convergence turns out to be too restrictive. We will need a more flexible notion of convergence, which we will discuss more in depth in the next section. Let us first give some more examples to motivate subsequent definitions.

***Example* 4.1.7** (Quasirandom graphs)**.** Let G_n be a sequence of quasirandom graphs with edge density approaching $1/2$, and $v(G_n) \to \infty$. The constant graphon $W \equiv 1/2$ seems like a reasonable candidate for its limit, and later we will see that this is indeed the case.

***Example* 4.1.8** (Stochastic block model). Consider an n vertex graph with two types of vertices: red and blue. Half of the vertices are red, and half of the vertices are blue. Two red vertices are adjacent with probability p_r, two blue vertices are adjacent with probability p_b, and finally, a red vertex and a blue vertex are adjacent with probability p_{rb}, all independently. Then as $n \to \infty$, the graphs converge to the step graphon shown here.

The preceding examples suggest that the limiting graphon looks like a blurry image of the adjacency matrix. However, there is an important caveat as illustrated in the next example.

***Example* 4.1.9** (Checkerboard). Consider the $2n \times 2n$ "checkerboard" graphon shown here (for $n = 4$).

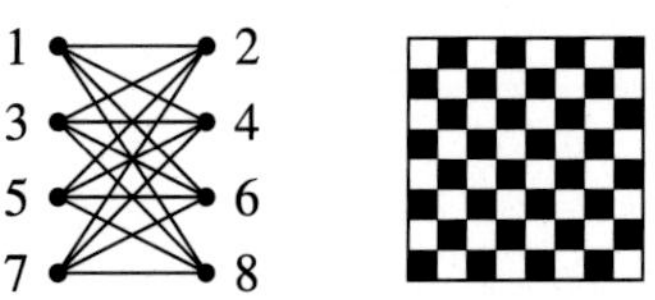

Since the 0's and 1's in the adjacency matrix are evenly spaced, one might suspect that this sequence converges to the constant $1/2$ graphon. However, this is not so. The checkerboard graphon is associated to the complete bipartite graph $K_{n,n}$, with the two vertex parts interleaved. By relabeling the vertices, we see that following is another representation of the associated graphon of the same graph.

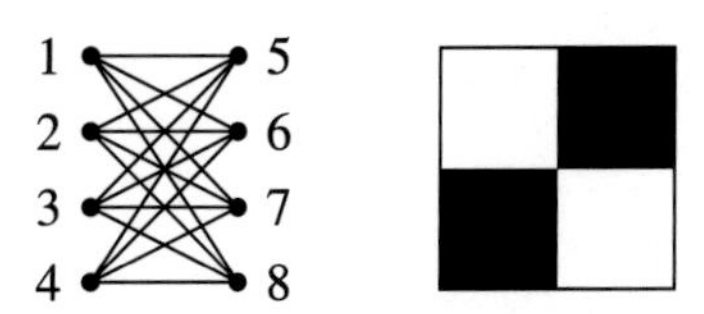

So the graphon is the same for all n. So the graphon shown on the right, which is also W_{K_2}, must be the limit of the sequence, and not the constant $1/2$ graphon. This example tells us that we must be careful about the possibility of rearranging vertices when studying graph limits.

A graphon is an infinite-dimensional object. We would like some ways to measure the *similarity* between two graphons. We will explain two different approaches:

- cut distance, and
- homomorphism densities.

One of the main results in the theory of graph limits is that these two approaches are equivalent – we will show this later in the chapter.

4.2 Cut Distance

There are many ways to measure the distance between two graphs. Different methods may be useful for different applications. For example, we can consider the ***edit distance*** between two graphs (say on the same set of vertices), defined to be the number of edges needed to be added/deleted to obtain one graph from the other. The notion of edit distance arose when discussing the induced graph removal lemmas in Section 2.8. However, edit distance is not suitable for graph limits, since it is incompatible with (quasi)random graphs. For example, given two n-vertex random graphs, independently generated with edge-probability $1/2$, we would like to say that they are similar, as these graphs will end up converging to the constant $1/2$ graphon as $n \to \infty$ (e.g., Example 4.1.7). However, two independent random graphs typically only agree on around half of their edges (even if we allow permuting vertices), and so it takes $(1/4 + o(1))n^2$ edge additions/deletions to obtain one from the other.

A more suitable notion of distance is motivated by the discrepancy condition from Theorem 3.1.1 on quasirandom graphs. Inspired by the condition **DISC**, we would like to say that a graph G is ε-close to the constant p graphon if

$$|e_G(X,Y) - p\,|X|\,|Y|| \leq \varepsilon\,|V(G)|^2 \quad \text{for all } X,Y \subseteq V(G).$$

Inspired by this notion, we now compare a pair of graphs G and G' on a common vertex set $V = V(G) = V(G')$. We say that ***G and G' are ε-close in cut norm*** if

$$|e_G(X,Y) - e_{G'}(X,Y)| \leq \varepsilon\,|V|^2 \quad \text{for all } X,Y \subseteq V. \tag{4.1}$$

(This term "cut" is often used to refer to the set of edges in a graph G between some $X \subseteq V(G)$ and its complement. The cut norm builds on this concept.) With this notion, two independent n-vertex random graphs with the same edge-probability are $o(1)$-close in cut norm as $n \to \infty$.

As illustrated in Example 4.1.9, we also need to consider possible relabelings of vertices. Intuitively, the cut distance between two graphs will come from the relabeling of vertices that gives the greatest alignment. The actual definition will be a bit more subtle, allowing vertex fractionalization. The general definition of cut distance will allow us to compare graphs with different numbers of vertices. It is conceptually easier to define cut distance using graphons.

The edit distance of graphs corresponds to the L^1 distance for graphons. For every $p \geq 1$, we define the ***L^p norm*** of a function $W\colon [0,1]^2 \to \mathbb{R}$ by

$$\|W\|_p := \left(\int_{[0,1]^2} |W(x,y)|^p \, dxdy \right)^{1/p},$$

and the ***L^∞ norm*** by

$$\|W\|_\infty := \sup\left\{t : W^{-1}([t,\infty)) \text{ has positive measure}\right\}.$$

(This is not simply the supremum of W; the definition should be invariant under measure zero changes of W.)

Definition 4.2.1 (Cut norm)
The ***cut norm*** of a measurable $W\colon [0,1]^2 \to \mathbb{R}$ is defined as

$$\|W\|_\square := \sup_{S,T\subseteq[0,1]} \left| \int_{S\times T} W \right|,$$

where S and T are measurable sets.

Let G and G' be two graphs sharing a common vertex set. Let W_G and $W_{G'}$ be their associated graphons (using the same ordering of vertices when constructing the graphons). Then G and G' are ε-close in cut norm (see (4.1)) if and only if

$$\|W_G - W_{G'}\|_\square \le \varepsilon.$$

(There is a subtlety in this claim that is worth thinking about: should we be worried about sets $S, T \subseteq [0,1]$ in Definition 4.2.1 of cut norm that contain fractions of some intervals that represent vertices? See Lemma 4.5.3 for a reformulation of the cut norm that may shed some light.)

We need a concept for an analog of a vertex set permutation for graphons. We write

$$\boldsymbol{\lambda(A)} := \text{the Lebesgue measure of } A.$$

Intuitively, this is the "length" or "area" of A. We will always be referring to Lebesgue measurable sets. (Measure theoretic technicalities are not central to the discussions here, so feel free to ignore them.)

Definition 4.2.2 (Measure preserving map)
We say that $\phi\colon [0,1] \to [0,1]$ is a ***measure preserving map*** if

$$\lambda(A) = \lambda(\phi^{-1}(A)) \quad \text{for all measurable } A \subseteq [0,1].$$

We say that ϕ is an ***invertible*** measure preserving map if there is another measure preserving map $\psi\colon [0,1] \to [0,1]$ such that $\phi \circ \psi$ and $\psi \circ \phi$ are both identity maps outside sets of measure zero.

***Example* 4.2.3.** For any constant $\alpha \in \mathbb{R}$, the function $\phi(x) = x + \alpha \bmod 1$ is measure preserving (this map rotates the circle $\mathbb{R}/\mathbb{Z}$ by α).

A more interesting example is $\phi(x) = 2x \bmod 1$, illustrated here.

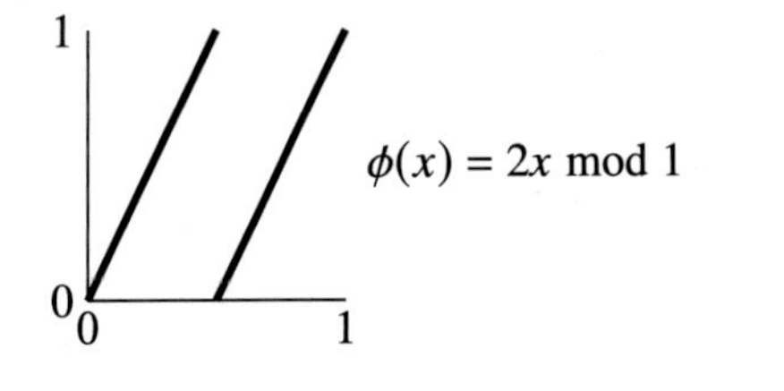

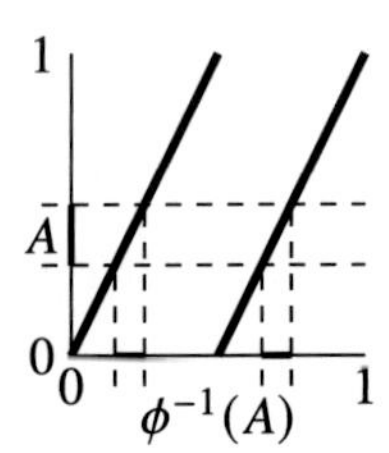

This map is also measure preserving. This might not seem to be the case at first, since ϕ seems to shrink some intervals by half. However, the definition of measure preserving actually says $\lambda(\phi^{-1}(A)) = \lambda(A)$ and not $\lambda(\phi(A)) = \lambda(A)$. For any interval $[a,b] \subseteq [0,1]$, we have $\phi^{-1}([a,b]) = [a/2, b/2] \cup [1/2 + a/2, 1/2 + b/2]$, which does have the same measure as $[a,b]$. This map is 2-to-1, and it is not invertible.

Given $W: [0,1]^2 \to \mathbb{R}$ and an invertible measure preserving map $\phi: [0,1] \to [0,1]$, we write

$$\boldsymbol{W^\phi(x,y)} := W(\phi(x), \phi(y)).$$

Intuitively, this operation relabels the vertex set.

Definition 4.2.4 (Cut metric)
Given two symmetric measurable functions $U, W: [0,1]^2 \to \mathbb{R}$, we define their ***cut distance*** (or ***cut metric***) to be

$$\begin{aligned}\boldsymbol{\delta_\square(U,W)} &:= \inf_\phi \left\| U - W^\phi \right\|_\square \\ &= \inf_\phi \sup_{S,T \subseteq [0,1]} \left| \int_{S \times T} (U(x,y) - W(\phi(x), \phi(y)))\, dx dy \right|,\end{aligned}$$

where the infimum is taken over all invertible measure preserving maps $\phi: [0,1] \to [0,1]$. Define the cut distance between two graphs G and G' by the cut distance of their associated graphons:

$$\boldsymbol{\delta_\square(G,G')} := \delta_\square(W_G, W_{G'}).$$

Likewise, we can also define the cut distance between a graph and a graphon U:

$$\boldsymbol{\delta_\square(G,U)} := \delta_\square(W_G, U).$$

Definition 4.2.5 (Convergence in cut metric)
We say that a sequence of graphs or graphons ***converges in cut metric*** if they form a Cauchy sequence with respect to $\delta_\square$. Furthermore, we say that W_n ***converges to W in cut metric*** if $\delta_\square(W_n, W) \to 0$ as $n \to \infty$.

Note that in $\delta_\square(G,G')$, we are doing more than just permuting vertices. A measure preserving map on $[0,1]$ is also allowed to split a single node into fractions.

It is possible for two different graphons to have cut distance zero. For example, they could differ on a measure-zero set, or they could be related via measure preserving maps.

Space of Graphons

We can form a metric space by identifying graphons with measure zero (i.e., treating such two graphs with cut distance zero as the same point).

Definition 4.2.6 (Graphon space)
Let $\widetilde{\mathcal{W}}_0$ be the set of graphons (i.e., symmetric measurable functions $[0,1]^2 \to [0,1]$) where any pair of graphons with cut distance zero are considered the same point in the space. This is a metric space under cut distance $\delta_\square$.

We view every graph G as a point in $\widetilde{\mathcal{W}}_0$ via its associated graphon (note that several graphs can be identified as the same point in $\widetilde{\mathcal{W}}_0$).

(The subscript 0 in $\widetilde{\mathcal{W}}_0$ is conventional. Sometimes, without the subscript, $\widetilde{\mathcal{W}}$ is used to denote the space of symmetric measurable functions $[0,1]^2 \to \mathbb{R}$.)

Here is a central theorem in the theory of graph limits, proved by Lovász and Szegedy (2007).

Theorem 4.2.7 (Compactness of graphon space)
The metric space $(\widetilde{\mathcal{W}}_0, \delta_\square)$ is compact.

One of the main goals of this chapter is to prove this theorem and show its applications.

The compactness of graphon space is related to the graph regularity lemma. In fact, we will use the regularity method to prove compactness. Both compactness and the graph regularity lemma tell us that despite the infinite variability of graphs, every graph can be ε-approximated by a graph from a finite set of templates.

We close this section with the following observation.

Theorem 4.2.8 (Graphs are dense in the space of graphons)
The set of graphs is dense in $(\widetilde{\mathcal{W}}_0, \delta_\square)$.

Proof. Let $\varepsilon > 0$. It suffices to show that for every graphon W there exists a graph G such that $\delta_\square(G, W) < \varepsilon$.

We approximate W in several steps, illustrated here.

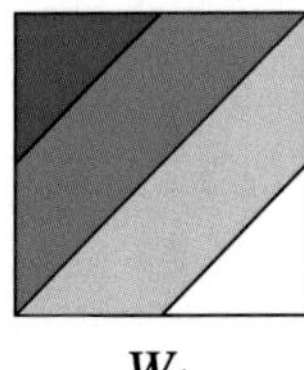
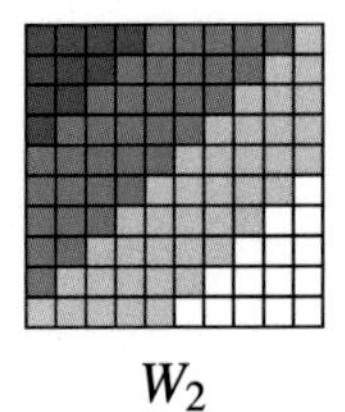

W $\qquad$ W_1 $\qquad$ W_2

First, by rounding down the values of $W(x, y)$, we construct a graphon W_1 whose values are all integer multiples of $\varepsilon/3$, such that

$$\|W - W_1\|_\infty \le \varepsilon/3.$$

Next, since every Lebesgue measurable subset of $[0, 1]^2$ can be arbitrarily well approximated using a union of boxes, we can find a step graphon W_2 approximating W_1 in L^1 norm:

$$\|W_1 - W_2\|_1 \le \varepsilon/3.$$

Finally, by replacing each block of W_2 by a sufficiently large quasirandom (bipartite) graph of edge density equal to the value of W_2, we find a graph G so that

$$\|W_2 - W_G\|_\square \le \varepsilon/3.$$

Then $\delta_\square(W, G) < \varepsilon$. $\square$

***Remark* 4.2.9.** In the preceding proof, to obtain $\|W_1 - W_2\|_1 \le \varepsilon/3$, the number of steps of W_2 cannot be uniformly bounded as a function of ε (i.e., it must depend on W as well – think about what happens for a random graph). Consequently the number of vertices of the final graph G produced by this proof is not bounded by a function of ε.

Later on, we will see a different proof showing that for every $\varepsilon > 0$, there is some $N(\varepsilon)$ so that every graphon lies within cut distance ε of some graph with $\leq N(\varepsilon)$ vertices (Proposition 4.8.1).

Since every compact metric space is complete, we have the following corollary.

Corollary 4.2.10 (Graphons complete graphs)
The graphon space $(\widetilde{\mathcal{W}}_0, \delta_\square)$ is the completion of the space of graphs with respect to the cut metric.

Exercise 4.2.11 (Zero-one valued graphons). Let W be a $\{0,1\}$-valued graphon. Suppose graphons W_n satisfy $\|W_n - W\|_\square \to 0$ as $n \to \infty$. Show that $\|W_n - W\|_1 \to 0$ as $n \to \infty$.

4.3 Homomorphism Density

Subgraph densities give another way of measuring graphs. It will be technically more convenient to work with graph homomorphisms instead of subgraphs.

Definition 4.3.1 (Homomorphism density)
A ***graph homomorphism*** from F to G is a map $\phi\colon V(F) \to V(G)$ such that if $uv \in E(F)$ then $\phi(u)\phi(v) \in E(G)$ (i.e., ϕ maps edges to edges). Define

$$\mathbf{Hom}(F, G) := \{\text{homomorphisms from } F \text{ to } G\}$$

and

$$\mathbf{hom}(F, G) := |\mathrm{Hom}(F, G)|\,.$$

Define the ***F-homomorphism density in G*** (or ***F-density in G*** for short) as

$$t(F, G) := \frac{\hom(F, G)}{v(G)^{v(F)}}.$$

This is also the probability that a uniformly random map $V(F) \to V(G)$ induces a graph homomorphism from F to G.

Example **4.3.2** (Homomorphism counts).

- $\hom(K_1, G) = v(G)$.
- $\hom(K_2, G) = 2e(G)$.
- $\hom(K_3, G) = 6 \cdot \#\text{triangles in } G$.
- $\hom(G, K_3)$ is the number of proper colorings of G using three labeled colors such as $\{\text{red, green, blue}\}$ (corresponding to the vertices of K_3).

Remark **4.3.3** (Subgraphs vs. homomorphisms). Note that homomorphisms from F to G do not quite correspond to copies of subgraphs F inside G, because these homomorphisms can be noninjective. Define the ***injective homomorphism density***:

$$t_{\mathrm{inj}}(F, G) := \frac{\#\text{injective homomorphisms from } F \text{ to } G}{v(G)(v(G)-1)\cdots(v(G)-v(F)+1)}.$$

Equivalently, this is the fraction of injective maps $V(F) \to V(G)$ that are graph homomorphisms (i.e., send edges to edges). The fraction of maps $V(F) \to V(G)$ that are noninjective is $\le \binom{v(F)}{2}/v(G)$. (For every fixed pair of vertices of F, the probability that they collide is exactly $1/v(G)$.) So,

$$\left|t(F,G) - t_{\text{inj}}(F,G)\right| \le \frac{1}{v(G)}\binom{v(F)}{2}.$$

If F is fixed, the right-hand side tends to zero as $v(G) \to \infty$. So all but a negligible fraction of such homomorphisms correspond to subgraphs. This is why we often treat subgraph densities interchangeably with homomorphism densities as they agree in the limit.

Now we define the corresponding notion of homomorphism density in graphons. We first give an example and then the general formula.

***Example* 4.3.4** (Triangle density in graphons)**.** The following quantity is the triangle density in a graphon W:

$$t(K_3, W) = \int_{[0,1]^3} W(x,y)W(y,z)W(z,x)\, dxdydz.$$

This definition agrees with Definition 4.3.1 for the triangle density in graphs. Indeed, for every graph G, the triangle density in G equals the triangle density in the associated graphon W_G; that is, $t(K_3, W_G) = t(K_3, G)$.

Definition 4.3.5 (Homomorphism density in graphon)
Let F be a graph and W a graphon. The ***F-density in W*** is defined to be

$$t(F,W) = \int_{[0,1]^{V(F)}} \prod_{ij \in E(F)} W(x_i, x_j) \prod_{i \in V(F)} dx_i.$$

We also use the same formula when W is a symmetric measurable function.

Note that for all graphs F and G, letting W_G be the graphon associated to G,

$$t(F,G) = t(F, W_G). \tag{4.2}$$

So, the two definitions of F-density agree.

Definition 4.3.6 (Left convergence)
We say that a sequence of graphons W_n is ***left-convergent*** if for every graph F, $t(F, W_n)$ converges as $n \to \infty$. We say that this sequence ***left-converges*** to a graphon W if $\lim_{n\to\infty} t(F, W_n) = t(F, W)$ for every graph F.

For a sequence of graphs, we say that it is ***left-convergent*** if the sequence of associated graphons $W_n = W_{G_n}$ is left-convergent, and that it ***left-converges*** to W if W_n does.

One usually has $v(G_n) \to \infty$, but it is not strictly necessary for this definition. Note that when $v(G_n) \to \infty$, homomorphism densities and subgraph densities coincide. (See Remark 4.3.3.)

It turns out that left-convergence is equivalent to convergence in cut metric. This foundational result in graph limits is due to Borgs, Chayes, Lovász, Sós, and Vesztergombi (2008).

Theorem 4.3.7 (Equivalence of convergence)
A sequence of graphons is left-convergent if and only if it is a Cauchy sequence with respect to the cut metric $\delta_\square$.

The sequence left-converges to some graphon W if and only if it converges to W in cut metric.

The implication that convergence in cut metric implies left-convergence is easier; it follows from the counting lemma (Section 4.5). The converse is more difficult, and we will establish it at the end of the chapter.

This allows us to talk about ***convergent sequences*** of graphs or graphons without specifying whether we are referring to left-convergence or convergence in cut metric. However, since a major goal of this chapter is to prove the equivalence between these two notions, we will be more specific about the notion of convergence.

From the compactness of the space of graphons and the equivalence of convergence (actually only needing the easier implication), we will be able to quickly deduce the existence of limit for a left-convergent sequence, which was first proved by Lovász and Szegedy (2006). Note that the following statement does not require knowledge of the cut metric.

Theorem 4.3.8 (Existence of limit for left-convergence)
Every left-convergent sequence of graphs or graphons left-converges to some graphon.

***Remark* 4.3.9.** One can artificially define a metric that coincides with left-convergence. Let $(F_n)_{n\geq 1}$ enumerate over all graphs. One can define a distance between graphons U and W by

$$\sum_{k\geq 1} 2^{-k} \left|t(F_k, W) - t(F_k, U)\right|.$$

We see that a sequence of graphons converges under this notion of distance if and only if it is left-convergent. This shows that left-convergence defines a metric topology on the space of graphons, but in practice the preceding distance is pretty useless.

Exercise 4.3.10 (Counting Eulerian orientations). Define $W\colon [0,1]^2 \to \mathbb{R}$ by $W(x,y) = 2\cos(2\pi(x-y))$. Let F be a graph. Show that $t(F,W)$ is the number of ways to orient all edges of F so that every vertex has the same number of incoming edges as outgoing edges.

4.4 W-Random Graphs

In this section, we explain how to use a graphon to create a random graph model. This hopefully gives more intuition about graphons.

The most common random graph model is the Erdős–Rényi random graph $\mathbf{G}(n,p)$, which is an n-vertex graph with every edge chosen with probability p.

Stochastic Block Model

The ***stochastic block model*** is a random graph model that generalizes the Erdős–Rényi random graph. We already saw an example in Example 4.1.8. Let us first illustrate the ***two-block model***, which has several parameters:

	q_r	q_b
q_r	p_{rr}	p_{rb}
q_b	p_{rb}	p_{bb}

with all the numbers lying in $[0, 1]$, and subject to $q_r + q_b = 1$. We form an n-vertex random graph as follows:

(1) Color each vertex red with probability q_r and blue with probability q_b, independently at random. These vertex colors are "hidden states" and are not part of the data of the output random graph. (This step is slightly different from Example 4.1.8 in an unimportant way.)

(2) For every pair of vertices, independently place an edge between them with probability
- p_{rr} if both vertices are red,
- p_{bb} if both vertices are blue, and
- p_{rb} if one vertex is red and the other is blue.

One can easily generalize the preceding to a ***k*-block model**, where vertices have k hidden states, with $q_1, \ldots, q_k$ (adding up to 1) being the vertex state probabilities, and a symmetric $k \times k$ matrix $(p_{ij})_{1 \le i,j \le k}$ of edge probabilities for pairs of vertices between various states.

W-Random Graph

The W-random graph is a further generalization. The stochastic block model corresponds to step graphons W.

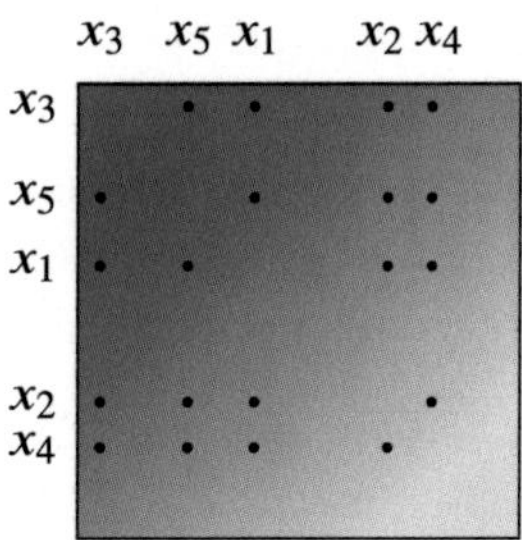

Definition 4.4.1 (W-random graph)

Let W be a graphon. The n-vertex ***W*-random graph** $\mathbf{G}(n, W)$ denotes the n-vertex random graph (with vertices labeled $1, \ldots, n$) obtained by first picking $x_1, \ldots, x_n$ uniformly at random from $[0, 1]$, and then putting an edge between vertices i and j with probability $W(x_i, x_j)$, independently for all $1 \le i < j \le n$.

Let us show that these W-random graphs left-converge to W with probability 1.

Theorem 4.4.2 (W-random graphs left-converge to W)

Let W be a graphon. For each n, let G_n be a random graph distributed as $\mathbf{G}(n, W)$. Then G_n left-converges to W with probability 1.

***Remark* 4.4.3.** The theorem does not require each G_n to be sampled independently. For example, we can construct the sequence of random graphs, with G_n distributed as $\mathbf{G}(n, W)$, by revealing one vertex at a time without resampling the previous vertices and edges. In this case, each G_n is a subgraph of the next graph G_{n+1}.

We will need the following standard result about concentration of Lipschitz functions. This can be proved using Azuma's inequality (e.g., see Chapter 7 of *The Probabilistic Method* by Alon and Spencer).

Theorem 4.4.4 (Bounded differences inequality)
Let $X_1 \in \Omega_1, \dots, X_n \in \Omega_n$ be independent random variables. Suppose $f \colon \Omega_1 \times \cdots \times \Omega_n \to \mathbb{R}$ is L-Lipschitz for some constant L in the sense of satisfying

$$\left|f(x_1, \dots, x_n) - f(x_1', \dots, x_n')\right| \le L \tag{4.3}$$

whenever $(x_1, \dots, x_n)$ and $(x_1', \dots, x_n')$ differ on exactly one coordinate. Then the random variable $Z = f(X_1, \dots, X_n)$ satisfies, for every $\lambda \ge 0$,

$$\mathbb{P}(Z - \mathbb{E}Z \ge \lambda L) \le e^{-2\lambda^2/n} \quad \text{and} \quad \mathbb{P}(Z - \mathbb{E}Z \le -\lambda L) \le e^{-2\lambda^2/n}.$$

Let us show that the F-density in a W-random graph rarely differs significantly from $t(F, W)$.

Theorem 4.4.5 (Sample concentration for graphons)
For every $\varepsilon > 0$, positive integer n, graph F, and graphon W, we have

$$\mathbb{P}\left(|t(F, \mathbf{G}(n, W)) - t(F, W)| > \varepsilon\right) \le 2\exp\left(\frac{-\varepsilon^2 n}{8v(F)^2}\right). \tag{4.4}$$

Proof. Recall from Remark 4.3.3 that the injective homomorphism density $t_{\text{inj}}(F, G)$ is defined to be the fraction of injective maps $V(F) \to V(G)$ that carry every edge of F to an edge of G. We will first prove that

$$\mathbb{P}\left(\left|t_{\text{inj}}(F, \mathbf{G}(n, W)) - t(F, W)\right| > \varepsilon\right) \le 2\exp\left(\frac{-\varepsilon^2 n}{2v(F)^2}\right). \tag{4.5}$$

Let $y_1, \dots, y_n$, and z_{ij} for each $1 \le i < j \le n$, be independent uniform random variables in $[0, 1]$. Let G be the graph on vertices $\{1, \dots, n\}$ with an edge between i and j if and only if $z_{ij} \le W(y_i, y_j)$, for every $i < j$. Then G has the same distribution as $\mathbf{G}(n, W)$. Let us group variables y_i, z_{ij} into $x_1, x_2, \dots, x_n$ where

$$x_1 = (y_1), \quad x_2 = (y_2, z_{12}), \quad x_3 = (y_3, z_{13}, z_{23}), \quad x_4 = (y_4, z_{14}, z_{24}, z_{34}), \quad \dots.$$

This amounts to exposing the graph G one vertex at a time. Define the function $f(x_1, \dots, x_n) = t_{\text{inj}}(F, G)$. Note that $\mathbb{E}f = \mathbb{E}\, t_{\text{inj}}(F, \mathbf{G}(n, W)) = t(F, W)$ by linearity of expectations. (In this step, it is important that we are using the injective variant of homomorphism densities.) Note that changing a single coordinate of f changes the value of the function by at most $v(F)/n$, since exactly a $v(F)/n$ fraction of injective maps $V(F) \to V(G)$ includes a fixed $v \in V(G)$ in the image. Then (4.5) follows from the bounded differences inequality, Theorem 4.4.4.

To deduce the theorem from (4.5), recall from Remark 4.3.3 that

$$\left|t(F,G) - t_{\mathrm{inj}}(F,G)\right| \le v(F)^2/(2v(G)).$$

If $\varepsilon < v(F)^2/n$, then the right-hand side of (4.4) is at least $2e^{-\varepsilon/8} \ge 1$, and so the inequality trivially holds. Otherwise, $|t(F,\mathbf{G}(n,W)) - t(F,W)| > \varepsilon$ implies $\left|t_{\mathrm{inj}}(F,\mathbf{G}(n,W)) - t(F,W)\right| > \varepsilon - v(F)^2/(2n) \ge \varepsilon/2$, and then we can apply (4.5) to conclude. □

Theorem 4.4.2 then follows from the Borel–Cantelli lemma, stated in Theorem 4.4.6, applied to Theorem 4.4.5 for all rational $\varepsilon > 0$.

Theorem 4.4.6 (Borel–Cantelli lemma)
Given a sequence of events $E_1, E_2, \ldots$, if $\sum_n \mathbb{P}(E_n) < \infty$, then with probability 1, only finitely many of them occur.

4.5 Counting Lemma

In Chapter 2 on the graph regularity lemma, we proved a counting lemma that gave a lower bound on the number of copies of some fixed graph H in a regularity partition. The same techniques can be modified to give a similar upper bound. Here we prove another graph counting lemma. The proof is more analytic, whereas the previous proofs in Chapter 2 were more combinatorial (embedding one vertex at a time).

Theorem 4.5.1 (Counting lemma)
Let F be a graph. Let W and U be graphons. Then

$$|t(F,W) - t(F,U)| \le |E(F)|\, \delta_\square(W,U).$$

Qualitatively, the counting lemma tells us that for every graph F, the function $t(F,\cdot)$ is continuous in $(\widetilde{\mathcal{W}}_0, \delta_\square)$, the graphon space with respect to the cut metric. It implies the easier direction of the equivalence in Theorem 4.3.7, namely that convergence in cut metric implies left-convergence.

Corollary 4.5.2 (Cut metric convergence implies left-convergence)
Every Cauchy sequence of graphons with respect to the cut metric is left-convergent.

In the rest of this section, we prove Theorem 4.5.1. It suffices to prove that

$$|t(F,W) - t(F,U)| \le |E(F)|\, \|W - U\|_\square\,. \tag{4.6}$$

Indeed, for every invertible measure preserving map $\phi\colon [0,1] \to [0,1]$, we have $t(F,U) = t(F,U^\phi)$. By considering the above inequality with U replaced by U^ϕ, and taking the infimum over all U^ϕ, we obtain Theorem 4.5.1.

The following reformulation of the cut norm is often quite useful.

Lemma 4.5.3 (Reformulation of cut norm)
For every measurable $W\colon [0,1]^2 \to \mathbb{R}$,

$$\|W\|_\square = \sup_{\substack{u,v:[0,1]\to[0,1]\\ \text{measurable}}} \left| \int_{[0,1]^2} W(x,y)u(x)v(y)\, dx dy \right|.$$

Proof. We want to show (left-hand side is how we defined the cut norm in Definition 4.2.1)

$$\sup_{\substack{S,T\subseteq[0,1]\\ \text{measurable}}} \left| \int_{[0,1]^2} W(x,y)1_S(x)1_T(y)\,dxdy \right| = \sup_{\substack{u,v:[0,1]\to[0,1]\\ \text{measurable}}} \left| \int_{[0,1]^2} W(x,y)u(x)v(y)\,dxdy \right|.$$

The right-hand side is at least as large as the left-hand side since we can take $u = 1_S$ and $v = 1_T$. On the other hand, the integral on the right-hand side is bilinear in u and v, and so it is always possible to change u and v to $\{0,1\}$-valued functions without decreasing the value of the integral (e.g., think about what is the best choice for v with u held fixed, and vice versa). If u and v are restricted to $\{0,1\}$-valued functions, then the two sides are identical. □

As a warm-up, let us illustrate the proof of the triangle counting lemma, which has all the ideas of the general proof but with simpler notation. As shown in the following illustration, the main idea to "replace" the W's by U's on the triangle one at a time using the cut norm.

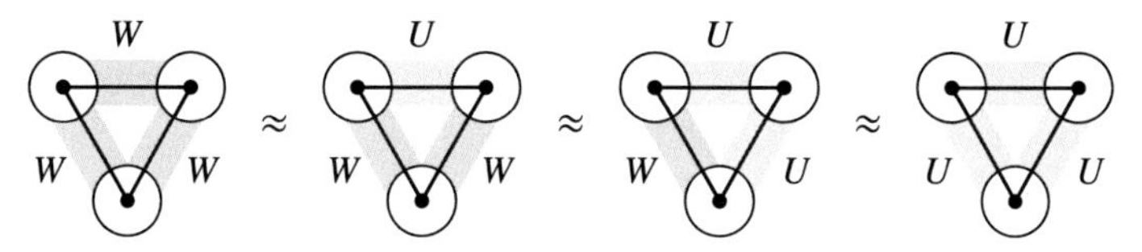

Proposition 4.5.4 (Triangle counting lemma)
Let W and U be graphons. Then

$$|t(K_3,W) - t(K_3,U)| \le 3\,\|W-U\|_\square\,.$$

Proof. Given three graphons W_{12}, W_{13}, W_{23}, define

$$t(W_{12},W_{13},W_{23}) = \int_{[0,1]^3} W_{12}(x,y)W_{13}(x,z)W_{23}(y,z)\,dxdydz.$$

So

$$t(K_3,W) = t(W,W,W) \quad \text{and} \quad t(K_3,U) = t(U,U,U).$$

Observe that $t(W_{12},W_{13},W_{23})$ is trilinear in W_{12},W_{13},W_{23}. We have

$$t(W,W,W) - t(U,W,W) = \int_{[0,1]^3} (W-U)(x,y)W(x,z)W(y,z)\,dxdydz.$$

For any fixed z, note that $x \mapsto W(x,z)$ and $y \mapsto W(y,z)$ are both measurable functions $[0,1] \to [0,1]$. So, applying Lemma 4.5.3 gives

$$\left| \int_{[0,1]^2} (W-U)(x,y)W(x,z)W(y,z)\,dxdy \right| \le \|W-U\|_\square$$

for every z. Now integrate over all z and applying the triangle inequality, we obtain

$$|t(W,W,W) - t(U,W,W)| \le \|W-U\|_\square\,.$$

We have similar inequalities in the other two coordinates. We can write

$$t(W,W,W) - t(U,U,U) = t(W,W,W-U) + t(W,W-U,U) + t(W-U,U,U).$$

Each term on the right-hand side is at most $\|W - U\|_\square$ in absolute value. So, the result follows. □

The preceding proof generalizes in a straightforward way to a general graph counting lemma.

Proof of the counting lemma (Theorem 4.5.1). Given a collection of graphons W_e indexed by the edges e of F, define

$$t_F(W_e : e \in E(F)) = \int_{[0,1]^{V(F)}} \prod_{ij \in E(F)} W_{ij}(x_i, x_j) \prod_{i \in V(H)} dx_i.$$

In particular, this quantity equals $t(F, W)$ if $W_e = W$ for all $e \in E(F)$. A straightforward generalization of the triangle case shows that if we change exactly one argument in the preceding function from W to U, then its value changes by at most $\|W - U\|_\square$ in absolute value. Thus, starting with $t_F(W_e : e \in E(F))$ with every $W_e = W$, we can change each argument from W to U, one by one, resulting in a total change of at most $e(F) \|W - U\|_\square$. This proves (4.6), and hence the theorem. □

4.6 Weak Regularity Lemma

In Chapter 2, we defined an ε-regular vertex partition of a graph to be a partition such that all but ε-fraction of pairs of vertices lie between ε-regular pairs of vertex parts. The number of parts is at most an exponential tower of height $O(\varepsilon^{-5})$.

The goal of this section is to introduce a weaker version of the regularity lemma, requiring substantially fewer parts for the partition. The guarantee provided by the partition can be captured by the cut norm.

Let us first state this notion for a graph and then for a graphon.

Definition 4.6.1 (Weak regular partition for graphs)
Given graph G, a partition $\mathcal{P} = \{V_1, \dots, V_k\}$ of $V(G)$ is called ***weak ε-regular*** if, for all $A, B \subseteq V(G)$,

$$\left| e(A, B) - \sum_{i,j=1}^{k} d(V_i, V_j) |A \cap V_i| |B \cap V_j| \right| \le \varepsilon v(G)^2.$$

***Remark* 4.6.2** (Interpreting weak regularity). Given $A, B \subseteq V(G)$, suppose we only knew how many vertices from A and B lie in each part of the partition (and not specifically which vertices), and we are asked to predict the number of edges between A and B. Then the sum above is the number of edges between A and B that one would naturally expect based on the edge densities between vertex parts. Being weak regular says that this prediction is roughly correct.

Weak regularity is more "global" compared to the notion of an ε-regular partition from Chapter 2. The edge densities between certain pairs $A \cap V_i$ and $B \cap V_j$ could differ significantly from that of V_i and V_j. All we ask is that on average these discrepancies mostly cancel out.

The following weak regularity lemma was proved by Frieze and Kannan (1999), initially motivated by algorithmic applications that we will mention in Remark 4.6.11.

Theorem 4.6.3 (Weak regularity lemma for graphs)
Let $0 < \varepsilon < 1$. Every graph has a weak ε-regular partition into at most $4^{1/\varepsilon^2}$ vertex parts.

Now let us state the corresponding notions for graphons.

Definition 4.6.4 (Stepping operator)
Given a symmetric measurable function $W\colon [0,1]^2 \to \mathbb{R}$, and a measurable partition $\mathcal{P} = \{S_1, \dots, S_k\}$ of $[0,1]$, define a symmetric measurable function $W_{\mathcal{P}}\colon [0,1]^2 \to \mathbb{R}$ by setting its value on each $S_i \times S_j$ to be the average value of W over $S_i \times S_j$. (Since we only care about functions up to measure zero sets, we can ignore all parts S_i with measure zero.)

In other words, $W_{\mathcal{P}}$ is a step graphon with steps given by $\mathcal{P}$ and values given by averaging W over the steps.

***Remark* 4.6.5.** The stepping operator is the orthogonal projection in the Hilbert space $L^2([0,1]^2)$ onto the subspace of functions constant on each step $S_i \times S_j$. It can also be viewed as the conditional expectation with respect to the σ-algebra generated by $S_i \times S_j$.

Definition 4.6.6 (Weak regular partition for graphons)
Given graphon W, we say that a measurable partition $\mathcal{P}$ of $[0,1]$ into finitely many parts is ***weak ε-regular*** if

$$\|W - W_{\mathcal{P}}\|_\square \le \varepsilon.$$

Theorem 4.6.7 (Weak regularity lemma for graphons)
Let $0 < \varepsilon < 1$. Then every graphon has a weak ε-regular partition into at most $4^{1/\varepsilon^2}$ parts.

***Remark* 4.6.8.** Technically speaking, Theorem 4.6.3 does not follow from Theorem 4.6.7 since the partition of $[0,1]$ for W_G could split intervals corresponding to individual vertices of G. However, the proofs of the two claims are exactly the same. Alternatively, one can allow a more flexible definition of a graphon as a symmetric measurable function $W\colon \Omega\times\Omega \to [0,1]$, and then take Ω to be the discrete probability space $V(G)$ endowed with the uniform measure.

Like the proof of the regularity lemma in Section 2.1, we use an energy increment strategy. Recall from Definition 2.1.10 that the energy of a vertex partition is the mean-squared edge-density between parts. Given a graphon W, we define the ***energy*** of a measurable partition $\mathcal{P} = \{S_1, \dots, S_k\}$ of $[0,1]$ by

$$\|W_{\mathcal{P}}\|_2^2 = \int_{[0,1]^2} W_{\mathcal{P}}(x,y)^2 \, dx dy = \sum_{i,j=1}^{k} \lambda(S_i)\lambda(S_j)(\text{avg of } W \text{ on } S_i \times S_j)^2.$$

Given $W, U : [0,1]^2 \to \mathbb{R}$, we write

$$\langle \boldsymbol{W}, \boldsymbol{U} \rangle := \int WU = \int_{[0,1]^2} W(x,y)U(x,y) \, dx dy.$$

Lemma 4.6.9 (L^2 energy increment)
Let W be a graphon. Let $\mathcal{P}$ be a finite measurable partition of $[0,1]$ that is not weak ε-regular for W. Then there is a measurable refinement $\mathcal{P}'$ of $\mathcal{P}$, dividing each part of $\mathcal{P}$ into at most four parts, such that

$$\|W_{\mathcal{P}'}\|_2^2 > \|W_P\|_2^2 + \varepsilon^2.$$

Proof. Because $\|W - W_{\mathcal{P}}\|_\square > \varepsilon$, there exist measurable subsets $S, T \subseteq [0,1]$ such that

$$|\langle W - W_{\mathcal{P}}, 1_{S\times T}\rangle| > \varepsilon.$$

Let $\mathcal{P}'$ be the refinement of $\mathcal{P}$ by introducing S and T, dividing each part of $\mathcal{P}$ into ≤ 4 subparts. We know that

$$\langle W_{\mathcal{P}}, W_{\mathcal{P}}\rangle = \langle W_{\mathcal{P}'}, W_{\mathcal{P}}\rangle$$

because $W_{\mathcal{P}}$ is constant on each step of $\mathcal{P}$, and $\mathcal{P}'$ is a refinement of $\mathcal{P}$. Thus,

$$\langle W_{\mathcal{P}'} - W_{\mathcal{P}}, W_{\mathcal{P}}\rangle = 0.$$

By the Pythagorean Theorem (in the Hilbert space $L^2([0,1]^2)$),

$$\|W_{\mathcal{P}'}\|_2^2 = \|W_{\mathcal{P}}\|_2^2 + \|W_{\mathcal{P}'} - W_{\mathcal{P}}\|_2^2. \tag{4.7}$$

Note that $\langle W_{\mathcal{P}'}, 1_{S\times T}\rangle = \langle W, 1_{S\times T}\rangle$ since S and T are both unions of parts of the partition $\mathcal{P}'$. So, by the Cauchy–Schwarz inequality,

$$\|W_{\mathcal{P}'} - W_{\mathcal{P}}\|_2 \geq |\langle W_{\mathcal{P}'} - W_{\mathcal{P}}, 1_{S\times T}\rangle| = |\langle W - W_{\mathcal{P}}, 1_{S\times T}\rangle| > \varepsilon.$$

So, by (4.7), we have $\|W_{\mathcal{P}'}\|_2^2 > \|W_{\mathcal{P}}\|_2^2 + \varepsilon^2$, as claimed. □

We will prove the following slight generalization of Theorem 4.6.7, allowing an arbitrary starting partition. (This will be useful later).

Theorem 4.6.10 (Weak regularity lemma for graphons)
Let $0 < \varepsilon < 1$. Let W be a graphon. Let $\mathcal{P}_0$ be a finite measurable partition of $[0,1]$. Then every graphon has a weak ε-regular partition $\mathcal{P}$, such that $\mathcal{P}$ refines $\mathcal{P}_0$, and each part of $\mathcal{P}_0$ is partitioned into at most $4^{1/\varepsilon^2}$ parts under $\mathcal{P}$.

This proposition specifically tells us that starting with any given partition, the regularity argument still works.

Proof. Starting with $i = 0$:

(1) If $\mathcal{P}_i$ is weak ε-regular, then STOP.
(2) Else, by Lemma 4.6.9, there exists a measurable partition $\mathcal{P}_{i+1}$ refining each part of $\mathcal{P}_i$ into at most four parts, such that $\|W_{\mathcal{P}_{i+1}}\|_2^2 > \|W_{\mathcal{P}_i}\|_2^2 + \varepsilon^2$.
(3) Increase i by 1 and go back to Step (1).

Since $0 \leq \|W_{\mathcal{P}}\|_2 \leq 1$ for every $\mathcal{P}$, the process terminates with $i < 1/\varepsilon^2$, resulting in a terminal $\mathcal{P}_i$ with the desired properties. □

Remark* 4.6.11** (Additive approximation of maximum cut). One of the initial motivations for developing the weak regularity lemma was to develop a general efficient algorithm for estimating the maximum cut in a dense graph. The ***maximum cut problem is a central problem in algorithms and combinatorial optimization:

MAX CUT: Given a graph S, find a $S \subseteq V(G)$ that maximizes $e(S, V(G) \setminus S)$.

Goemans and Williamson (1995) found an efficient 0.878-approximation algorithm. (This means that the algorithm outputs some S with $e(S, V(G) \setminus S)$ at least a factor 0.878 of the optimum.) Their seminal algorithm uses a semidefinite relaxation. The Unique Games Conjecture (currently still open) would imply that it would be NP-hard to obtain a better approximation than the Goemans–Williamson algorithm (Khot, Kindler, Mossel, and O'Donnell 2007). It is also known that approximating beyond $16/17 \approx 0.941$ is NP-hard (Håstad 2001).

On the other hand, an algorithmic version of the weak regularity lemma gives us an efficient algorithm to approximate the maximum cut for dense graphs with an additive error. This means, given $\varepsilon > 0$, we wish to find a cut whose number of edges is within εn^2 of the optimum. The basic idea is to find a weak regular partition $V(G) = V_1 \cup \cdots \cup V_k$, and then do a brute-force search through all possible sizes $|S \cap V_i|$. See Frieze and Kannan (1999) for more details. These ideas have been further developed into efficient sampling algorithms, sampling only $\mathrm{poly}(1/\varepsilon)$ random vertices, for estimating the maximum cut in a dense graph (e.g., Alon, Fernandez de la Vega, Kannan, and Karpinski (2003b)).

The following exercise offers another approach to the weak regularity lemma. It gives an approximation of a graphon as a linear combination of $\leq \varepsilon^{-2}$ indicator functions of boxes. The polynomial dependence of ε^{-2} is important for designing efficient approximation algorithms.

Exercise 4.6.12 (Weak regularity decomposition).

(a) Let $\varepsilon > 0$. Show that for every graphon W, there exist measurable $S_1, \ldots, S_k, T_1, \ldots, T_k \subseteq [0,1]$ and reals $a_1, \ldots, a_k \in \mathbb{R}$, with $k < \varepsilon^{-2}$, such that

$$\left\| W - \sum_{i=1}^{k} a_i \mathbf{1}_{S_i \times T_i} \right\|_{\square} \leq \varepsilon.$$

The rest of the exercise shows how to recover a regularity partition from the preceding approximation.

(b) Show that the stepping operator is contractive with respect to the cut norm, in the sense that, if $W \colon [0,1]^2 \to \mathbb{R}$ is a measurable symmetric function, then $\|W_{\mathcal{P}}\|_{\square} \leq \|W\|_{\square}$.

(c) Let $\mathcal{P}$ be a partition of $[0,1]$ into measurable sets. Let U be a graphon that is constant on $S \times T$ for each $S, T \in \mathcal{P}$. Show that for every graphon W, one has

$$\|W - W_{\mathcal{P}}\|_{\square} \leq 2\,\|W - U\|_{\square}.$$

(d) Use (a) and (c) to give a different proof of the weak regularity lemma (with slightly worse bounds than the one given in class): show that, for every $\varepsilon > 0$ and every graphon W, there exists a partition $\mathcal{P}$ of $[0,1]$ into $2^{O(1/\varepsilon^2)}$ measurable sets such that $\|W - W_{\mathcal{P}}\|_{\square} \leq \varepsilon$.

Exercise 4.6.13* (Second neighborhood distance). Let $0 < \varepsilon < 1/2$. Let W be a graphon. Define $\tau_{W,x} \colon [0,1] \to [0,1]$ by

$$\tau_{W,x}(z) = \int_{[0,1]} W(x,y)W(y,z)\,dy.$$

(This models the second neighborhood of x.) Prove that if a finite set $S \subseteq [0,1]$ satisfies

$$\|\tau_{W,s} - \tau_{W,t}\|_1 > \varepsilon \qquad \text{for all distinct } s,t \in S,$$

then $|S| \le (1/\varepsilon)^{C/\varepsilon^2}$, where C is some absolute constant.

Exercise 4.6.14 (Strong regularity lemma). Let $\boldsymbol{\varepsilon} = (\varepsilon_1, \varepsilon_2, \dots)$ be a sequence of positive reals. By repeatedly applying the weak regularity lemma, show that there is some $M = M(\boldsymbol{\varepsilon})$ such that for every graphon W, there is a pair of partitions $\mathcal{P}$ and Q of $[0,1]$ into measurable sets, such that Q refines $\mathcal{P}$, $|Q| \le M$ (here $|Q|$ denotes the number of parts of Q),

$$\|W - W_Q\|_\square \le \varepsilon_{|\mathcal{P}|} \qquad \text{and} \qquad \|W_Q\|_2^2 \le \|W_{\mathcal{P}}\|_2^2 + \varepsilon_1^2.$$

Furthermore, deduce the strong regularity lemma in the following form:

$$W = W_{\mathrm{str}} + W_{\mathrm{psr}} + W_{\mathrm{sml}},$$

where W_{str} is a k-step graphon with $k \le M$, $\|W_{\mathrm{psr}}\|_\square \le \varepsilon_k$, and $\|W_{\mathrm{sml}}\|_1 \le \varepsilon_1$. State your bounds on M explicitly in terms of $\boldsymbol{\varepsilon}$. (Note: the parameter choice $\varepsilon_k = \varepsilon/k^2$ roughly corresponds to Szemerédi's regularity lemma, in which case your bound on M should be an exponential tower of 2's of height $\varepsilon^{-O(1)}$; if not, then you are doing something wrong.)

4.7 Martingale Convergence Theorem

In this section we prove a result about martingales that will be used in the proof of the compactness of the graphon space.

Martingales are a standard notion in probability theory. It is a stochastic sequence where the expected change at each step is zero, even conditioned on all prior values of the sequence.

Definition 4.7.1 (Discrete time martingale)
A ***martingale*** is a random real sequence $X_0, X_1, X_2, \dots$ such that, for all $n \ge 0$, $\mathbb{E}\,|X_n| < \infty$, and

$$\mathbb{E}[X_{n+1} | X_0, \dots, X_n] = X_n.$$

***Remark* 4.7.2.** The preceding definition is sufficient for our purposes. In order to give a more formal definition of a martingale, we need to introduce the notion of a *filtration*. See any standard measure theory-based introduction to probability. (Williams [1991, chapters 10–11]) has a particularly lucid discussion of martingales and their convergence theorem discussed in what follows.) This martingale is indexed by integers and hence called "discrete-time." There are also continuous-time martingales (e.g., Brownian motion), which we will not discuss here.

Example 4.7.3 (Partial sum of independent mean zero random variables). Let $Z_1, Z_2, \ldots$ be a sequence of independent mean zero random variables (e.g., ± 1 with equal probability). Then $X_n = Z_1 + \cdots + Z_n$, $n \geq 0$, is a martingale.

Example 4.7.4 (Betting strategy). Consider any betting strategy in a "fair" casino, where the expected value of each bet is zero. Let X_n be the balance after n rounds of betting. Then X_n is a martingale regardless of the betting strategy. So every betting strategy has zero expected gain after n rounds. Also see the **optional stopping theorem** for a more general statement (e.g., Williams [1991, chapter 10]).

The original meaning of the word "martingale" refers to the following better strategy on a sequence of fair coin tosses. Each round the better is allowed to bet an arbitrary amount Z: if heads, the better gains Z dollars, and if tails, the better loses Z dollars.

Start betting 1 dollar. If one wins, stop. If one loses, then double one's bet for the next coin. And then repeat (i.e., keep doubling one's bet until the first win, at which point one stops).

A "fallacy" is that this strategy always results in a final net gain of \$1, the supposed reason being that with probability 1 one eventually sees a head. This initially appears to contradict the earlier claim that all betting strategies have zero expected gain. Thankfully there is no contradiction. In real life, one starts with a finite budget and could possibly go bankrupt with this betting strategy, thereby leading to a forced stop. In the optional stopping theorem, there are some boundedness hypotheses that are violated by the above strategy.

The following construction of martingales is most relevant for our purposes.

Example 4.7.5 (Doob martingale). Let X be some "hidden" random variable. Partial information is revealed about X gradually over time. For example, X is some fixed function of some random inputs. So, the exact value of X is unknown, but its distribution can be derived from the distribution of the inputs. Initially, one does not know any of the inputs. Over time, some of the inputs are revealed. Let

$$X_n = \mathbb{E}[X \mid \text{all information revealed up to time } n].$$

Then $X_0, X_1, \ldots$ is a martingale. (Why?) Informally, X_n is the best guess (in expectation) of X based on all the information available up to time n. We have $X_0 = \mathbb{E}X$ (when no information is revealed). All information is revealed as $n \to \infty$, and the martingale X_n converges to the random variable X with probability 1.

Here is a real-life example. Let $X \in \{0, 1\}$ be whether a candidate wins in a presidential election. Let X_n be the inferred probability that the candidate wins, given all the information known at time t_n. Then X_n converges to the "truth," a $\{0, 1\}$-value, eventually becoming deterministic when the election result is finalized.

Then X_n is a martingale. At time t_n, knowing X_n, if the expectation for X_{n+1} (conditioned on everything known at time t_n) were different from X_n, then one should have adjusted X_n accordingly in the first place.

The precise notion of "information" in the preceding formula can be formalized using the notion of *filtration* in probability theory.

Here is the main result of this section.

Theorem 4.7.6 (Martingale convergence theorem)
Every bounded martingale converges with probability 1.

In other words, if $X_0, X_1, \ldots$ is a martingale with $X_n \in [0,1]$ for every n, then the sequence is convergent with probability 1.

***Remark* 4.7.7.** The proof actually shows that the boundedness condition can be replaced by the weaker L^1-boundedness condition $\sup_n \mathbb{E}|X_n| < \infty$. Even more generally, a hypothesis called "uniform integrability" is enough.

Some boundedness condition is necessary. For example, in Example 4.7.3, a running sum of independent uniform ± 1 is a nonbounded martingale and never converges.

Proof. If a sequence $X_0, X_1, \cdots \in [0,1]$ does not converge, then there exist a pair of rational numbers $0 < a < b < 1$ such that X_n "up-crosses" $[a,b]$ infinitely many times, meaning that there is an infinite sequence $s_1 < t_1 < s_2 < t_2 < \cdots$ such that $X_{s_i} < a < b < X_{t_i}$ for all i.

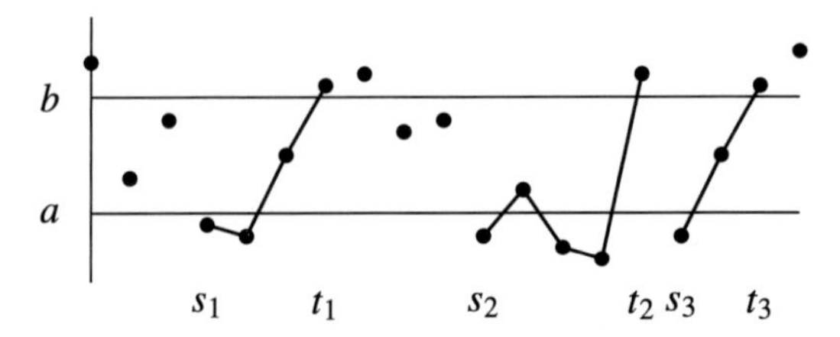

We will show that for each $a < b$, the probability that a bounded martingale $X_0, X_1, \cdots \in [0,1]$ up-crosses $[a,b]$ infinitely many times is zero. Then, by taking a union of all countably many such pairs (a,b) of rationals, we deduce that the martingale converges with probability 1.

Consider the following betting strategy. Imagine that X_n is a stock price. At any time, if X_n dips below a, we buy and hold one share until X_n reaches above b, at which point we sell this share. (Note that we always hold either zero or one share. We do not buy more until we have sold the currently held share.) Start with a budget of $Y_0 = 1$ (so we will never go bankrupt). Let Y_n be the value of our portfolio (cash on hand plus the value of the share if held) at time n. Then Y_n is a martingale. (Why?) So $\mathbb{E}Y_n = Y_0 = 1$. Also, $Y_n \geq 0$ for all n. If one buys and sells at least k times up to time n, then $Y_n \geq k(b-a)$. (This is only the net profit from buying and selling; the actual Y_n may be higher due to the initial cash balance and the value of the current share held.) So, by Markov's inequality, for every n,

$$\mathbb{P}(\geq k \text{ up-crossings up to time } n) \leq \mathbb{P}(Y_n \geq k(b-a)) \leq \frac{\mathbb{E}Y_n}{k(b-a)} = \frac{1}{k(b-a)}.$$

By the monotone convergence theorem,

$$\mathbb{P}(\geq k \text{ up-crossings}) = \lim_{n\to\infty} \mathbb{P}(\geq k \text{ up-crossings up to time } n) \leq \frac{1}{k(b-a)}.$$

Letting $k \to \infty$, the probability of having infinitely many up-crossings is zero. □

4.8 Compactness of the Graphon Space

Using the weak regularity lemma and the martingale convergence theorem, let us prove that the space of graphons is compact with respect to the cut metric.

Proof of compactness of the graphon space (Theorem 4.2.7). As $\widetilde{\mathcal{W}}_0$ is a metric space, it suffices to prove sequential compactness. Fix a sequence $W_1, W_2, \ldots$ of graphons. We want to show that there is a subsequence that converges (with respect to $\delta_\square$) to some limit graphon.

Step 1. Regularize.

For each n, apply the weak regularity lemma (Theorem 4.6.7) repeatedly, to obtain a sequence of partitions $\mathcal{P}_{n,1}, \mathcal{P}_{n,2}, \mathcal{P}_{n,3}, \ldots$ (everything in this proof is measurable, and we will stop repeatedly mentioning it) such that

(a) $\mathcal{P}_{n,k+1}$ refines $\mathcal{P}_{n,k}$ for all n, k,

(b) $|\mathcal{P}_{n,k}| = m_k$ where m_k is a function of only k, and

(c) $\|W_n - W_{n,k}\|_\square \le 1/k$ where $W_{n,k} = (W_n)_{\mathcal{P}_{n,k}}$.

The weak regularity lemma only guarantees that $|\mathcal{P}_{n,k}| \le m_k$, but if we allow empty parts, then we can achieve equality in (b).

Step 2. Passing to a subsequence.

Initially, each $\mathcal{P}_{n,k}$ partitions $[0,1]$ into arbitrary measurable sets. By restricting to a subsequence, we may assume that

- For each k and $i \in [m_k]$, the measure of the ith part of $\mathcal{P}_{n,k}$ converges to some value $\alpha_{k,i}$ as $n \to \infty$.
- For each k and $i, j \in [m_k]$, the value of $W_{n,k}$ on the product of the ith and jth parts of $\mathcal{P}_{n,k}$ converges to some value $\beta_{k,i,j}$ as $n \to \infty$.

Now construct, for each k, the following limiting objects as $n \to \infty$ along the preceding subsequence:

- Let $\mathcal{P}_k = \{I_{k,1}, \ldots, I_{k,m_k}\}$ denote a partition of $[0,1]$ into intervals with lengths $\lambda(I_{k,i}) = \alpha_{k,i}$ for each $i \in [m_k]$.
- Let U_k denote a step graphon with steps $\mathcal{P}_k$, and whose value on $I_{k,i} \times I_{k,j}$ is $\beta_{k,i,j}$ for each $i, j \in [m_k]$.

Then, for each k,

$$\delta_\square(W_{n,k}, U_k) \to 0, \quad \text{as } n \to \infty. \tag{4.8}$$

(In fact, some rearrangement of the step graphon $W_{n,k}$ converges pointwise almost everywhere to the step graphon U_k.)

For each k, since $W_{n,k} = (W_{n,k+1})_{\mathcal{P}_{n,k}}$ for every n, we have

$$U_k = (U_{k+1})_{\mathcal{P}_k}.$$

.368

.5 .3
.3 .4

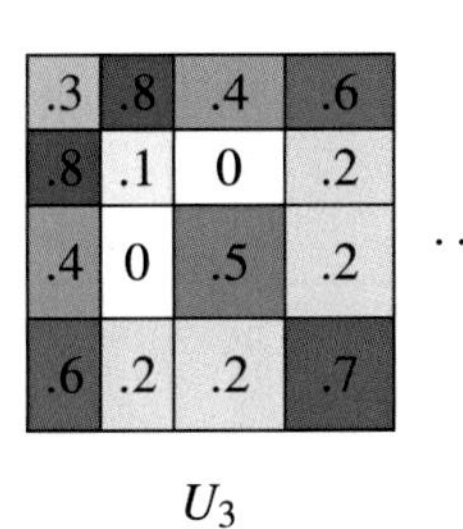

U_1 U_2 U_3

Step 3. Finding the limit.

Now each U_k can be thought of as a random variable on probability space $[0,1]^2$ (i.e., $U_k(X,Y)$ with $(X,Y) \sim \text{Uniform}([0,1]^2)$). The condition $U_k = (U_{k+1})_{\mathcal{P}_k}$ implies that the sequence $U_1, U_2, \ldots$ is a martingale. Since each U_k is bounded between 0 and 1, by the martingale convergence theorem (Theorem 4.7.6), there exists a graphon U such that $U_k \to U$ pointwise almost everywhere as $k \to \infty$.

We claim that $W_1, W_2, \ldots$ (which is a relabeled subsequence of the original sequence) converges to U in cut metric.

Let $\varepsilon > 0$. Then there exists some $k > 3/\varepsilon$ such that $\|U - U_k\|_1 < \varepsilon/3$, by pointwise convergence and the dominated convergence theorem. Then $\delta_\square(U, U_k) < \varepsilon/3$. By (4.8), there exists some $n_0 \in \mathbb{N}$ such that $\delta_\square(W_{n,k}, U_k) < \varepsilon/3$ for all $n > n_0$. Finally, since we chose $k > 3/\varepsilon$, we already know that $\delta_\square(W_n, W_{n,k}) < \varepsilon/3$ for all n. We conclude that

$$\delta_\square(U, W_n) \le \delta_\square(U, U_k) + \delta_\square(U_k, W_{n,k}) + \delta_\square(W_{n,k}, W_n) \le \varepsilon/3 + \varepsilon/3 + \varepsilon/3 = \varepsilon.$$

Since $\varepsilon > 0$ can be chosen to be arbitrarily small, we find that the subsequence W_n converges to U in cut metric. □

Quick Applications

The compactness of $(\widetilde{\mathcal{W}}_0, \delta_\square)$ is a powerful statement. We will use it to prove the equivalence of cut metric convergence and left-convergence in the next section. Right now, let us show how to use compactness to deduce the existence of limits for a left-convergent sequence of graphons.

Proof of Theorem 4.3.8 (existence of limit for a left-convergent sequence of graphons).
Let $W_1, W_2, \ldots$ be a sequence of graphons such that the sequence of F-densities $\{t(F, W_n)\}_n$ converges for every graph F. Since $(\widetilde{\mathcal{W}}_0, \delta_\square)$ is a compact metric space by Theorem 4.2.7, it is also sequentially compact, and so there is a subsequence $(n_i)_{i=1}^\infty$ and a graphon W such that $\delta_\square(W_{n_i}, W) \to 0$ as $i \to \infty$. Fix any graph F. By the counting lemma, Theorem 4.5.1, it follows that $t(F, W_{n_i}) \to t(F, W)$. But by assumption, the sequence $\{t(F, W_n)\}_n$ converges. Therefore $t(F, W_n) \to t(F, W)$ as $n \to \infty$. Thus, W_n left-converges to W. □

Let us now examine a different aspect of compactness. Recall that by definition, a set is compact if every open cover has a finite subcover.

Recall from Theorem 4.2.8 that the set of graphs is dense in the space of graphons with respect to the cut metric. This was proved by showing that for every $\varepsilon > 0$ and graphon W, one can find a graph G such that $\delta_\square(G, W) < \varepsilon$. However, the size of G produced by this proof depends on both ε and W, since the proof proceeds by first taking a discrete L^1 approximation of W, which could involve an unbounded number of steps to approximate. In contrast, we show in what follows that the number of vertices of G needs to depend only on ε and not on W.

Proposition 4.8.1 (Uniform approximation of graphons by graphs)
For every $\varepsilon > 0$ there is some positive integer $N = N(\varepsilon)$ such that every graphon lies within cut distance ε of a graph on at most N vertices.

Proof. Let $\varepsilon > 0$. For a graph G, define the open ε-ball (with respect to the cut metric) around G:

$$B_\varepsilon(G) = \{W \in \widetilde{\mathcal{W}}_0 : \delta_\square(G, W) < \varepsilon\}.$$

Since every graphon lies within cut distance ε from some graph (Theorem 4.2.8), the balls $B_\varepsilon(G)$ cover $\widetilde{\mathcal{W}}_0$ as G ranges over all graphs. By compactness, this open cover has a finite subcover, and let N be the maximum number of vertices in graphs G of this subcover. Then every graphon lies within cut distance ε of a graph on at most N vertices. □

The following exercise asks to make the preceding proof quantitative.

Exercise 4.8.2. Show that for every $\varepsilon > 0$, every graphon lies within cut distance at most ε from some graph on at most C^{1/ε^2} vertices, where C is some absolute constant.

Hint: Use the weak regularity lemma.

Remark* 4.8.3** (Ineffective bounds from compactness). Arguments using compactness usually do not generate quantitative bounds, meaning, for example, the proof of Proposition 4.8.1 does not give any specific function $n(\varepsilon)$, only that such a function always exists. In case where one does not have an explicit bound, we call the bound ***ineffective. Ineffective bounds also often arise from arguments involving ergodic theory and nonstandard analysis. Sometimes a different argument can be found that generates a quantitative bound (e.g., Exercise 4.8.2), but it is not always known how to do this. Here we illustrate a simple example of a compactness application (unrelated to dense graph limits) that gives an ineffective bound, but it remains an open problem to make the bound effective.

This example concerns bounded degree graphs. It is sometimes called a "regularity lemma" for bounded degree graphs, but it is very different from the regularity lemmas we have encountered so far.

A ***rooted graph*** (G, v) consists of a graph G with a vertex $v \in v(G)$ designated as the ***root***. Given a graph G and positive integer r, we can obtain a random rooted graph by first picking a vertex v of G as the root uniformly at random, and then removing all vertices more than distance r from v. We define the ***r-neighborhood-profile*** of G to be the probability distribution on rooted graphs generated by this process.

Recall that the ***total variation distance*** between two probability distributions μ and λ is defined by

$$d_{TV}(\mu, \lambda) = \sup_E |\mu(E) - \lambda(E)|,$$

where E ranges over all events. In the case of two discrete discrete random distributions μ and λ, the preceding definition can be written as half the ℓ^1 distance between the two probability distributions:

$$d_{TV}(\mu, \lambda) = \frac{1}{2} \sum_x |\mu(x) - \lambda(x)|.$$

The following is an unpublished observation of Alon.

Theorem 4.8.4 ("Regularity lemma" for bounded degree graphs)
For every $\varepsilon > 0$ and positive integers Δ and r there exists a positive integer $N = N(\varepsilon, \Delta, r)$ such that for every graph G with maximum degree at most Δ, there exists a graph G' with at most N vertices, so that the total variation distance between the r-neighborhood-profiles of G and G' is at most ε.

Proof. Let $\mathcal{G} = \mathcal{G}_{\Delta,r}$ be the set of all possible rooted graphs with maximum degree Δ and radius at most r around the root. Then $|\mathcal{G}| < \infty$. The r-neighborhood-profile p_G of any rooted graph G can be represented as a point $p_G \in [0,1]^{\mathcal{G}}$ with coordinate sum 1, and let $A = \{p_G : \text{graph } G\} \subseteq [0,1]^{\mathcal{G}}$ be the set of all points that can arise this way. Since $[0,1]^{\mathcal{G}}$ is compact, the closure of A is compact. Since the union of the open ε-neighborhoods (with respect to d_{TV}) of p_G, ranging over all graphs G, covers the closure of A, by compactness there is some finite subcover. This subcover is a finite collection $\mathcal{X}$ of graphs so that for every graph G, p_G lies within ε total variance distance to some $p_{G'}$ with $G' \in \mathcal{X}$. We conclude by letting N be the maximum number of vertices of a graph from $\mathcal{X}$. □

Despite the short proof using compactness, it remains an open problem to make the preceding result quantitative.

Open Problem 4.8.5 (Effective "regularity lemma" for bounded degree graphs)
Find some specific $N(\varepsilon, \Delta, r)$ so that Theorem 4.8.4 holds.

4.9 Equivalence of Convergence

In this section, we prove Theorem 4.3.7, that left-convergence is equivalent to convergence in cut metric. The counting lemma (Theorem 4.5.1) already showed that cut metric convergence implies left-convergence. It remains to show the converse. In other words, we need to show that if $W_1, W_2, \ldots$ is a sequence of graphons such that $t(F, W_n)$ converges as $n \to \infty$ for every graph F, then W_n is a Cauchy sequence in $(\widetilde{\mathcal{W}}_0, \delta_\square)$.

By the compactness of the graphon space, there is always some (subsequential) limit point W of the sequence W_n under the cut metric. We want to show that this limit point is unique. Suppose U is another limit point. It remains to show that W and U are in fact the same point in $\widetilde{\mathcal{W}}_0$.

Let $(n_i)_{i=1}^\infty$ be a subsequence such that $W_{n_i} \to W$. By the counting lemma, $t(F, W_{n_i}) \to t(F, W)$ for all graphs F, and by convergence of F-densities, $t(F, W_n) \to t(F, W)$ for all graphs F. Similarly, $t(F, W_n) \to t(F, U)$ for all F. Hence, $t(F, U) = t(F, W)$ for all F. All that remains is to prove the following claim.

Theorem 4.9.1 (Uniqueness of moments)
Let U and W be graphons such that $t(F, W) = t(F, U)$ for all graphs F. Then $\delta_\square(U, W) = 0$.

***Remark* 4.9.2.** The result is reminiscent of results from probability theory on the uniqueness of moments, which roughly says that if two "sufficiently well-behaved" real random variables X and Y share the same moments (i.e., $\mathbb{E}[X^k] = \mathbb{E}[Y^k]$ for all nonnegative integers k), then X and Y must be identically distributed. One needs some technical conditions for the conclusion to hold. For example, Carleman's condition says that if the moments of X satisfy $\sum_{k=1}^\infty \mathbb{E}[X^{2k}]^{-1/(2k)} = \infty$, then the distribution of X is uniquely determined by its moments. This sufficient condition holds as long as the kth moment of X does not grow too quickly with k. It holds for many distributions in practice.

We need some preparation before proving the uniqueness of moments theorem.

Lemma 4.9.3 (Tail bounds for U-statistics)
Let $U\colon [0,1]^2 \to [-1,1]$ be a symmetric measurable function. Let $x_1, \ldots, x_k \in [0,1]$ be chosen independently and uniformly at random. Let $\varepsilon > 0$. Then

$$\mathbb{P}\left(\left|\frac{1}{\binom{k}{2}}\sum_{i<j} U(x_i, x_j) - \int_{[0,1]^2} U\right| \geq \varepsilon\right) \leq 2e^{-k\varepsilon^2/8}.$$

Proof. Let $f(x_1, \ldots, x_n)$ denote the expression inside the absolute value. So $\mathbb{E}f = 0$. Also f changes by at most $2(k-1)/\binom{k}{2} = 4/k$ whenever we change exactly one coordinate of f. By the bounded differences inequality, Theorem 4.4.4, we obtain

$$\mathbb{P}(|f| \geq \varepsilon) \leq 2\exp\left(\frac{-2\varepsilon^2}{(4/k)^2 k}\right) = 2e^{-k\varepsilon^2/8}. \qquad \square$$

Let us now consider a variation of the W-random graph model from Section 4.4. Let $x_1, \ldots, x_k \in [0,1]$ be chosen independently and uniformly at random. Let $\mathbf{H}(k, W)$ be an edge-weighted random graph on vertex set $[k]$ with edge ij having weight $W(x_i, x_j)$, for each $1 \leq i < j \leq n$. Note that this definition makes sense for any symmetric measurable $W\colon [0,1]^2 \to \mathbb{R}$. Furthermore, when W is a graphon, the W-random graph $\mathbf{G}(k, W)$ can be obtained by independently sampling each edge of $\mathbf{H}(k, W)$ with probability equal to its edge weight. We shall study the joint distributions of $\mathbf{G}(k, W)$ and $\mathbf{H}(k, W)$ coupled through the preceding two-step process.

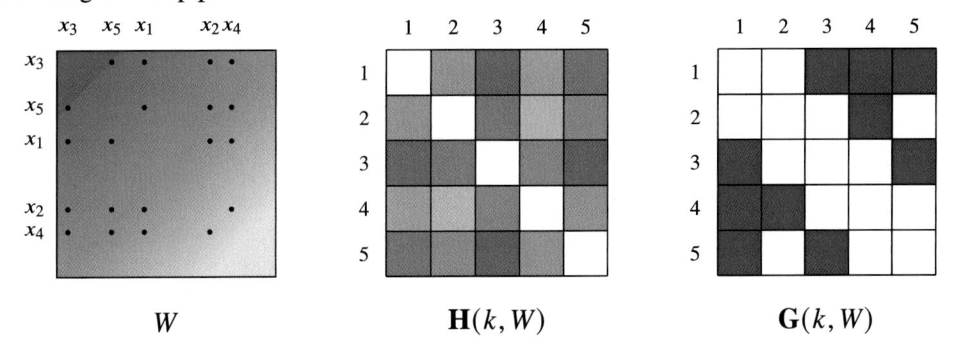

W $\qquad$ $\mathbf{H}(k, W)$ $\qquad$ $\mathbf{G}(k, W)$

Similar to Definition 4.2.4 of the cut distance $\delta_\square$, define the distance based on the L^1 norm:

$$\boldsymbol{\delta_1(W, U)} := \inf_\phi \|W - U^\phi\|_1$$

where the infimum is taken over all invertible measure preserving maps $\phi\colon [0,1] \to [0,1]$. Since $\|\cdot\|_\square \leq \|\cdot\|_1$, we have $\delta_\square \leq \delta_1$.

Lemma 4.9.4 (1-norm convergence for $\mathbf{H}(k, W)$)
Let W be a graphon. Then $\delta_1(\mathbf{H}(k, W), W) \to 0$ as $k \to \infty$ with probability 1.

Proof. First, we prove the result for step graphons W. In this case, with probability 1 the fraction of vertices of $\mathbf{H}(k, W)$ that fall in each step of W converges to the length of each step by the law of large numbers. If so, then after sorting the vertices of $\mathbf{H}(k, W)$, the associated graphon $\mathbf{H}(k, W)$ is obtained from W by changing the step sizes by $o(1)$ as $k \to \infty$, and then zeroing out the diagonal blocks, as illustrated in what follows. Then $\mathbf{H}(k, W)$ converges to W pointwise almost everywhere as $k \to \infty$. In particular, $\delta_1(\mathbf{H}(k, W), W) \to 0$.

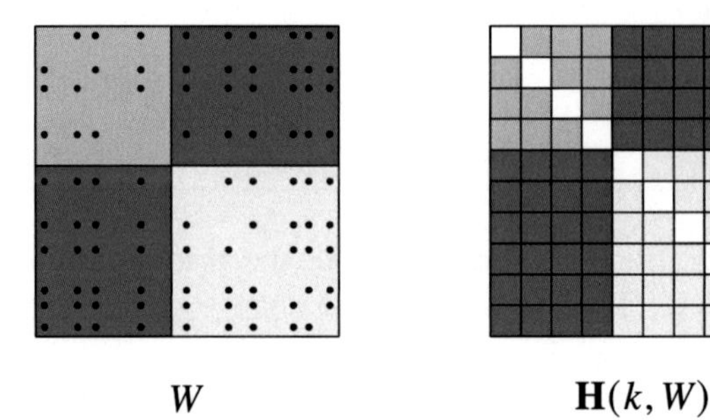

W $\qquad$ $\mathbf{H}(k, W)$

Now let W be any graphon. For any other graphon W', by using the same random vertices for $\mathbf{H}(k, W)$ and $\mathbf{H}(k, W')$, the two random graphs are coupled so that with probability 1,

$$\|\mathbf{H}(k, W) - \mathbf{H}(k, W')\|_1 = \|\mathbf{H}(k, W - W')\|_1 = \|W - W'\|_1 + o(1) \quad \text{as } k \to \infty$$

by Lemma 4.9.3 applied to $U(x, y) = |W(x, y) - W'(x, y)|$.

For every $\varepsilon > 0$, we can find some step graphon W' so that $\|W - W'\|_1 \le \varepsilon$ (by approximating the Lebesgue measure using boxes). We saw earlier that $\delta_1(\mathbf{H}(k, W'), W') \to 0$. It follows that with probability 1,

$$\begin{aligned}\delta_1(\mathbf{H}(k, W), W) &\le \|\mathbf{H}(k, W) - \mathbf{H}(k, W')\|_1 + \delta_1(\mathbf{H}(k, W'), W') + \|W' - W\|_1 \\ &= 2\,\|W' - W\|_1 + o(1) \le 2\varepsilon + o(1)\end{aligned}$$

as $k \to \infty$. Since $\varepsilon > 0$ can be chosen to be arbitrarily small, we have $\delta_1(\mathbf{H}(k, W), W) \to 0$ with probability 1. □

Proof of Theorem 4.9.1 (uniqueness of moments). By inclusion-exclusion, for any k-vertex labeled graph F,

$$\begin{aligned}&\Pr[\mathbf{G}(k, W) \cong F \text{ as labeled graphs}] \\ &\qquad = \sum_{F' \supseteq F} (-1)^{e(F') - e(F)} \Pr[\mathbf{G}(k, W) \supseteq F' \text{ as labeled graphs}],\end{aligned}$$

where the sum ranges over all graphs F' with $V(F') = V(F)$ and $E(F') \supseteq E(F)$. Since

$$t(F', W) = \Pr[\mathbf{G}(k, W) \supseteq F' \text{ as labeled graphs}],$$

we see that the distribution of $\mathbf{G}(k, W)$ is determined by the values of $t(F, W)$ over all F. Since $t(F, W) = t(F, U)$ for all F, $\mathbf{G}(k, W)$ and $\mathbf{G}(k, U)$ are identically distributed.

Our strategy is to prove

$$W \overset{\delta_1}{\approx} \mathbf{H}(k, W) \overset{\delta_\square}{\approx} \mathbf{G}(k, W) \overset{D}{\equiv} \mathbf{G}(k, U) \overset{\delta_\square}{\approx} \mathbf{H}(k, U) \overset{\delta_1}{\approx} U.$$

By Lemma 4.9.4, $\delta_1(\mathbf{H}(k, W), W) \to 0$ with probability 1.

By coupling $\mathbf{H}(k, W)$ and $\mathbf{G}(k, W)$ using the same random vertices as noted earlier, so that $\mathbf{G}(k, W)$ is generated from $\mathbf{H}(k, W)$ by independently sampling each edge with probability equal to the edge weight, we have

$$\mathbb{P}(\delta_\square(\mathbf{G}(k, W), \mathbf{H}(k, W)) \geq \varepsilon) \to 0 \quad \text{as } k \to 0 \text{ for every fixed } \varepsilon > 0.$$

We leave the details of this claim as the exercise that follows. It can be proved via the Chernoff bound and the union bound. We need to be a bit careful about the definition of the cut norm, as one needs to consider fractional vertices.

Exercise 4.9.5 (Edge-sampling an edge-weighted graph and cut norm). Let H be an edge-weighted graph on k vertices, with edge weights in $[0, 1]$, and let G be a random graph obtained from H by independently keeping each edge with probability equal to its edge-weight. Prove that for every $\varepsilon > 0$ and $\delta > 0$, there exists k_0 such that $\delta_\square(G, H) < \varepsilon$ with probability $> 1 - \delta$, provided that $k \geq k_0$.

So with probability 1,

$$\delta_\square(\mathbf{H}(k, W), \mathbf{G}(k, W)) \to 0 \qquad \text{as } k \to \infty.$$

Since $\delta_\square \leq \delta_1$, we have, with probability 1,

$$\delta_\square(W, \mathbf{G}(k, W)) \leq \delta_1(W, \mathbf{H}(k, W)) + \delta_\square(\mathbf{H}(k, W), \mathbf{G}(k, W)) = o(1).$$

Likewise, $\delta_\square(U, \mathbf{G}(k, U)) = o(1)$ with probability 1. Since $\mathbf{G}(k, W)$ and $\mathbf{G}(k, U)$ are identically distributed as noted earlier, we deduce that $\delta_\square(W, U) = 0$. □

This finishes the proof of the equivalence between left-convergence and cut metric convergence. This equivalence can be recast as counting and inverse counting lemmas. We state the inverse counting lemma in what follows and leave the proof as an instructive exercise in applying the compactness of the graphon space. (One need not invoke anything from the proof of the uniqueness of moments theorem. You may wish to review the discussions on applying compactness at the end of the previous section and the beginning of this section.)

Corollary 4.9.6 (Inverse counting lemma)
For every $\varepsilon > 0$ there is some $\eta > 0$ and integer $k > 0$ such that, if U and W are graphons with

$$|t(F, U) - t(F, W)| \leq \eta \quad \text{whenever } v(F) \leq k,$$

then $\delta_\square(U, W) \leq \varepsilon$.

Exercise 4.9.7. Prove the inverse counting lemma Corollary 4.9.6 using the compactness of the graphon space (Theorem 4.2.7) and the uniqueness of moments (Theorem 4.9.1).

Hint: Consider a hypothetical sequence of counterexamples.

***Remark* 4.9.8.** The inverse counting lemma was first proved by Borgs, Chayes, Lovász, Sós, and Vesztergombi (2008) in the following quantitative form:

Theorem 4.9.9 (Inverse counting lemma)
Let k be a positive integer. Let U and W be graphons with

$$|t(F,U) - t(F,W)| \le 2^{-k^2} \quad \text{whenever } v(F) \le k.$$

Then (here C is some absolute constant)

$$\delta_\square(U,W) \le \frac{C}{\sqrt{\log k}}.$$

Exercise 4.9.10. Prove that there exists a function $f\colon (0,1] \to (0,1]$ such that for all graphons U and W, there exists a graph F with

$$\frac{|t(F,U) - t(F,W)|}{e(F)} \ge f(\delta_\square(U,W)).$$

Exercise 4.9.11* (Generalized maximum cut). For symmetric measurable functions $W, U\colon [0,1]^2 \to \mathbb{R}$, define

$$C(W,U) := \sup_\phi \langle W, U^\phi \rangle = \sup_\phi \int W(x,y)U(\phi(x),\phi(y))\,dxdy,$$

where ϕ ranges over all invertible measure preserving maps $[0,1] \to [0,1]$. Extend the definition of $C(\cdot,\cdot)$ to graphs via $C(G,\cdot) := C(W_G,\cdot)$ and so on.

(a) Is $C(U,W)$ continuous jointly in (U,W) with respect to the cut norm? Is it continuous in U if W is held fixed?
(b) Show that if W_1 and W_2 are graphons such that $C(W_1,U) = C(W_2,U)$ for all graphons U, then $\delta_\square(W_1,W_2) = 0$.
(c) Let $G_1, G_2, \ldots$ be a sequence of graphs such that $C(G_n,U)$ converges as $n \to \infty$ for every graphon U. Show that $G_1, G_2, \ldots$ is convergent.
(d) Can the hypothesis in (c) be replaced by "$C(G_n,H)$ converges as $n \to \infty$ for every graph H"?

Exercise 4.9.12 (Characterizing graphs in terms of homomorphism counts).

(a) Let G_1 and G_2 be two graphs such that $\hom(F,G_1) = \hom(F,G_2)$ for every graph F. Show that G_1 and G_2 are isomorphic.
(b) Let G_1 and G_2 be two graphs such that $\hom(G_1,H) = \hom(G_2,H)$ for every graph H. Show that G_1 and G_2 are isomorphic.

Further Reading

The book *Large Networks and Graph Limits* by Lovász (2012) is the authoritative reference on the subject. His survey article titled *Very Large Graphs* (2009) also gives an excellent overview.

One particularly striking application of the theory of dense graph limits is to large deviations for random graphs by Chatterjee and Varadhan (2011). See the survey article *An Introduction to Large Deviations for Random Graphs* by Chatterjee (2016) as well as his book (Chatterjee 2017).

Chapter Summary

- A **graphon** is a symmetric measurable function $W\colon [0,1]^2 \to [0,1]$.
  - Every graph G can be turned into an associated graphon W_G.
  - A graphon can be turned into a random graph model known a W-**random graph**, generalizing the **stochastic block model**.
- The **cut metric** of two graphons U and W is defined by

$$\begin{aligned}\delta_\square(U,W) &= \inf_\phi \|U - W^\phi\|_\square \\ &= \inf_\phi \sup_{S,T\subseteq[0,1]} \left|\int_{S\times T} (U(x,y) - W(\phi(x),\phi(y)))\, dxdy\right|,\end{aligned}$$

  where the infimum is taken over all invertible measure preserving maps $\phi\colon [0,1] \to [0,1]$.
- Given a sequence of graphons (or graphs) $W_1, W_2, \ldots$, we say that it
  - **converges in cut metric** if it is a Cauchy sequence with respect to the cut metric $\delta_\square$;
  - **left-converges** if the homomorphism density $t(F, W_n)$ converges for every fixed graph F as $n \to \infty$.
- The **graphon space is compact** under the cut metric.
  - Proof uses the weak regularity lemma and the martingale convergence theorem.
  - Compactness has powerful consequences.
- Convergence in cut metric and left-convergence are **equivalent** for a sequence of graphons.
  - ($\Rightarrow$) follows from a counting lemma.
  - ($\Leftarrow$) was proved here using compactness.

5

Graph Homomorphism Inequalities

Chapter Highlights

- A suite of techniques for proving inequalities between subgraph densities
- The maximum/minimum triangle density in a graph of given edge density.
- How to apply Cauchy–Schwarz and Hölder inequalities
- Lagrangian method (another proof of Turán's theorem, and linear inequalities between clique densities)
- Entropy method (and applications to Sidorenko's conjecture)

In this chapter, we study inequalities between graph homomorphism densities. Here is a typical example.

Question 5.0.1 (Linear inequality between homomorphism densities)
Given fixed graphs $F_1, \dots, F_k$ and reals $c_1, \dots, c_k$, does

$$c_1 t(F_1, G) + c_2 t(F_2, G) + \cdots + c_k t(F_k, G) \geq 0 \tag{5.1}$$

hold for all graphs G? Recall $t(F, G) = \hom(F, G)/v(G)^{v(F)}$.

Although the left-hand side is a linear combination of various graph homomorphism densities in G, polynomial combinations can also be written this way, as $t(F_1, G)t(F_2, G) = t(F_1 \sqcup F_2, G)$ where $F_1 \sqcup F_2$ is the disjoint union of the two graphs.

More generally, we would like to understand constrained optimization problems in terms of graph homomorphism density. Many problems in extremal graph theory can be cast in this framework. For example, Turán's theorem from Chapter 1 on the maximum edge density of a K_r-free graph can be phrased in terms of the optimization problem

$$\text{maximize } t(K_2, G) \quad \text{subject to } t(K_r, G) = 0.$$

Turán's theorem (Corollary 1.2.6) says that the answer is $1/(r-1)$, achieved by $G = K_{r-1}$. We will see another proof of Turán's theorem later in this chapter, in Section 5.4, using the method of Lagrangians.

***Remark* 5.0.2** (Undecidability). Perhaps surprisingly, Question 5.0.1 is *undecidable* as shown by Hatami and Norine (2011). This means that there is no algorithm that always correctly decides whether a given inequality is true for all graphs (although this fact does not prevent us from proving/disproving specific inequalities). This undecidability stands in stark contrast to the decidability of polynomial inequalities over the reals, which follows from a classic result of Tarski (1948) that the first-order theory of real numbers is decidable (via quantifier elimination). This undecidability of graph homomorphism inequalities is related to **Matiyasevich's theorem** (1970) (also known as the Matiyasevich–Robinson–Davis–Putnam theorem), giving a negative solution to **Hilbert's Tenth Problem**, showing that Diophantine equations are undecidable. (Equivalently: polynomial inequalities over the integers are undecidable.) In fact, the proof of the former proceeds by converting polynomial inequalities over the integers to inequalities between $t(F, G)$ for various F.

As in the case of Diophantine equations, the undecidability of graph homomorphism inequalities should be positively viewed as evidence of the richness of this space of problems. There are still many open problems, such as Sidorenko's inequality that we will see shortly.

***Remark* 5.0.3** (Graphs vs. graphons). In the space of graphons with respect to the cut norm, $W \mapsto t(F, W)$ is continuous (by the counting lemma, Theorem 4.5.1), and graphs are a dense subset (Theorem 4.2.8). It follows that any inequality for continuous functions of $t(F, G)$ over various F's (e.g., linear combinations as in Question 5.0.1) holds for all graphs G if and only if they hold for all graphons W in place of G. Furthermore, due to the compactness of the space of graphons, the extremum of continuous functions of F-densities is always attained at some graphon. The graphon formulation of the results can be often succinct and attractive.

For example, consider the following extremal problem (already mentioned in Chapter 4), where $p \in [0, 1]$ is a given constant:

$$\text{minimize } t(C_4, G) \quad \text{subject to } t(K_2, G) \geq p.$$

The infimum p^4 is not attained by any single graph, but rather by a sequence of quasirandom graphs (see Section 3.1). However, if we enlarge the space from graphs G to graphons W, then the minimizer is attained, in this case by the constant graphon p.

Sidorenko's Conjecture and Forcing Conjecture

There are many important open problems on graph homomorphism inequalities. A major conjecture in extremal combinatorics is Sidorenko's conjecture (1993) (an equivalent conjecture was given earlier by Erdős and Simonovits).

Definition 5.0.4 (Sidorenko graphs)
We say that a graph F is ***Sidorenko*** if, for every graph G,

$$t(F, G) \geq t(K_2, G)^{e(F)}.$$

Conjecture 5.0.5 (Sidorenko's conjecture)
Every bipartite graph is Sidorenko.

In other words, the conjecture says that for a fixed bipartite graph F, the F-density in a graph of a given edge density is asymptotically minimized by a random graph. We will develop techniques in this chapter to prove several interesting special cases of Sidorenko's conjecture.

Every Sidorenko graph is necessarily bipartite. Indeed, given a nonbipartite F, we can take a nonempty bipartite G to get $t(F,G) = 0$ while $t(K_2,G) > 0$.

A notable open case of Sidorenko's conjecture is $F = K_{5,5} \setminus C_{10}$ (left side of the following figure). This F is called the ***Möbius graph*** since it is the point-face incidence graph of a minimum simplicial decomposition of a Möbius strip (right side).

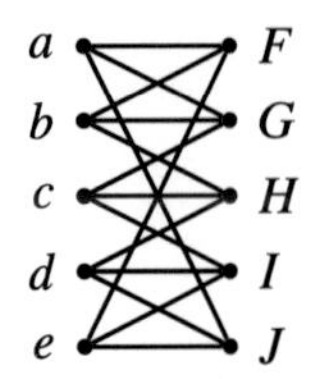

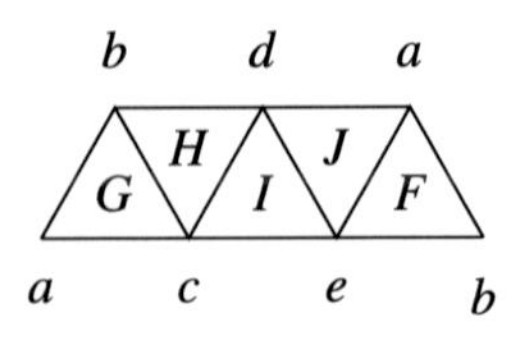

Sidorenko's conjecture has the equivalent graphon formulation: for every bipartite graph F and graphon W,

$$t(F,W) \geq t(K_2,W)^{e(F)}.$$

Note that equality occurs when $W \equiv p$, the constant graphon. One can think of Sidorenko's conjecture as a separate problem for each F, and asking to minimize $t(F,W)$ among graphons W with $\int W \geq p$. Whether the constant graphon is the unique minimizer is the subject of an even stronger conjecture known as the forcing conjecture.

Definition 5.0.6 (Forcing graphs)
We say that a graph F is ***forcing*** if every graphon W with $t(F,W) = t(K_2,W)^{e(F)}$ is a constant graphon (up to a set of measure zero).

By translating back and forth between graph limits and sequences of graphs, the forcing property is equivalent to a quasirandomness condition. Thus any forcing graph can play the role of C_4 in Theorem 3.1.1. This is what led Chung, Graham, and Wilson to consider forcing graphs. In particular, C_4 is forcing.

Proposition 5.0.7 (Forcing and quasirandomness)
A graph F is forcing if and only if for every constant $p \in [0,1]$, every sequence of graphs $G = G_n$ with

$$t(K_2,G) = p + o(1) \qquad \text{and} \qquad t(F,G) = p^{e(F)} + o(1)$$

is quasirandom in the sense of Definition 3.1.2.

Exercise 5.0.8. Prove Proposition 5.0.7.

The forcing conjecture states a complete characterization of forcing graphs (Skokan and Thoma 2004; Conlon, Fox, and Sudakov 2010).

Conjecture 5.0.9 (Forcing conjecture)
A graph is forcing if and only if it is bipartite and has at least one cycle.

Exercise 5.0.10. Prove the "only if" direction of the forcing conjecture.

Exercise 5.0.11. Prove that every forcing graph is Sidorenko.

Exercise 5.0.12 (Forcing and stability). Show that a graph F is forcing if and only if for every $\varepsilon > 0$, there exists $\delta > 0$ such that, if a graph G satisfies $t(F, G) \le t(K_2, G)^{e(F)} + \delta$, then $\delta_\square(G, p) \le \varepsilon$.

The following exercise shows that to prove a graph is Sidorenko, we do not lose anything by giving away a constant factor. The proof is a quick and neat application of the tensor power trick.

Exercise 5.0.13 (Tensor power trick). Let F be a bipartite graph. Suppose there is some constant $c > 0$ such that

$$t(F, G) \ge c\, t(K_2, G)^{e(F)} \quad \text{for all graphs } G.$$

Show that F is Sidorenko.

5.1 Edge versus Triangle Densities

What are all the pairs of edge and triangles densities that can occur in a graph (or graphon)? Since the set of graphs is dense in the space of graphons, the closure of $\{(t(K_2, G), t(K_3, G)) :$ graph $G\}$ is the

$$\textbf{\textit{edge-triangle region}} := \{(t(K_2, W), t(K_3, W)) : \text{graphon } W\} \subseteq [0,1]^2. \tag{5.2}$$

This is a closed subset of $[0,1]^2$, due to the compactness of the space of graphons. This set has been completely determined, and it is illustrated in Figure 5.1. We will discuss its features in this section.

The upper and lower boundaries of this region correspond to the answers of the following question.

Question 5.1.1 (Extremal triangle density given edge density)
Fix $p \in [0,1]$. What are the minimum and maximum possible $t(K_3, W)$ among all graphons with $t(K_2, W) = p$?

For a given $p \in [0,1]$, the set $\{t(K_3, W) : t(K_2, W) = p\}$ is a closed interval. Indeed, if W_0 achieves the minimum triangle density, and W_1 achieves the maximum, then their linear interpolation $W_t = (1-t)W_0 + tW_1$, ranging over $0 \le t \le 1$, must have triangle density continuously interpolating between those of W_0 and W_1, and therefore achieves every intermediate value.

Maximum Triangle Density

The maximization part of Question 5.1.1 is easier. The answer is $p^{3/2}$.

Theorem 5.1.2 (Max triangle density)
For every graph G,

$$t(K_3, G) \le t(K_2, G)^{3/2}.$$

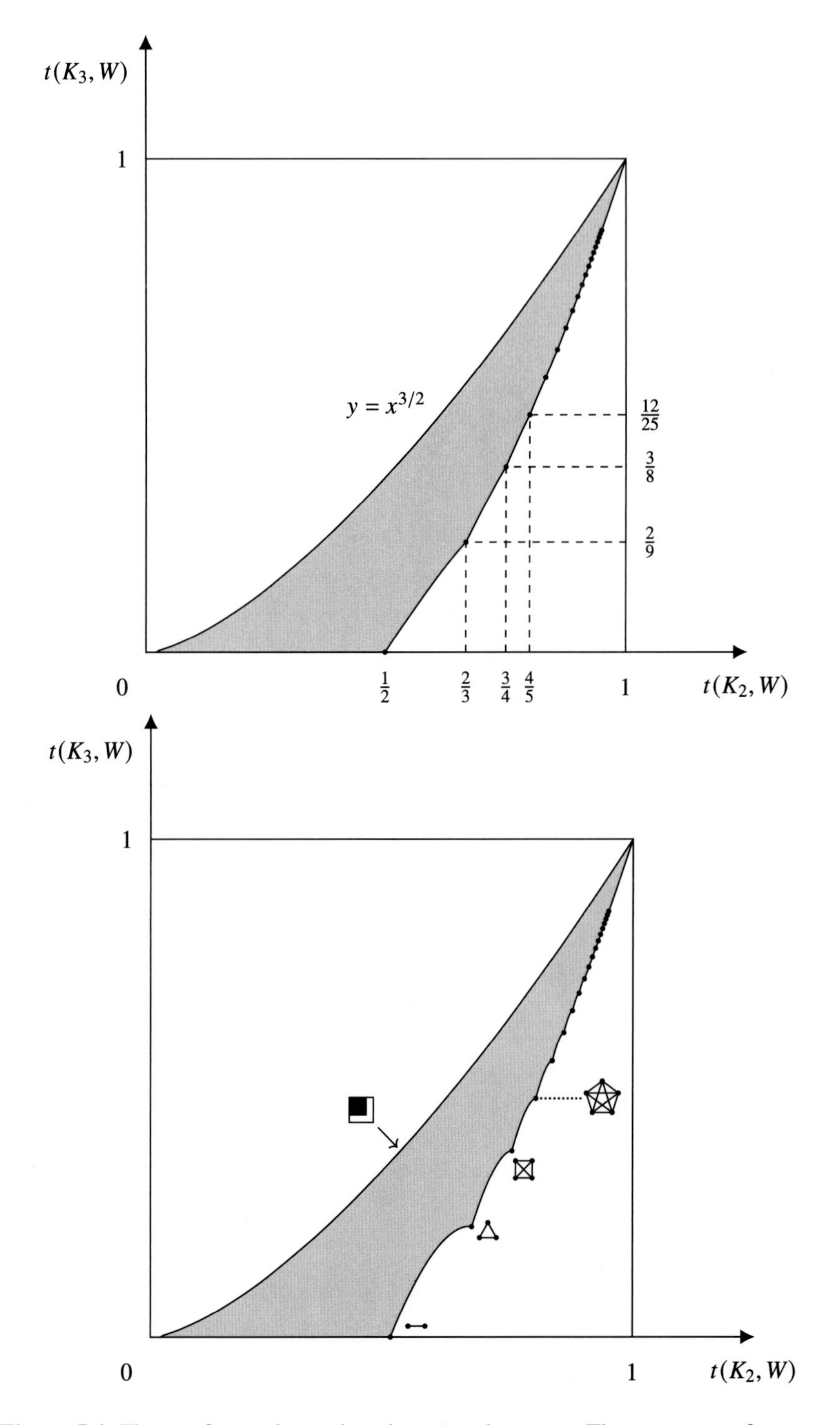

Figure 5.1 The top figure shows the edge-triangle region. This region is often depicted as in the bottom figure, which better highlights the concave scallops on the lower boundary but is a less accurate plot.

This inequality is asymptotically tight for G being a clique on a subset of vertices. The equivalent graphon inequality $t(K_3, W) \leq t(K_2, W)^{3/2}$ attains equality for the clique graphon:

$$W(x, y) = \begin{cases} 1 & \text{if } x, y \leq a, \\ 0 & \text{otherwise.} \end{cases} \tag{5.3}$$

For the preceding W, we have $t(K_3, G) = a^3$ while $t(K_2, G) = a^2$.

Proof. The quantities $\hom(K_3, G)$ and $\hom(K_2, G)$ count the number of closed walks in the graphs of length 3 and 2, respectively. Let $\lambda_1 \geq \cdots \geq \lambda_n$ be the eigenvalues of the adjacency matrix A_G of G. Then

$$\hom(K_3, G) = \operatorname{tr} A_G^3 = \sum_{i=1}^{k} \lambda_i^3 \qquad \text{and} \qquad \hom(K_2, G) = \operatorname{tr} A_G^2 = \sum_{i=1}^{k} \lambda_i^2.$$

Then (see Lemma 5.1.3)

$$\hom(K_3, G) = \sum_{i=1}^{n} \lambda_i^3 \leq \left(\sum_{i=1}^{n} \lambda_i^2 \right)^{3/2} = \hom(K_2, G)^{3/2}.$$

After dividing by $v(G)^3$ on both sides, the result follows. □

Lemma 5.1.3 (A power sum inequality)
Let $t \geq 1$, and $a_1, \cdots, a_n \geq 0$. Then,

$$a_1^t + \cdots + a_n^t \leq (a_1 + \cdots + a_n)^t.$$

Proof. Assume at least one a_i is positive, or else both sides equal to zero. Then

$$\frac{\text{LHS}}{\text{RHS}} = \sum_{i=1}^{n} \left(\frac{a_i}{a_1 + \cdots + a_n} \right)^t \leq \sum_{i=1}^{n} \frac{a_i}{a_1 + \cdots + a_n} = 1.$$ □

***Remark* 5.1.4.** We will see additional proofs of Theorem 5.1.2 not invoking eigenvalues later in Exercise 5.2.14 and in Section 5.3. Theorem 5.1.2 is an inequality in "physical space" (as opposed to going into the "frequency space" of the spectrum), and it is a good idea to think about how to prove it while staying in the physical space.

More generally, the clique graphon (5.3) also maximizes K_r-densities among all graphons of given edge density.

Theorem 5.1.5 (Maximum clique density)
For any graphon W and integer $k \geq 3$,

$$t(K_k, W) \leq t(K_2, W)^{k/2}.$$

Proof. There exist integers $a, b \geq 0$ such that $k = 3a + 2b$ (e.g., take $a = 1$ if k is odd and $a = 0$ if k is even). Then $aK_3 + bK_2$ (a disjoint union of a triangles and b isolated edges) is a subgraph of K_k. So

$$t(K_k, W) \leq t(aK_3 + bK_2, W) = t(K_3, W)^a t(K_2, W)^b \leq t(K_2, W)^{3a/2+b} = t(K_2, W)^{k/2}. \qquad \square$$

Remark 5.1.6 (Kruskal–Katona theorem). Thanks to a theorem of Kruskal (1963) and Katona (1968), the exact answer to the following nonasymptotic question is completely known:

What is the maximum number of copies of K_k's in an n-vertex graph with m edges?

When $m = \binom{a}{2}$ for some integer a, the optimal graph is a clique on a vertices. More generally, for any value of m, the optimal graph is obtained by adding edges in *colexicographic order*:

$$12, 13, 23, 14, 24, 34, 15, 25, 35, 45, \ldots.$$

This is stronger than Theorem 5.1.5, which only gives an asymptotically tight answer as $n \to \infty$. The full Kruskal–Katona theorem also answers this question:

What is the maximum number of k-cliques in an r-graph with n vertices and m edges?

When $m = \binom{a}{r}$, the optimal r-graph is a clique on a vertices. (An asymptotic version of this statement can be proved using techniques in Section 5.3.) More generally, the optimal r-graph is obtained by adding the edges in colexicographic order. For example, for 3-graphs, the edges should be added in the following order:

$$123, 124, 134, 234, 125, 135, 235, 145, 245, 345, \ldots.$$

Here $a_1 \ldots a_r < b_1 \ldots b_r$ in colexicographic order if $a_i < b_i$ at the last i where $a_i \neq b_i$ (i.e., dictionary order when read from right to left). Here we sort the elements of each r-tuple in increasing order.

The Kruskal–Katona theorem can be proved by a compression/shifting argument. The idea is to repeatedly modify the graph so that we eventually end up at the optimal graph. At each step, we "push" all the edges toward a clique along some "direction" in a way that does not reduce the number of k-cliques in the graph.

Minimum Triangle Density

Now we turn to the lower boundary of the edge-triangle region. What is the minimum triangle density in a graph of given edge density p?

For $p \leq 1/2$, we can have complete bipartite graphs of density $p + o(1)$, which are triangle-free. For $p > 1/2$, the triangle density must be positive due to Mantel's theorem (Theorem 1.1.1) and supersaturation (Theorem 1.3.4). It turns out that among graphs with edge density $p + o(1)$, the triangle density is asymptotically minimized by certain complete multipartite graphs, although this is not easy to prove.

For each positive integer k, we have

$$t(K_2, K_k) = 1 - \frac{1}{k} \quad \text{and} \quad t(K_3, K_k) = \left(1 - \frac{1}{k}\right)\left(1 - \frac{2}{k}\right).$$

As k ranges over all positive integers, these pairs form special points on the lower boundary of the edge-triangle region, as illustrated in Figure 5.1 on page 162. (Recall that K_k is associated to the same graphon as a complete k-partite graph with equal parts.)

Now suppose the given edge density p lies strictly between $1 - 1/(k-1)$ and $1 - 1/k$ for some integer $k \geq 2$. To obtain the graphon with edge density p and minimum triangle density, we first start with K_k with all vertices having equal weight. And then we shrink the relative weight of exactly one of the k vertices (while keeping the remaining $k - 1$ vertices to have the same vertex weight). For example, the graphon illustrated here is obtained by starting with K_4 and shrinking the weight on one vertex.

	I_1	I_2	I_3	I_4
I_1	0	1	1	1
I_2	1	0	1	1
I_3	1	1	0	1
I_4	1	1	1	0

During this process, the total edge density (account for vertex weights) decreases continuously from $1 - 1/k$ to $1 - 1/(k-1)$. At some point, the edge density is equal to p. It turns out that this vertex-weighted k-clique W minimizes triangle density among all graphons with edge density p.

The preceding claim is much more difficult to prove than the maximum triangle density result. This theorem, stated here, due to Razborov (2008), was proved using an involved Cauchy–Schwarz calculus that he coined *flag algebra*. We will say a bit more about this method in Section 5.2.

Theorem 5.1.7 (Minimum triangle density)
Fix $0 \leq p \leq 1$ and $k = \lceil 1/(1-p) \rceil$. The minimum of $t(K_3, W)$ among graphons W with $t(K_2, W) = p$ is attained by the stepfunction W associated to a k-clique with node weights $a_1, a_2, \cdots, a_k$ with sum equal to 1, $a_1 = \cdots = a_{k-1} \geq a_k$, and $t(K_2, W) = p$.

We will not prove this theorem in full here. See Lovász (2012, Section 16.3.2) for a proof of Theorem 5.1.7. Later in this chapter, we give lower bounds that match the edge-triangle region at the cliques. In particular, Theorem 5.4.4 will allow us to determine the convex hull of the region.

The graphon described in Theorem 5.1.7 turns out to be not unique unless $p = 1 - 1/k$ for some positive integer k. Indeed, suppose $1 - 1/(k-1) < p < 1 - 1/k$. Let $I_1, \ldots, I_k$ be the partition of $[0, 1]$ into the intervals corresponding to the vertices of the vertex-weighted k-clique, with $I_1, \ldots, I_{k-1}$ all having equal length, and I_k strictly smaller length. Now replace the graphon on $I_{k-1} \cup I_k$ by an arbitrary triangle-free graphon of the same edge density.

	I_1	I_2	I_3	I_4
I_1	0	1	1	1
I_2	1	0	1	1
I_3	1	1	any triangle-free graphon	
I_4	1	1		

This operation does not change the edge-density or the triangle-density of the graphon. (Check!) The nonuniqueness of the minimizer hints at the difficulty of the result.

This completes our discussion of the edge-triangle region (Figure 5.1 on page 162).

Theorem 5.1.7 was generalized from K_3 to K_4 (Nikiforov 2011), and then to all cliques K_r (Reiher 2016). The construction for the minimizing graphon is the same as for the triangle case.

Theorem 5.1.8 (Minimum clique density)
Fix $0 \le p \le 1$ and $k = \lceil 1/(1-p) \rceil$. The minimum of $t(K_r, W)$ among graphons W with $t(K_2, W) = p$ is attained by the stepfunction W associated to a k-clique with node weights $a_1, a_2, \cdots, a_k$ with sum equal to 1, $a_1 = \cdots = a_{k-1} \ge a_k$, and $t(K_2, W) = p$.

Exercise 5.1.9. Prove that C_6 is Sidorenko.

Hint: Write $\hom(C_6, G)$ and $\hom(K_2, G)$ in terms of the spectrum of G.

5.2 Cauchy–Schwarz

We will apply the Cauchy–Schwarz inequality in the following form: given real-valued functions f and g on the same space (always assuming the usual measurability assumptions without further comments), we have

$$\left(\int_X fg\right)^2 \le \left(\int_X f^2\right)\left(\int_X g^2\right).$$

It is one of the most versatile inequalities in combinatorics.

We write the variables being integrated underneath the integral sign. The domain of integration (usually $[0, 1]$ for each variable) is omitted to avoid clutter. We write

$$\int_{x,y,\ldots} f(x, y, \ldots) \qquad \text{for} \qquad \int f(x, y, \ldots)\, dx dy \cdots .$$

In practice, we will often apply the Cauchy–Schwarz inequality by changing the order of integration and separating an integral into an outer integral and an inner integral. A typical application of the Cauchy–Schwarz inequality is demonstrated in the following calculation (here one should think of x, y, z each as collections of variables):

$$\begin{aligned}\int_{x,y,z} f(x,y)g(x,z) &= \int_x \left(\int_y f(x,y)\right)\left(\int_z g(x,z)\right)\\ &\le \left(\int_x\left(\int_y f(x,y)\right)^2\right)^{1/2}\left(\int_x\left(\int_z g(x,z)\right)^2\right)^{1/2}\\ &= \left(\int_{x,y,y'} f(x,y)f(x,y')\right)^{1/2}\left(\int_{x,z,z'} g(x,z)g(x,z')\right)^{1/2}.\end{aligned}$$

Note that in the final step, "expanding a square" has the effect of "duplicating a variable." It is useful to recognize expressions with duplicated variables that can be folded back into a square.

Let us warm up by proving that $K_{2,2}$ is Sidorenko. We actually already proved this statement in Proposition 3.1.14 in the context of the Chung–Graham–Wilson theorem on quasirandom graphs. We repeat the same calculations here to demonstrate the integral notation.

Theorem 5.2.1 ($K_{2,2}$ is Sidorenko)

$$t(K_{2,2}, W) \ge t(K_2, W)^4$$

The theorem follows from the next two lemmas.

Lemma 5.2.2

$$t(K_{1,2}, W) \ge t(K_2, W)^2$$

Proof. By rewriting as a square and then applying the Cauchy–Schwarz inequality,

$$\begin{aligned}t(K_{1,2}, W) = \int_{x,y,y'} W(x,y)W(x,y') &= \int_x\left(\int_y W(x,y)\right)^2\\ &\ge \left(\int_{x,y} W(x,y)\right)^2 = t(K_2,W)^2.\end{aligned}$$ □

Lemma 5.2.3

$$t(K_{2,2}, W) \ge t(K_{1,2}, W)^2$$

Proof. Similar to the previous proof, we have

$$\begin{aligned}t(K_{2,2}, W) &= \int_{x,y,z,z'} W(x,z)W(x,z')W(y,z)W(y,z')\\ &= \int_{x,y}\left(\int_z W(x,z)W(y,z)\right)^2 \ge \left(\int_{x,y,z} W(x,z)W(y,z)\right)^2 = t(K_{1,2},W)^2.\end{aligned}$$ □

Proofs involving Cauchy–Schwarz are sometimes called sum-of-squares proofs. The Cauchy–Schwarz inequality can be proved by writing the difference between the two sides as a sum-of-squares quantity:

$$\left(\int f^2\right)\left(\int g^2\right) - \left(\int fg\right)^2 = \frac{1}{2}\int_{x,y}(f(x)g(y) - f(y)g(x))^2.$$

Commonly, $g = 1$, in which case we can also write

$$\left(\int f^2\right) - \left(\int f\right)^2 = \int_x \left(f(x) - \int_y f(y)\right)^2.$$

For example, we can write the proof of Lemma 5.2.3 as

$$t(K_{1,2}, W) - t(K_2, W)^2 \geq \int_x \left(\int_y W(x,y) - t(K_2, W)\right)^2.$$

Exercise 5.2.4. Write $t(K_{2,2}, W) - t(K_2, W)^4$ as a single sum-of-squares expression.

The next inequality tells us that if we color the edges of K_n using two colors, then at least $1/4 + o(1)$ fraction of all triangles are monochromatic (Goodman 1959). Note that this $1/4$ constant is tight since it is obtained by a uniform random coloring. In the graphon formulation below, the graphons W and $1 - W$ correspond to edges of each color. We have equality for the constant $1/2$ graphon.

Theorem 5.2.5 (Triangle is common)

$$t(K_3, W) + t(K_3, 1 - W) \geq 1/4$$

Proof. Expanding, we have

$$\begin{aligned} t(K_3, 1 - W) &= \int (1 - W(x,y))(1 - W(x,z))(1 - W(y,z))\, dxdydz \\ &= 1 - 3t(K_2, W) + 3t(K_{1,2}, W) - t(K_3, W). \end{aligned}$$

So

$$\begin{aligned} t(K_3, W) + t(K_3, 1 - W) &= 1 - 3t(K_2, W) + 3t(K_{1,2}, W) \\ &\geq 1 - 3t(K_2, W) + 3t(K_2, W)^2 \\ &= \frac{1}{4} + 3\left(t(K_2, W) - \frac{1}{2}\right)^2 \geq \frac{1}{4}. \end{aligned}$$ □

Which graphs, other than triangles, have the preceding property? We do not know the full answer.

Definition 5.2.6 (Common graphs)
We say that a graph F is ***common*** if, for all graphons W,

$$t(F, W) + t(F, 1 - W) \geq 2^{-e(F)+1}.$$

In other words, the left-hand side is minimized by the constant $1/2$ graphon.

Although it was initially conjectured that all graphs are common, this turns out to be false. In particular, K_t is not common for all $t \geq 4$ (Thomason 1989).

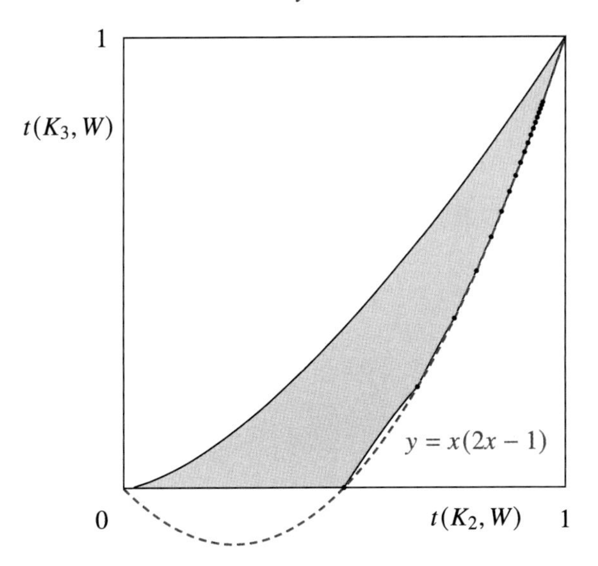

Figure 5.2 The Goodman lower bound on the triangle density from Theorem 5.2.8 plotted on top of the edge-triangle region (Figure 5.1 on page 162).

Proposition 5.2.7

Every Sidorenko graph is common.

Proof. Suppose F is Sidorenko. Let $p = t(K_2, W)$. Then $t(F, W) \geq p^{e(F)}$ and $t(F, 1 - W) \geq t(K_2, 1 - W)^{e(F)} = (1 - p)^{e(F)}$. Adding up and using convexity,

$$t(F, W) + t(F, 1 - W) \geq p^{e(F)} + (1 - p)^{e(F)} \geq 2^{-e(F)+1}. \qquad \square$$

The converse is false. The triangle is common but not Sidorenko. (Recall that every Sidorenko graph is bipartite.)

We also have the following lower bound on the minimum triangle density, given edge density (Goodman 1959). See Figure 5.2 for a plot.

Theorem 5.2.8 (Lower bound on triangle density)

$$t(K_3, W) \geq t(K_2, W)(2t(K_2, W) - 1)$$

The inequality is tight whenever $W = K_n$, in which case $t(K_2, W) = 1 - 1/n$ and $t(K_3, W) = \binom{n}{3}/n^3 = (1 - 1/n)(1 - 2/n)$. In particular, Goodman's bound implies that $t(K_3, W) > 0$ whenever $t(K_2, W) > 1/2$, which we saw from Mantel's theorem.

Proof. Since $0 \leq W \leq 1$, we have $(1 - W(x, z))(1 - W(y, z)) \geq 0$, and so

$$W(x, z)W(y, z) \geq W(x, z) + W(y, z) - 1.$$

Thus

$$\begin{aligned} t(K_3, G) &= \int_{x,y,z} W(x,y)W(x,z)W(y,z) \\ &\geq \int_{x,y,z} W(x,y)(W(x,z) + W(y,z) - 1) \\ &= 2t(K_{1,2}, W) - t(K_2, W) \\ &\geq 2t(K_2, W)^2 - t(K_2, W). \end{aligned}$$

□

Finally, let us demonstrate an application of the Cauchy–Schwarz inequality in the following form, for nonnegative functions f and g:

$$\left(\int f^2 g\right)\left(\int g\right) \geq \left(\int fg\right).$$

Recall that a graph F is Sidorenko if $t(F, W) \geq t(K_2, W)^{e(F)}$ for all graphons W (Definition 5.0.4).

Theorem 5.2.9 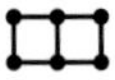
is Sidorenko.

Proof. The idea is to "fold" the graph F in the theorem statement in half along the middle using the Cauchy–Schwarz inequality. Using w and x to indicate the two vertices in the middle, we have

$$t(F, W) = \int_{w,x} \left(\int_{y,z} W(w,y)W(y,z)W(z,x)\right)^2 W(w,x).$$

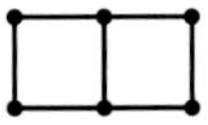

So,

$$\begin{aligned} t(F, W)t(K_2, W) &\geq \left(\int_{w,x,y,z} W(w,y)W(y,z)W(z,x)W(w,x)\right)^2 \\ &= t(C_4, W)^2 \geq t(K_2, W)^8, \end{aligned}$$

with the last step due to Theorem 5.2.1. Therefore $t(F, W) \geq t(K_2, W)^7$ and hence F is Sidorenko. □

Remark* 5.2.10** (Flag algebra). The preceding examples were all simple enough to be found by hand. As mentioned earlier, every application of the Cauchy–Schwarz inequality can be rewritten in the form of a sum of a squares. One could actually search for these sum-of-squares proofs more systematically using a computer program. This idea, first introduced by Razborov (2007), can be combined with other sophisticated methods to determine the lower boundary of the edge-triangle region (Razborov 2008). Razborov coined the term ***flag algebra to describe a formalization of such calculations. The technique is also sometimes called ***graph algebra***, ***Cauchy–Schwarz calculus***, ***sum-of-squares proof***.

Conceptually, the idea is that we are looking for all the ways to obtain nonnegative linear combinations of squared expressions. In a typical application, one is asked to solve an extremal problem of the form

$$\begin{array}{lll} \text{Minimize} & t(F_0, W) & \\ \text{Subject to} & t(F_1, W) = q_1, \quad \ldots, & t(F_\ell, W) = q_\ell, \\ & W \text{ a graphon.} & \end{array}$$

The technique is very flexible. The objectives and constraints could be any linear combinations of densities. It could be maximization instead of minimization. Extensions of the techniques can handle wider classes of extremal problems, such as for hypergraphs, directed graphs, edge-colored graphs, permutations, and more.

Let us illustrate the technique. The nonnegativity of squares implies inequalities such as

$$\int_{x,y,z} W(x,y)W(x,z)\left(\int_{u,w} (aW(x,u)W(y,u) - bW(x,w)W(w,u)W(u,z) + c)\right)^2 \geq 0.$$

Here $a, b, c \in \mathbb{R}$ are constants (to be chosen). We can expand the preceding expression, and then, for instance,

$$\text{replace} \quad \left(\int_{u,w} G_{x,y,z}(u,w)\right)^2 \quad \text{by} \quad \int_{u,w,u',w'} G_{x,y,z}(u,w)G_{x,y,z}(u',w').$$

We obtain a nonnegative linear combination of $t(F, W)$ over various F with undetermined real coefficients.

The idea is to now consider all such nonnegative expressions. (In practice, on a computer, we consider a large but finite set of such inequalities.) Then we try to optimize the previously undetermined real coefficients (a, b, c in the previous inequality). By adding together an optimized nonnegative linear combination of all such inequalities, and combining with the given constraints, we aim to obtain an inequality $t(F_0, W) \geq \alpha$ for some real α. This would prove a bound on the minimization problem stated earlier. We can find such coefficient and nonnegative combinations efficiently using a **semidefinite program (SDP)** solver. If we also happen to have an example of W satisfying the constraints and matching the bound (i.e., $t(F_0, W) = \alpha$), then we would have solved the extremal problem.

The flag algebra method, with computer assistance, has successfully solved many interesting extremal problems in graph theory. For example, a conjecture of Erdős (1984) on the maximum pentagon density in a triangle-free graph was solved using flag algebra methods; the extremal construction is a blow-up of a 5-cycle (Grzesik 2012; Hatami, Hladký, Kráľ, Norine, and Razborov 2013).

Theorem 5.2.11 (Maximum number 5-cycles in a triangle-free graph)
Every n-vertex triangle-free graph has at most $(n/5)^5$ cycles of length 5.

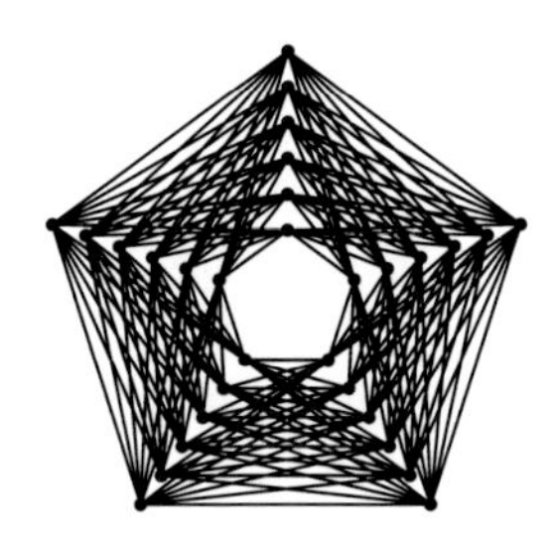

Let us mention another nice result obtained using the flag algebra method: What is the maximum possible number of induced copies of a given graph H among all n-vertex graphs (Pippenger and Golumbic 1975)?

The optimal limiting density (as a fraction of $\binom{n}{v(H)}$, as $n \to \infty$) is called the ***inducibility*** of graph H. They conjectured that for every $k \geq 5$, the inducibility of a k-cycle is $k!/(k^k - k)$, obtained by an *iterated blow-up* of a k-cycle. ($k = 5$ is illustrated here; in the limit there should be infinitely many fractal-like iterations).

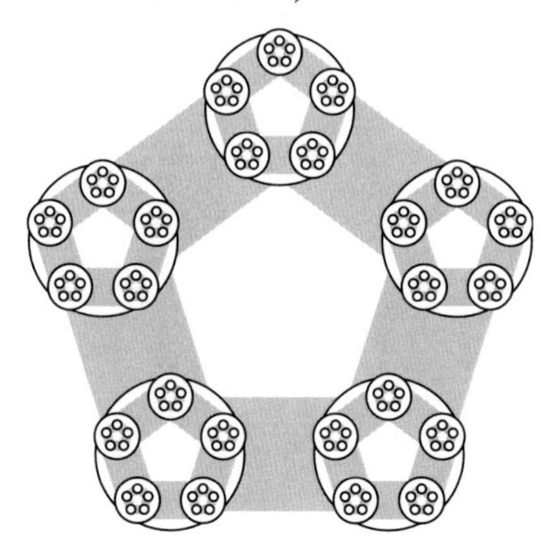

The conjecture for 5-cycles was proved by using flag algebra methods combined with additional "stability" methods (Balogh, Hu, Lidický, and Pfender 2016). The constant factor in the following theorem is tight.

Theorem 5.2.12 (Inducibility of the 5-cycle)
Every n-vertex graph has at most $n^5/(5^5 - 5)$ induced 5-cycles.

Although the flag algebra method has successfully solved several extremal problems, in many interesting cases, the method does not give a tight bound. Nevertheless, for many open extremal problems, such as the tetrahedron hypergraph Turán problem, the best known bound comes from this approach.

***Remark* 5.2.13** (Incompleteness). Can every true linear inequality for graph homomorphism densities be proved via Cauchy–Schwarz/sum-of-squares?

Before giving the answer, we first discuss classical results about real polynomials. Suppose $p(x_1, \ldots, x_n)$ is a real polynomial such that $p(x_1, \ldots, x_n) \geq 0$ for all $x_1, \ldots, x_n \in \mathbb{R}$. Can such a nonnegative polynomial always be written as a sum of squares? Hilbert (1888; 1893) proved that the answer is yes for $n \leq 2$ and no in general for $n \geq 3$. The first explicit counterexample was given by Motzkin (1967):

$$p(x, y) = x^4y^2 + x^2y^4 + 1 - 3x^2y^2$$

is always nonnegative due to the AM-GM inequality, but it cannot be written as a nonnegative sum of squares. Solving Hilbert's 17th problem, Artin (1927) proved that every $p(x_1, \ldots, x_n) \geq 0$ can be written as a sum of squares of *rational* functions, meaning that there is some nonzero polynomial q such that pq^2 can be written as a sum of squares of polynomials. For the earlier example,

$$p(x, y) = \frac{x^2y^2(x^2 + y^2 + 1)(x^2 + y^2 - 2)^2 + (x^2 - y^2)^2}{(x^2 + y^2)^2}.$$

Let us return to the topic of inequalities between graph homomorphism densities. If $f(W) = \sum_i c_i t(F_i, W)$ is nonnegative for every graphon W, can f always be written as a nonnegative sum of squares of rational functions in $t(F, W)$? In other words, can every true inequality be proved using a finite number of Cauchy–Schwarz inequalities (i.e., via vanilla flag algebra calculations).

It turns out that the answer is no (Hatami and Norine 2011). Indeed, if there were always a sum-of-squares proof, then we could obtain an algorithm for deciding whether $f(W) \geq 0$ (with rational coefficients, say) holds for all graphons W, thereby contradicting the undecidability of the problem (Remark 5.0.2). Consider the algorithm that enumerates over all possible forms of sum-of-squares expressions (with undetermined coefficients that can then be solved for) and in parallel enumerates over all graphs G and checks whether $f(G) \geq 0$. If every true inequality had a sum-of-squares proof, then this algorithm would always terminate and tell us whether $f(W) \geq 0$ for all graphons W.

Exercise 5.2.14 (Another proof of maximum triangle density). Let $W\colon [0,1]^2 \to \mathbb{R}$ be a symmetric measurable function. Write W^2 for the function taking value $W^2(x, y) = W(x, y)^2$.

(a) Show that $t(C_4, W) \leq t(K_2, W^2)^2$.

(b) Show that $t(K_3, W) \leq t(K_2, W^2)^{1/2} t(C_4, W)^{1/2}$.

Combining the two inequalities, we deduce $t(K_3, W) \leq t(K_2, W^2)^{3/2}$, which is somewhat stronger than Theorem 5.1.2. We will see another proof later in Corollary 5.3.10.

Exercise 5.2.15. Prove that the skeleton of the 3-cube is Sidorenko.

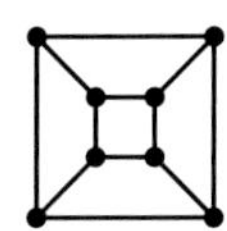

Exercise 5.2.16. Prove that K_4^- is common, where K_4^- is K_4 with one edge removed.

Exercise 5.2.17. Prove that every path is Sidorenko, by extending the proof of Theorem 5.3.4.

Exercise 5.2.18 (A lower bound on clique density). Show that, for every positive integer $r \geq 3$, and graphon W, writing $p = t(K_2, W)$,

$$t(K_r, W) \geq p(2p - 1)(3p - 2) \cdots ((r - 1)p - (r - 2)).$$

Note that this inequality is tight when W is the associated graphon of a clique.

Exercise 5.2.19 (Triangle vs. diamond). Prove there is a function $f\colon [0,1] \to [0,1]$ with $f(x) \geq x^2$ and $\lim_{x \to 0} f(x)/x^2 = \infty$ such that

$$t(K_4^-, W) \geq f(t(K_3, W))$$

for all graphons W. Here K_4^- is K_4 with one edge removed.

Hint: Apply the triangle removal lemma.

5.3 Hölder

Hölder's inequality is a generalization of the Cauchy–Schwarz inequality. It says that, given $p_1, \dots, p_k \geq 1$ with $1/p_1 + \cdots + 1/p_k = 1$, and real-valued functions $f_1, \dots, f_k$ on a common space, we have

$$\int f_1 f_2 \cdots f_k \leq \|f_1\|_{p_1} \cdots \|f_k\|_{p_k},$$

where the ***p*-norm** of a function f is defined by

$$\|f\|_p := \left(\int |f|^p \right)^{1/p}.$$

In practice, the case $p_1 = \cdots = p_k = k$ of Hölder's inequality is used often.

We can apply Hölder's inequality to show that $K_{s,t}$ is Sidorenko. The proof is essentially verbatim to the proof of Theorem 5.2.1 that $t(K_{2,2}, W) \geq t(K_2, W)^4$ from the previous section, except that we now apply Hölder's inequality instead of the Cauchy–Schwarz inequality. We outline the steps in what follows and leave the details as an exercise.

Theorem 5.3.1 (Complete bipartite graphs are Sidorenko)

$$t(K_{s,t}, W) \geq t(K_2, W)^{st}$$

Lemma 5.3.2

$$t(K_{s,1}, W) \geq t(K_2, W)^s$$

Lemma 5.3.3

$$t(K_{s,t}, W) \geq t(K_{s,1}, W)^t$$

Sidorenko's Conjecture for 3-Edge Path

It is already quite a nontrivial fact that all paths are Sidorenko (Mulholland and Smith 1959; Atkinson, Watterson, and Moran 1960; Blakley and Roy 1965). You are encouraged to try it yourself before looking at the next proof.

Theorem 5.3.4
The 3-edge path is Sidorenko.

Let us give two short proofs that both appeared as answers to a MathOverflow question `https://mathoverflow.net/q/189222`. Later in Section 5.5 we will see another proof using the entropy method.

The first proof is a special case of a more general technique by Sidorenko (1991).

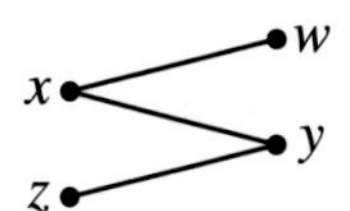

First Proof That the 3-Edge Path Is Sidorenko. Let P_4 be the 3-edge path. Let W be a graphon. Let $g(x) = \int_y W(x, y)$, representing the "degree" of vertex x. We have

$$t(P_4, W) = \int_{w,x,y,z} W(x,w)W(x,y)W(z,y) = \int_{x,y,z} g(x)W(x,y)W(z,y).$$

By relabeling, we can also write it as

$$t(P_4, W) = \int_{x,y,z} W(x,y)W(z,y)g(z).$$

Applying the Cauchy–Schwarz inequality twice, followed by Hölder's inequality,

$$\begin{aligned} t(P_4, W) &= \sqrt{\int_{x,y,z} g(x)W(x,y)W(z,y)}\sqrt{\int_{x,y,z} W(x,y)W(z,y)g(z)} \\ &\geq \int_{x,y,z} \sqrt{g(x)}W(x,y)W(z,y)\sqrt{g(z)} \\ &= \int_y \left(\int_x \sqrt{g(x)}W(x,y)\right)^2 \\ &\geq \left(\int_{x,y} \sqrt{g(x)}W(x,y)\right)^2 \\ &= \left(\int_x g(x)^{3/2}\right)^2 \geq \left(\int_x g(x)\right)^3 = \left(\int_{x,y} W(x,y)\right)^3. \end{aligned}$$

□

The second proof is due to Lee (2019).

Second Proof That the 3-Edge Path Is Sidorenko. Define $g(x) = \int_y W(x,y)$ as earlier. We have

$$t(P_4, W) = \int_{w,x,y,z} W(x,w)W(x,y)W(z,y) = \int_{x,y} g(x)W(x,y)g(y).$$

Note that

$$\int_{x,y} \frac{W(x,y)}{g(x)} = \int_x \frac{g(x)}{g(x)} = 1.$$

Similarly we have

$$\int_{x,y} \frac{W(x,y)}{g(y)} = 1.$$

So, by Hölder's inequality

$$\begin{aligned} t(P_4, W) &= \left(\int_{x,y} g(x)W(x,y)g(y)\right)\left(\int_{x,y} \frac{W(x,y)}{g(x)}\right)\left(\int_{x,y} \frac{W(x,y)}{g(y)}\right) \\ &\geq \left(\int_{x,y} W(x,y)\right)^3. \end{aligned}$$

□

A Generalization of Hölder's Inequality

Now we discuss a powerful variant of Hölder's inequality due to Finner (1992), which is related more generally to Brascamp–Lieb inequalities. Here is a representative example.

Theorem 5.3.5 (Generalized Hölder inequality for a triangle)
Let X, Y, Z be measure spaces. Let $f\colon X \times Y \to \mathbb{R}$, $g\colon X \times Z \to \mathbb{R}$, and $h\colon Y \times Z \to \mathbb{R}$ be measurable functions (assuming integrability whenever needed). Then

$$\int_{x,y,z} f(x,y)g(x,z)h(y,z) \le \|f\|_2 \, \|g\|_2 \, \|h\|_2 \,.$$

Note that a straightforward application of Hölder's inequality, when X, Y, Z are probability spaces (so that $\int_{x,y,z} f(x,y) = \int_{x,y} f(x,y)$), would yield

$$\int_{x,y,z} f(x,y)g(x,z)h(y,z) \le \|f\|_3 \, \|g\|_3 \, \|h\|_3 \,.$$

This is weaker than Theorem 5.3.5. Indeed, in a probability space, $\|f\|_2 \le \|f\|_3$ by Hölder's inequality.

Proof of Theorem 5.3.5. We apply the Cauchy–Schwarz inequality three times. First to the integral over x (this affects f and g while leaving h intact):

$$\int_{x,y,z} f(x,y)g(x,z)h(y,z) \le \int_{y,z} \left(\int_x f(x,y)^2\right)^{1/2} \left(\int_x g(x,z)^2\right)^{1/2} h(y,z).$$

Next, we apply the Cauchy–Schwarz inequality to the variable y. (This affects f and h while leaving g intact.) Continuing the preceding inequality,

$$\le \int_z \left(\int_{x,y} f(x,y)^2\right)^{1/2} \left(\int_x g(x,z)^2\right)^{1/2} \left(\int_y h(y,z)^2\right)^{1/2}.$$

Finally, we apply the Cauchy–Schwarz inequality to the variable z. (This affects g and h while leaving x intact.) Continuing the preceding inequality,

$$\le \left(\int_{x,y} f(x,y)^2\right)^{1/2} \left(\int_{x,z} g(x,z)^2\right)^{1/2} \left(\int_{y,z} h(y,z)^2\right)^{1/2}.$$

This completes the proof of Theorem 5.3.5. □

***Remark* 5.3.6** (Projection inequalities). What is the maximum volume of a body $K \subseteq \mathbb{R}^3$ whose projection on each coordinate plane is at most 1? A unit cube has volume 1, but is this the largest possible?

Letting $|\cdot|$ denote both volume and area (depending on the dimension) and $\pi_{xy}(K)$ denote the projection of K onto the xy-plane, and likewise with the other planes. Using $1_K(x,y,z) \le f(x,y)g(x,z)h(y,z)$, Theorem 5.3.5 implies

$$|K|^2 \le |\pi_{xy}(K)||\pi_{xz}(K)||\pi_{yz}(K)|. \tag{5.4}$$

In particular, if all three projections have volume at most 1, then $|K| \le 1$.

The inequality (5.4), which holds more generally in higher dimensions, is due to Loomis and Whitney (1949). See Exercise 5.3.9 in what follows. It has important applications in combinatorics. A powerful generalization known as **Shearer's entropy inequality** will be discussed in Section 5.5. Also see Exercise 5.5.19 for a strengthening of the projection inequalities.

Now let us state a more general form of Theorem 5.3.5, which can be proved using the same techniques. The key point of the inequality in Theorem 5.3.5 is that each variable (i.e., x, y, and z) is contained in exactly two of the factors (i.e., $f(x,y)$, $g(x,z)$, and $h(y,z)$). Everything works the same way as long as each variable is contained in exactly k factors, as long as we use L^k norms on the right-hand side.

For example,

$$\int_{u,v,w,x,y,z} f_1(u,v)f_2(v,w)f_3(w,z)f_4(x,y) \cdot f_5(y,z)f_6(z,u)f_7(u,x)f_8(u,z)f_9(w,y) \le \prod_{i=1}^{9} \|f_i\|_3 .$$

Here the factors in the integral correspond to edges of a 3-regular graph shown. In particular, every variable lies in exactly three factors.

More generally, each function f_i can take as input any number of variables, as long as every variable appears in exactly k functions. For example,

$$\int_{w,x,y,z} f(w,x,y)g(w,y,z)h(x,z) \le \|f\|_2\|g\|_2\|h\|_2.$$

The inequality is stated more generally in what follows. Given $x = (x_1,\dots,x_m) \in X_1 \times\cdots\times X_m$ and $I \subseteq [m]$, we write $\pi_I(x) = (x_i)_{i\in I} \in \prod_{i\in I} X_i$ for the projection onto the coordinate subspace of I.

Theorem 5.3.7 (Generalized Hölder inequality)
Let $X_1,\dots,X_m$ be measure spaces. Let $I_1,\dots,I_\ell \subseteq [m]$ be such that each element of $[m]$ appears in exactly k different I_i's. For each $i \in [\ell]$, let $f_i \colon \prod_{j\in I_i} X_j \to \mathbb{R}$. Then

$$\int_{X_1\times\cdots\times X_m} f_1(\pi_{I_1}(x))\cdots f_\ell(\pi_{I_\ell}(x))\,dx \le \|f_1\|_k \cdots \|f_\ell\|_k .$$

Furthermore, if every X_i is a probability space, then we can relax the hypothesis to "each element of $[m]$ appears in *at most* k different I_i's."

Exercise 5.3.8. Prove Theorem 5.3.7 by generalizing the proof of Theorem 5.3.5.

The next exercise generalizes the projection inequality from Remark 5.3.6. Also see Exercise 5.5.19 for a strengthening.

Exercise 5.3.9 (Projection inequalities). Let $I_1,\dots,I_\ell \subseteq [d]$ such that each element of $[d]$ appears in exactly k different I_i's. Prove that for any compact body $K \subseteq \mathbb{R}^d$, with $|\cdot|$ denoting volume in the appropriate dimension,

$$|K|^k \le |\pi_{I_1}(K)|\cdots|\pi_{I_\ell}(K)|.$$

The version of Theorem 5.3.7 with each X_i being a probability space is useful for graphons.

Corollary 5.3.10 (Upper bound on F-density)
For any graph F with maximum degree at most k, and graphon W,

$$t(F,W) \le \|W\|_k^{e(F)}.$$

In particular, since

$$\|W\|_k^k = \int W^k \le t(K_2, W),$$

the inequality implies that

$$t(F,W) \le t(K_2, W)^{e(F)/k}.$$

This implies the upper bound on clique densities (Theorems 5.1.2 and 5.1.5). The stronger statement of Corollary 5.3.10 with the L^k norm of W on the right-hand side has no direct interpretations for subgraph densities, but it is important for certain applications such as understanding large deviation rates in random graphs (Lubetzky and Zhao 2017).

More generally, using different L^p norms for different factors in Hölder's inequality, we have the following statement (Finner 1992).

Theorem 5.3.11 (Generalized Hölder inequality)
Let $X_1, \dots, X_m$ be measure spaces. For each $i \in [\ell]$, let $p_i \ge 1$, let $I_i \subseteq [m]$, and $f_i \colon \prod_{j \in I_i} X_j \to \mathbb{R}$. If either

(a) $\sum_{i: j \in I_i} 1/p_i = 1$ for each $j \in [m]$,
OR
(b) each X_i is a probability space and $\sum_{i: j \in I_i} 1/p_i \le 1$ for each $j \in [m]$,

then

$$\int_{X_1 \times \cdots \times X_\ell} f_1(\pi_{I_1}(x)) \cdots f_\ell(\pi_{I_\ell}(x))\, dx \le \|f_1\|_{p_1} \cdots \|f_\ell\|_{p_\ell}.$$

Exercise 5.3.12. Prove Theorem 5.3.11.

An Application of Generalized Hölder Inequalities

Now we turn to another graph inequality where the preceding generalization of Hölder's inequality plays a key role.

Question 5.3.13 (Maximum number of independent sets in a regular graph)
Fix d. Among d-regular graphs, which graph G maximizes $i(G)^{1/v(G)}$, where $i(G)$ denotes the number of independent sets of G?

The answer turns out to be $G = K_{d,d}$. We can also take G to be a disjoint union of copies of $K_{d,d}$'s, and this would not change $i(G)^{1/v(G)}$. This result was shown by Kahn (2001) for bipartite regular graphs G, and later extended by Zhao (2010) to all regular graphs G.

Theorem 5.3.14 (Maximum number of independent sets in a regular graph)
For every n-vertex d-regular graph G,

$$i(G) \le i(K_{d,d})^{n/(2d)} = (2^{d+1} - 1)^{n/(2d)}.$$

The set of independent sets of G is in bijection with the set of graph homomorphisms from G to the following graph:

Indeed, a map between their vertex sets forms a graph homomorphism if and only if the vertices of G that map to the nonlooped vertex is an independent set of G.

Let us first prove Theorem 5.3.14 for bipartite regular G. The following more general inequality was shown by Galvin and Tetali (2004). It implies the bipartite case of Theorem 5.3.14 by the preceding discussion.

Theorem 5.3.15 (The maximum number of H-colorings in a regular graph)
For every n-vertex d-regular bipartite graph G, and any graph H (allowing looped vertices on H)

$$\hom(G, H) \le \hom(K_{d,d}, H)^{n/(2d)}.$$

This is equivalent to the following statement.

Theorem 5.3.16
For any d-regular bipartite graph F,

$$t(F, W) \le t(K_{d,d}, W)^{e(F)/d^2}.$$

Let us prove this theorem in the case $F = C_6$ to illustrate the technique more concretely. The general proof is basically the same. Let

$$f(x_1, x_2) = \int_y W(x_1, y)W(x_2, y).$$

This function should be thought of as the codegree of vertices x_1 and x_2. Then, grouping the factors in the integral according to their right endpoint, we have

x_1 y_1
x_2 y_2
x_3 y_3

$$\begin{aligned} t(C_6, W) &= \int_{x_1,x_2,x_3,y_1,y_2,y_3} W(x_1, y_1)W(x_2, y_1)W(x_1, y_2)W(x_3, y_2)W(x_2, y_3)W(x_3, y_3) \\ &= \int_{x_1,x_2,x_3} \left(\int_{y_1} W(x_1, y_1)W(x_2, y_1)\right)\left(\int_{y_2} W(x_1, y_2)W(x_3, y_2)\right) \\ &\qquad \cdot \left(\int_{y_3} W(x_2, y_3)W(x_3, y_3)\right) \\ &= \int_{x_1,x_2,x_3} f(x_1, x_2)f(x_1, x_3)f(x_2, x_3) \\ &\le \|f\|_2^3 \qquad \text{[by generalized Hölder, Theorem 5.3.5 / 5.3.7].} \end{aligned}$$

On the other hand, we have

$$\begin{aligned}
\|f\|_2^2 &= \int_{x_1,x_2} f(x_1,x_2)^2 \\
&= \int_{x_1,x_2} \left(\int_{y_1} W(x_1,y_1)W(x_2,y_1) \right) \left(\int_{y_2} W(x_1,y_2)W(x_2,y_2) \right) \\
&= \int_{x_1,x_2,y_1,y_2} W(x_1,y_1)W(x_2,y_1)W(x_1,y_2)W(x_2,y_2) \\
&= t(C_4, W).
\end{aligned}$$

x_1 y_1
x_2 y_2

This proves Theorem 5.3.16 in the case $F = C_6$. The theorem in general can be proved via a similar calculation.

Exercise 5.3.17. Complete the proof of Theorem 5.3.16 by generalizing the preceding argument.

***Remark* 5.3.18.** Kahn (2001) first proved the bipartite case of Theorem 5.3.14 using Shearer's entropy inequality, which we will see in Section 5.5. His technique was extended by Galvin and Tetali (2004) to prove Theorem 5.3.15. The proof using the generalized Hölder's inequality presented here was given by Lubetzky and Zhao (2017).

So far we proved Theorem 5.3.14 for bipartite regular graphs. To prove it for all regular graphs, we apply the following inequality by Zhao (2010). Here $G \times K_2$ (tensor product) is the bipartite double cover of G. An example is illustrated here:

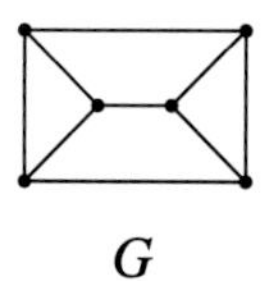

G

$G \times K_2$

The vertex set of $G \times K_2$ is $V(G) \times \{0,1\}$. Its vertices are labeled v_i with $v \in V(G)$ and $i \in \{0,1\}$. Its edges are u_0v_1 for all $uv \in E(G)$. Note that $G \times K_2$ is always a bipartite graph.

Theorem 5.3.19 (Bipartite double cover for independent sets)
For every graph G,

$$i(G)^2 \le i(G \times K_2).$$

Assuming Theorem 5.3.19, we can now prove Theorem 5.3.14 by reducing the statement to the bipartite case, which we proved earlier. Indeed, for every d-regular graph G,

$$i(G) \le i(G \times K_2)^{1/2} \le i(K_{d,d})^{n/(2d)},$$

where the last step follows from applying Theorem 5.3.14 to the bipartite graph $G \times K_2$.

Proof of Theorem 5.3.19. Let $2G$ denote a disjoint union of two copies of G. Label its vertices by v_i with $v \in V$ and $i \in \{0,1\}$ so that its edges are u_iv_i with $uv \in E(G)$ and $i \in \{0,1\}$. We will give an injection $\phi\colon I(2G) \to I(G \times K_2)$. Recall that $I(G)$ is the set of independent sets of G. The injection would imply $i(G)^2 = i(2G) \le i(G \times K_2)$ as desired.

Fix an arbitrary order on all subsets of $V(G)$. Let S be an independent set of $2G$. Let

$$E_{\text{bad}}(S) := \{uv \in E(G) : u_0, v_1 \in S\}.$$

Note that $E_{\text{bad}}(S)$ is a bipartite subgraph of G, since each edge of E_{bad} has exactly one endpoint in $\{v \in V(G) : v_0 \in S\}$ but not both (or else S would not be independent). Let A denote the first subset (in the previously fixed ordering) of $V(G)$ such that all edges in $E_{\text{bad}}(S)$ have one vertex in A and the other outside A. Define $\phi(S)$ to be the subset of $V(G) \times \{0,1\}$ obtained by "swapping" the pairs in A. That is, for all $v \in A$, $v_i \in \phi(S)$ if and only if $v_{1-i} \in S$ for each $i \in \{0,1\}$, and for all $v \notin A$, $v_i \in \phi(S)$ if and only if $v_i \in S$ for each $i \in \{0,1\}$. It is not hard to verify that $\phi(S)$ is an independent set in $G \times K_2$. The swapping procedure fixes the "bad" edges.

It remains to verify that ϕ is an injection. For every $S \in I(2G)$, once we know $T = \phi(S)$, we can recover S by first setting

$$E'_{\text{bad}}(T) = \{uv \in E(G) : u_i, v_i \in T \text{ for some } i \in \{0,1\}\},$$

so that $E_{\text{bad}}(S) = E'_{\text{bad}}(T)$, and then finding A as earlier and swapping the pairs of A back. (Remark: it follows that $T \in I(G \times K_2)$ lies in the image of ϕ if and only if $E'_{\text{bad}}(T)$ is bipartite.) □

***Remark* 5.3.20** (Reverse Sidorenko). Does Theorem 5.3.15 generalize to all regular graphs G like Theorem 5.3.14? Unfortunately, no. For example, when $H =$ ⚲⚲ consists of two isolated loops, $\hom(G,H) = 2^{c(G)}$, with $c(G)$ being the number of connected components of G. So $\hom(G,H)^{1/v(G)}$ is minimized among d-regular graphs G for $G = K_{d+1}$, which is the connected d-regular graph with the fewest vertices.

Theorem 5.3.15 actually extends to every triangle-free regular graph G. Furthermore, for every regular graph G that contains a triangle, there is some graph H for which the inequality in Theorem 5.3.15 fails.

There are several interesting families of graphs H where Theorem 5.3.15 is known to extend to all regular graphs G. Notably, this is true for $H = K_q$, which is significant since $\hom(G, K_q)$ is the number of proper q-colorings of G.

There are also generalizations to nonregular graphs. For example, for a graph G without isolated vertices, letting d_u denote the degree of $u \in V(G)$, we have

$$i(G) \le \prod_{uv \in E(G)} i(K_{d_u,d_v})^{1/(d_u d_v)},$$

and similarly for the number of proper q-colorings. In fact, the results mentioned in this remark about regular graphs are proved by induction on vertices of G and thus require considering the larger family of not necessarily regular graphs G.

The results discussed in this remark are due to Sah, Sawhney, Stoner, and Zhao (2019; 2020). The term *reverse Sidorenko inequalities* was introduced to describe inequalities such as $t(F,W)^{1/e(F)} \le t(K_{d,d},W)^{1/d^2}$, which mirror the inequality $t(F,W)^{1/e(F)} \ge t(K_2,W)$ in Sidorenko's conjecture. Also see the earlier survey by Zhao (2017) for discussions of related results and open problems.

We already know through the quasirandom graph equivalences (Theorem 3.1.1) that C_4 is forcing. The next exercise generalizes this fact.

Exercise 5.3.21. Prove that $K_{s,t}$ is forcing whenever $s,t \ge 2$.

Exercise 5.3.22. Let F be a bipartite graph with vertex bipartition $A \cup B$ such that every vertex in B has degree d. Let d_u denote the degree of u in F. Prove that for every graphon W,

$$t(F,W) \le \prod_{uv \in E(F)} t(K_{d_u,d_v},W)^{1/(d_u d_v)}.$$

Exercise 5.3.23 (Sidorenko for 3-edge path with vertex weights). Let $W\colon [0,1]^2 \to [0,\infty)$ be a measurable function (not necessarily symmetric). Let $p,q,r,s\colon [0,1] \to [0,\infty)$ be measurable functions. Prove that

$$\int_{w,x,y,z} p(w)q(x)r(y)s(z)W(x,w)W(x,y)W(z,y)$$
$$\ge \left(\int_{x,y} (p(x)q(x)r(y)s(y))^{1/3} W(x,y)\right)^3.$$

x, w, y, z

Exercise 5.3.24. For a graph G, let $f_q(G)$ denote the number of maps $V(G) \to \{0,1,\dots,q\}$ such that $f(u) + f(v) \le q$ for every $uv \in E(G)$. Prove that for every n-vertex d-regular graph G (not necessarily bipartite),

$$f_q(G) \le f_q(K_{d,d})^{n/(2d)}.$$

5.4 Lagrangian

Another Proof of Turán's Theorem

Here is another proof of Turán's theorem due to Motzkin and Straus (1965). It can be viewed as a continuous/analytic analogue of the Zykov symmetrization proof of Turán's theorem from Section 1.2 (the third proof there).

Theorem 5.4.1 (Turán's theorem)
The number of edges in an n-vertex K_{r+1}-free graph is at most

$$\left(1 - \frac{1}{r}\right)\frac{n^2}{2}.$$

Proof. Let G be a K_{r+1}-free graph on vertex set $[n]$. Consider the function

$$f(x_1, \ldots, x_n) = \sum_{ij \in E(G)} x_i x_j.$$

We want to show that

$$f\left(\frac{1}{n}, \ldots, \frac{1}{n}\right) \le \frac{1}{2}\left(1 - \frac{1}{r}\right).$$

In fact, we will show that

$$\max_{\substack{x_1,\ldots,x_n \ge 0 \\ x_1+\cdots+x_n=1}} f(x_1, \ldots, x_n) \le \frac{1}{2}\left(1 - \frac{1}{r}\right).$$

By compactness, the maximum is achieved at some $x = (x_1, \ldots, x_n)$. Let us choose such a maximizing vector with the minimum support size (i.e., the number of nonzero coordinates).

Suppose $ij \notin E(G)$ for some pair of distinct $x_i, x_j > 0$. If we replace (x_i, x_j) by $(s, x_i + x_j - s)$, then f changes linearly in s (since $x_i x_j$ does not come up as a summand in f), and since f is already maximized at x, it must not actually change with s. So we can replace (x_i, x_j) by $(x_i + x_j, 0)$, which keeps f the same while decreasing the number of nonzero coordinates of x.

Thus the support of x is a clique in G. By labeling vertices, suppose $x_1, \ldots, x_k > 0$ and $x_{k+1} = x_{k+2} = \cdots = x_n = 0$. Since G is K_{r+1}-free, this clique has size $k \le r$. So,

$$f(x) = \sum_{1 \le i < j \le k} x_i x_j \le \frac{1}{2}\left(1 - \frac{1}{k}\right)\left(\sum_{i=1}^{k} x_i\right)^2 = \frac{1}{2}\left(1 - \frac{1}{k}\right) \le \frac{1}{2}\left(1 - \frac{1}{r}\right). \qquad \square$$

Remark* 5.4.2** (Hypergraph Lagrangians). The ***Lagrangian of a hypergraph H with vertex set $[n]$ is defined to be

$$\boldsymbol{\lambda(H)} := \max_{\substack{x_1,\ldots,x_n \ge 0 \\ x_1+\cdots+x_n=1}} f(x_1, \ldots, x_n), \qquad \text{where } f(x_1, \ldots, x_n) = \sum_{e \in E(H)} \prod_{i \in e} x_i.$$

It is a useful tool for certain hypergraph Turán problems. The preceding proof of Turán's theorem shows that for every graph G, $\lambda(G) = (1 - 1/\omega(G))/2$, where $\omega(G)$ is the size of the largest clique in G. A maximizing x has coordinate $1/\omega(G)$ on vertices of the clique and zero elsewhere.

As an alternate but equivalent perspective, the above proof can rephrased in terms of maximizing the edge density among K_{r+1}-free vertex-weighted graphs (vertex weights are given by the vector x above). The proof shifts weights between nonadjacent vertices while not decreasing the edge density, and this process preserves K_{r+1}-freeness.

Linear Inequalities Between Clique Densities

The next theorem shows that to check whether a *linear* inequality in clique densities in G holds, it suffices to check it for G being cliques (Bollobás 1976; Schelp and Thomason 1998). The K_r density in a vertex-weighted clique can be expressed in terms of elementary symmetric polynomials, which we recall are given as follows:

$$
\begin{aligned}
e_0(x_1,\ldots,x_n) &= 1,\\
e_1(x_1,\ldots,x_n) &= x_1 + \cdots + x_n,\\
e_2(x_1,\ldots,x_n) &= \sum_{1\le i<j\le n} x_i x_j,\\
e_3(x_1,\ldots,x_n) &= \sum_{1\le i<j<k\le n} x_i x_j x_k,\\
&\vdots\\
e_n(x_1,\ldots,x_n) &= x_1\cdots x_n.
\end{aligned}
$$

Lemma 5.4.3 (Extreme points of a linear combination of symmetric polynomials)
Let $f(x_1,\ldots,x_n)$ be a real linear combination of elementary symmetric polynomials in $x_1,\ldots,x_n$. Suppose $x = (x_1,\ldots,x_n)$ minimizes $f(x)$ among all vectors $x \in \mathbb{R}^n$ with $x_1,\ldots,x_n \ge 0$ and $x_1+\cdots+x_n = 1$, and furthermore x has minimum support size among all such minimizers. Then, up to permuting the coordinates of x, there is some $1 \le k \le n$ so that

$$x_1 = \cdots = x_k = 1/k \quad\text{and}\quad x_{k+1} = \cdots = x_n = 0.$$

Proof. Suppose $x_1,\ldots,x_k > 0$ and $x_{k+1} = \cdots = x_n = 0$ with $k \ge 2$. Fixing $x_3,\ldots,x_n$, we see that as a function of (x_1,x_2), f has the form

$$Ax_1x_2 + Bx_1 + Bx_2 + C$$

where A, B, C depend on $x_3,\ldots,x_n$. Notably, the coefficients of x_1 and x_2 agree, since f is a symmetric polynomial. Holding $x_1 + x_2$ fixed, f has the form

$$Ax_1x_2 + C'.$$

If $A \ge 0$, then holding $x_1 + x_2$ fixed, we can set either x_1 or x_2 to be zero while not increasing f, which contradicts the hypothesis that the minimizing x has minimum support size. So $A < 0$, so that with $x_1 + x_2$ held fixed, $Ax_1x_2 + C'$ is minimized uniquely at $x_1 = x_2$. Thus, $x_1 = x_2$. Likewise, $x_1 = \cdots = x_k$, as claimed. □

Theorem 5.4.4 (Linear inequalities between clique densities)
Let $c_1,\cdots,c_\ell \in \mathbb{R}$. The inequality

$$\sum_{r=1}^{\ell} c_r t(K_r, G) \ge 0$$

is true for every graph G if and only if it is true with $G = K_n$ for every positive integer n.

More explicitly, the preceding inequality holds for all graphs G if and only if

$$\sum_{r=1}^{\ell} c_r \cdot \frac{n(n-1)\cdots(n-r+1)}{n^r} \ge 0 \qquad \text{for every } n \in \mathbb{N}.$$

Since this is a single variable polynomial in n, it is usually easy to check this inequality. We will see some examples right after the proof.

Proof. The only nontrivial direction is the "if" implication. Suppose the displayed inequality holds for all cliques G. Let G be an arbitrary graph with vertex set $[n]$. Let

$$f(x_1,\dots,x_n) = \sum_{r=1}^{\ell} r!c_r \sum_{\substack{\{i_1,\dots,i_r\} \\ r\text{-clique in } G}} x_{i_1}\cdots x_{i_r}.$$

So,

$$f\left(\frac{1}{n},\dots,\frac{1}{n}\right) = \sum_{r=1}^{\ell} c_r t(K_r, G).$$

It suffices to prove that

$$\min_{\substack{x_1,\dots,x_n \ge 0 \\ x_1+\cdots+x_n=1}} f(x_1,\dots,x_n) \ge 0.$$

By compactness, we can assume that the minimum is attained at some x. Among all minimizing x, choose one with the smallest support (i.e., the number of nonzero coordinates).

As in the previous proof, if $ij \notin E(G)$ for some pair of distinct $x_i, x_j > 0$, then, replacing (x_i, x_j) by $(s, x_i + x_j - s)$, f changes linearly in s. Since f is already minimized at x, it must stay constant as s changes. So we can replace (x_i, x_j) by $(x_i + x_j, 0)$, which keeps f the same while decreasing the number of nonzero coordinates of x. Thus, the support of x is a clique in G. Suppose x is supported on the first k coordinates. Then f is a linear combination of elementary symmetric polynomials in $x_1,\dots,x_k$. By Lemma 5.4.3, $x_1 = \cdots = x_k = 1/k$. Then $f(x) = \sum_{r=1}^{\ell} c_r t(K_r, K_k) \ge 0$ by hypothesis. □

***Remark* 5.4.5.** This proof technique can be adapted to show the stronger result that among all graphs G with a given number of vertices, the quantity $\sum_{r=1}^{\ell} c_r t(K_r, G)$ is minimized when G is a multipartite graph. Compare with the Zykov symmetrization proof of Turán's theorem (Theorem 1.2.4).

The theorem only considers linear inequalities between clique densities. The statement fails in general for inequalities with other graph densities. (Why?)

Theorem 5.4.4 can be equivalently stated in terms of the convex hull of the region of all possible clique density tuples.

Corollary 5.4.6 (Convex hull of feasible clique densities)
Let $\ell \ge 3$. In $\mathbb{R}^{\ell-1}$, the convex hull of

$$\{(t(K_2,W), t(K_3,W),\dots,t(K_\ell,W)) : \text{graphons } W\}$$

is the same as the convex hull of

$$\{(t(K_2,K_n), t(K_3,K_n),\dots,t(K_\ell,K_n)) : n \in \mathbb{N}\}.$$

For $\ell = 3$, the points

$$(t(K_2,K_n), t(K_3,K_n)) = \left(1 - \frac{1}{n}, \left(1 - \frac{1}{n}\right)\left(1 - \frac{2}{n}\right)\right), \quad n \in \mathbb{N},$$

are the extremal points of the convex hull of the edge-triangle region from (5.2). The actual region, illustrated in Figure 5.1, has a lower boundary consisting of concave curves connecting the points $(t(K_2, K_n), t(K_3, K_n))$.

Exercise 5.4.7 (Turán's theorem from the convex hull of feasible clique densities). Show that Corollary 5.4.6 implies the following version of Turán's theorem: $t(K_2, G) \le 1 - 1/r$ for every K_{r+1}-free graph G.

Exercise 5.4.8 (A generalization of Turán's theorem). In an n-vertex graph, assign weight $r/(r-1)$ to each edge, where r is the number of vertices in the largest clique containing that edge. Prove that the sum of all edge weights is at most $n^2/2$.

Exercise 5.4.9. For each graph F, let $c_F \in \mathbb{R}$ be such that $c_F \ge 0$ whenever F is not a clique (no restrictions when F is a clique). Assume that $c_F \ne 0$ for finitely many F's. Prove that the inequality

$$\sum_F c_F t_{\text{inj}}(F, G) \ge 0$$

is true for every graph G if and only if it is true with $G = K_n$ for every positive integer n.

Exercise 5.4.10 (Cliquey edges). Let n, r, t be nonnegative integers. Show that every n-vertex graph with at least $(1 - \frac{1}{r})\frac{n^2}{2} + t$ edges contains at least rt edges that belong to a K_{r+1}.

Hint: Rephrase the statement as a linear inequality between the number of edges and the number of cliquey edges in every graph.

Exercise 5.4.11 (A hypergraph Turán density). Let F be the 3-graph with 10 vertices and six edges illustrated here (lines denote edges). Prove that the hypergraph Turán density of F is $2/9$.

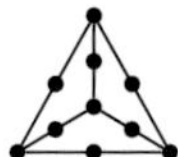

Exercise 5.4.12* (Maximizing $K_{1,2}$ density). Prove that, for every $p \in [0, 1]$, among all graphons W with $t(K_2, W) = p$, the maximum possible value of $t(K_{1,2}, W)$ is attained by either a "clique" or a "hub" graphon, illustrated here.

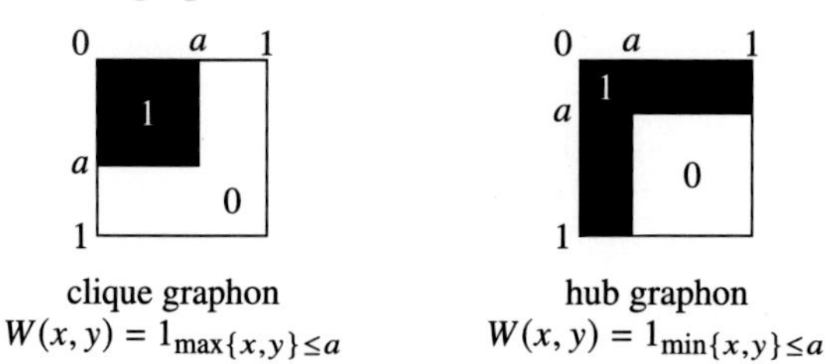

clique graphon $W(x, y) = 1_{\max\{x,y\} \le a}$

hub graphon $W(x, y) = 1_{\min\{x,y\} \le a}$

5.5 Entropy

In this section, we explain how to use entropy to prove certain graph homomorphism inequalities.

Entropy Basics

Definition 5.5.1 (Entropy)
Let X be a discrete random variable taking values in some set S. For each $s \in S$, let $p_s = \mathbb{P}(X = s)$. We define the ***(binary) entropy*** of X to be

$$\boldsymbol{H(X)} := \sum_{s \in S} -p_s \log_2 p_s.$$

(By convention, if $p_s = 0$, then the corresponding summand is set to zero.)

Exercise 5.5.2. Show that $H(X) \geq 0$ always.

Intuitively, $H(X)$ measures the amount of "surprise" in the randomness of X. A more rigorous interpretation of this intuition is given by the **Shannon noiseless coding theorem**, which says that the minimum number of bits needed to encode n independent copies of X is $nH(X) + o(n)$.

Here are some basic properties of entropy.

Lemma 5.5.3 (Uniform bound)
If X is a random variable supported on a finite set S, then

$$H(X) \leq \log_2 |S|.$$

Equality holds if and only if X is uniformly distributed on S.

Proof. Let function $f(x) = -x \log_2 x$ is concave for $x \in [0, 1]$. We have, by concavity,

$$H(X) = \sum_{s \in S} f(p_s) \leq |S| f\left(\frac{1}{|S|} \sum_{s \in S} p_s\right) = |S| f\left(\frac{1}{|S|}\right) = \log_2 |S|. \qquad \square$$

We write $\boldsymbol{H(X,Y)}$ for the entropy of the joint random variables (X,Y). This means that

$$\boldsymbol{H(X,Y)} := H(Z) = \sum_{(x,y)} -\mathbb{P}(X = x, Y = y) \log_2 \mathbb{P}(X = x, Y = y).$$

In particular,

$$H(X,Y) = H(X) + H(Y) \qquad \text{if } X \text{ and } Y \text{ are independent.}$$

We can similarly define $H(X,Y,Z)$, and so on.

Definition 5.5.4 (Conditional entropy)
Given jointly distributed discrete random variables X and Y, define

$$\boldsymbol{H(X|Y)} := \sum_{y} \mathbb{P}(Y = y) H(X|Y = y).$$

Here $H(X|Y = y) = \sum_x -\mathbb{P}(X = x|Y = y) \log_2 \mathbb{P}(X = x|Y = y)$ is entropy of the random variable X conditioned on the event $Y = y$.

Intuitively, $H(X|Y)$ measures the expected amount of new information or surprise in X after Y has already been revealed. For example:

- If X is completely determined by Y (i.e., $X = f(Y)$ for some function f), then $H(X|Y) = 0$.
- If X and Y are independent, then $H(X|Y) = H(X)$.
- If X and Y are conditionally independent on Z, then $H(X,Y|Z) = H(X|Z) + H(Y|Z)$ and $H(X|Y,Z) = H(X|Z)$.

Lemma 5.5.5 (Chain rule)

$$H(X,Y) = H(X) + H(Y|X)$$

Proof. Writing $p(x,y) = \mathbb{P}(X = x, Y = y)$ and so on, we have by Bayes' rule

$$p(x|y)p(y) = p(x,y),$$

and so (here we skip y if $p(y) = 0$):

$$\begin{aligned}
H(X|Y) &= \sum_y \mathbb{P}(Y = y)H(X|Y = y) \\
&= \sum_y -p(y)\sum_x p(x|y)\log_2 p(x|y) \\
&= \sum_{x,y} -p(x,y)\log_2 \frac{p(x,y)}{p(y)} \\
&= \sum_{x,y} -p(x,y)\log_2 p(x,y) + \sum_y p(y)\log_2 p(y) \\
&= H(X,Y) - H(Y).
\end{aligned}$$

□

Lemma 5.5.6 (Subadditivity)
$H(X,Y) \le H(X) + H(Y)$. More generally,

$$H(X_1,\ldots,X_n) \le H(X_1) + \cdots + H(X_n).$$

Proof. Let $f(t) = \log_2(1/t)$, which is convex. We have

$$\begin{aligned}
&H(X) + H(Y) - H(X,Y) \\
&\quad= \sum_{x,y}\big(-p(x,y)\log_2 p(x) - p(x,y)\log_2 p(y) + p(x,y)\log_2 p(x,y)\big) \\
&\quad= \sum_{x,y} p(x,y)\log_2 \frac{p(x,y)}{p(x)p(y)} \\
&\quad= \sum_{x,y} p(x,y) f\left(\frac{p(x)p(y)}{p(x,y)}\right) \\
&\quad\ge f\left(\sum_{x,y} p(x,y)\frac{p(x)p(y)}{p(x,y)}\right) = f(1) = 0.
\end{aligned}$$

More generally, by iterating the preceding inequality for two random variables, we have

$$\begin{aligned} H(X_1, \ldots, X_n) &\le H(X_1, \ldots, X_{n-1}) + H(X_n) \\ &\le H(X_1, \ldots, X_{n-2}) + H(X_{n-1}) + H(X_n) \\ &\le \cdots \le H(X_1) + \cdots + H(X_n). \end{aligned}$$

□

***Remark* 5.5.7.** The nonnegative quantity

$$I(X;Y) := H(X) + H(Y) - H(X,Y)$$

is called ***mutual information***. Intuitively, it measures the amount of common information between X and Y.

Lemma 5.5.8 (Dropping conditioning)
$H(X|Y) \le H(X)$. More generally,

$$H(X|Y,Z) \le H(X|Z).$$

Proof. By chain rule and subadditivity, we have

$$H(X|Y) = H(X,Y) - H(Y) \le H(X).$$

The inequality conditioning on Z follows, since the preceding implies that

$$H(X|Y,Z=z) \ge H(X|Z=z)$$

holds for every z, and taking expectation of z yields $H(X|Y,Z) \le H(X|Z)$. □

Remark* 5.5.9.** Another way to state the dropping condition inequality is the ***data processing inequality: $H(X|f(Y)) \ge H(X|Y)$ for any function f.

Applications to Sidorenko's Conjecture

Now let us use entropy to establish some interesting cases of Sidorenko's conjecture. Recall that a bipartite graph F is said to be ***Sidorenko*** if

$$t(F,G) \ge t(K_2,G)^{e(F)}$$

for every graph G. Sidorenko's conjecture says that every bipartite graph is Sidorenko.

The entropy approach to Sidorenko's conjecture was first introduced by Li and Szegedy (2011) and further developed in subsequent works (Szegedy (2015); Conlon, Kim, Lee, and Lee (2018)). Here we illustrate the entropy approach to Sidorenko's conjecture with several examples.

To show that F is Sidorenko, we need to show that for every graph G,

$$\frac{\hom(F,G)}{v(G)^{v(F)}} \ge \left(\frac{2e(G)}{v(G)^2}\right)^{e(F)}. \tag{5.5}$$

We write $\mathbf{Hom}(F,G)$ for the set of all maps $V(F) \to V(G)$ that give a graph homomorphism $F \to G$. This set has cardinality $\hom(F,G)$. Our strategy is to construct a random element $\Phi \in \mathrm{Hom}(F,G)$ whose entropy satisfies

$$H(\Phi) \geq e(F)\log_2(2e(G)) - (2e(F) - v(F))\log_2 v(G). \tag{5.6}$$

The uniform bound $H(\Phi) \leq \log_2 \hom(F, G)$ then implies (5.5).

Let us illustrate this technique for a three-edge path. We have already seen two proofs of the following inequality in Section 5.3. Now we present a different proof using the entropy method along with generalizations.

Theorem 5.5.10
The 3-edge path is Sidorenko.

Proof. Let P_4 denote the 3-edge path and G a graph. An element of $\mathrm{Hom}(P_4, G)$ is a walk of length three. We choose randomly a walk $XYZW$ in G as follows:

- XY is a uniform random edge of G (by this we mean first choosing an edge of G uniformly at random, and then let X be a uniformly chosen endpoint of this edge, and then Y the other endpoint);
- Z is a uniform random neighbor of Y;
- W is a uniform random neighbor of Z.

A key observation is that YZ is also distributed as a uniform random edge of G (pause and think about why). Indeed, conditioned on the choice of Y, the vertices X and Z are both independent and uniform neighbors of Y, so XY and YZ are identically distributed, and hence YZ is a uniform random edge of G.

Similarly, ZW is distributed as uniform random edge.

Also, since X and Z are conditionally independent given Y,

$$H(Z|X,Y) = H(Z|Y) \qquad \text{and similarly} \qquad H(W|X,Y,Z) = H(W|Z).$$

Furthermore,

$$H(Y|X) = H(Z|Y) = H(W|Z)$$

since XY, YZ, ZW are each identically distributed as a uniform random edge.

Thus,

$$\begin{aligned}
H(X,Y,Z,W) &= H(X) + H(Y|X) + H(Z|X,Y) + H(W|X,Y,Z) && \text{[chain rule]}\\
&= H(X) + H(Y|X) + H(Z|Y) + H(W|Z) && \text{[cond. indep.]}\\
&= H(X) + 3H(Y|X) && \text{[prev. paragraph]}\\
&= 3H(X,Y) - 2H(X) && \text{[chain rule]}\\
&= 3\log_2(2e(G)) - 2H(X) && \text{[XY uniform]}\\
&\geq 3\log_2(2e(G)) - 2\log_2 v(G) && \text{[uniform bound].}
\end{aligned}$$

This proves (5.5) and thus shows that P_4 is Sidorenko. Indeed, by the uniform bound,

$$\log_2 \hom(P_4, F) \geq H(X,Y,Z,W) \geq 3\log_2(2e(G)) - 2\log_2 v(G),$$

and hence

$$t(P_4, G) = \frac{\hom(P_4, G)}{v(G)^4} \geq \left(\frac{2e(G)}{v(G)^2}\right)^3 = t(K_2, G)^3. \qquad \square$$

Let us outline how to extend the preceding proof strategy from the 3-edge path to any tree T. Define a ***T-branching random walk*** in a graph G to be a random $\Phi \in \text{Hom}(T, G)$ defined by fixing an arbitrary root v of T (the choice of v will not matter in the end). Then set $\Phi(v)$ to be a random vertex of G with each vertex of G chosen proportional to its degree. Then extend Φ to a random homomorphism $T \to G$ one vertex at a time: if $u \in V(T)$ is already mapped to $\Phi(u)$ and its neighbor $w \in V(T)$ has not yet been mapped, then set $\Phi(w)$ to be a uniform random neighbor of $\Phi(u)$, independent of all previous choices. The resulting random $\Phi \in \text{Hom}(T, G)$ has the following properties:

- for each edge of T, its image under Φ is a uniform random edge of G and with the two possible edge orientations equally likely; and
- for each vertex v of T, conditioned on $\Phi(v)$, the neighbors of v in T are mapped by Φ to conditionally independent and uniform neighbors of $\Phi(v)$ in G.

Furthermore, as in the proof of Theorem 5.5.10,

$$\begin{aligned} H(\Phi) &= e(T)\log_2(2e(G)) - (e(T)-1)H(\Phi(v)) \\ &\geq e(T)\log_2(2e(G)) - (e(T)-1)\log_2 v(G). \end{aligned} \tag{5.7}$$

(Exercise: fill in the details.) Together with the uniform bound $H(\Phi) \leq \log_2 \hom(T, G)$, we proved the following.

Theorem 5.5.11

Every tree is Sidorenko.

We saw earlier that $K_{s,t}$ is Sidorenko, which can be proved by two applications of Hölder's inequality (see Section 5.3). Here let us give another proof using entropy. This entropy proof is subtler than the earlier Hölder's inequality proof, but it will soon lead us more naturally to the next generalization.

Theorem 5.5.12

Every complete bipartite graph is Sidorenko.

Let us demonstrate the proof for $K_{2,2}$ for concreteness. The same proof extends to all $K_{s,t}$.

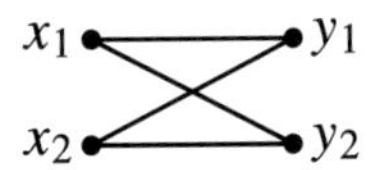

Proof that $K_{2,2}$ is Sidorenko. As earlier, we construct a random element of $\text{Hom}(K_{2,2}, G)$. Pick a random $(X_1, X_2, Y_1, Y_2) \in V(G)^4$ with $X_iY_j \in E(G)$ for all i, j as follows:

- X_1Y_1 is a uniform random edge;
- Y_2 is a uniform random neighbor of X_1;
- X_2 is a conditionally independent copy of X_1 given (Y_1, Y_2).

The last point deserves some attention. It does *not* say that we choose a uniform random common neighbor of Y_1 and Y_2, as one might naively attempt. Instead, one can think of the first two steps as defining the $K_{1,2}$-branching random walk for (X_1, Y_1, Y_2). Under this distribution, we can first sample (Y_1, Y_2) according to its marginal and then produce two conditionally independent copies of X_1 (with the second copy now called X_2).

We have

$$\begin{aligned}
&H(X_1,X_2,Y_1,Y_2)\\
&= H(Y_1,Y_2) + H(X_1,X_2|Y_1,Y_2) && \text{[chain rule]}\\
&= H(Y_1,Y_2) + 2H(X_1|Y_1,Y_2) && \text{[cond. indep.]}\\
&= 2H(X_1,Y_1,Y_2) - H(Y_1,Y_2) && \text{[chain rule]}\\
&\geq 2(2\log_2(2e(G)) - \log_2 v(G)) - H(Y_1,Y_2) && \text{[(5.7)]}\\
&\geq 2(2\log_2(2e(G)) - \log_2 v(G)) - 2\log_2 v(G) && \text{[uniform bound]}\\
&= 4\log(2e(G)) - 4\log_2 v(G).
\end{aligned}$$

Together with the uniform bound $H(X_1,X_2,Y_1,Y_2) \leq \log_2 \hom(K_{2,2},G)$, we deduce that $K_{2,2}$ is Sidorenko. □

Exercise 5.5.13. Complete the proof of Theorem 5.5.12 for general $K_{s,t}$.

The following result was first proved by Conlon, Fox, and Sudakov (2010) using the dependent random choice technique. The entropy proof was found later by Li and Szegedy (2011).

Theorem 5.5.14

Let F be a bipartite graph that has a vertex adjacent to all vertices in the other part. Then F is Sidorenko.

Let us illustrate the proof for the following graph F. The proof extends to the general case.

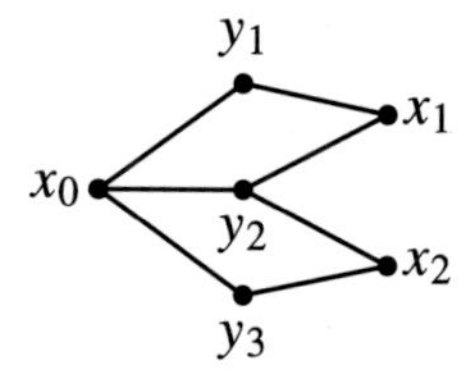

Proof that the preceding graph is Sidorenko. Pick $(X_0,X_1,X_2,Y_1,Y_2,Y_3) \in V(G)^6$ randomly as follows:

- X_0Y_1 is a uniform random edge;
- Y_2 and Y_3 are independent uniform random neighbors of X_0;
- X_1 is a conditionally independent copy of X_0 given (Y_1,Y_2);
- X_2 is a conditionally independent copy of X_0 given (Y_2,Y_3).

We have the following properties:

- X_0,X_1,X_2 are conditionally independent given (Y_1,Y_2,Y_3);
- X_1 and (X_0,Y_3,X_2) are conditionally independent given (Y_1,Y_2);
- The distribution of (X_0,Y_1,Y_2) is identical to the distribution of (X_1,Y_1,Y_2).

So (the first and fourth steps by chain rule, and the second and third steps by conditional independence),

$$\begin{aligned}
&H(X_0,X_1,X_2,Y_1,Y_2,Y_3)\\
&= H(X_0,X_1,X_2|Y_1,Y_2,Y_3) + H(Y_1,Y_2,Y_3)\\
&= H(X_0|Y_1,Y_2,Y_3) + H(X_1|Y_1,Y_2,Y_3) + H(X_2|Y_1,Y_2,Y_3) + H(Y_1,Y_2,Y_3)\\
&= H(X_0|Y_1,Y_2,Y_3) + H(X_1|Y_1,Y_2) + H(X_2|Y_2,Y_3) + H(Y_1,Y_2,Y_3)\\
&= H(X_0,Y_1,Y_2,Y_3) + H(X_1,Y_1,Y_2) + H(X_2,Y_2,Y_3) - H(Y_1,Y_2) - H(Y_2,Y_3).
\end{aligned}$$

By (5.7),

$$
\begin{aligned}
H(X_0,Y_1,Y_2,Y_3) &\geq 3\log_2(2e(G)) - 2\log_2 v(G),\\
H(X_1,Y_1,Y_2) &\geq 2\log_2(2e(G)) - \log_2 v(G),\\
\text{and}\quad H(X_2,Y_2,Y_3) &\geq 2\log_2(2e(G)) - \log_2 v(G).
\end{aligned}
$$

And by the uniform bound,

$$H(Y_1,Y_2) = H(Y_2,Y_3) \leq 2\log_2 v(G).$$

Putting everything together, we have

$$\log_2 \hom(F,G) \geq H(X_0,X_1,X_2,Y_1,Y_2,Y_3) \geq 7\log_2(2e(G)) - 8\log_2 v(G).$$

Thereby verifying (5.7), showing that F is Sidorenko. □

(Where did we use the assumption that F has vertex complete to the other part?)

Exercise 5.5.15. Complete the proof of Theorem 5.5.14.

Shearer's Inequality

Another important tool in the entropy method is Shearer's inequality, which is a powerful generalization of subadditivity. Before stating it in full generality, let us first see a simple instance of Shearer's lemma.

Theorem 5.5.16 (Shearer's entropy inequality, special case)

$$2H(X,Y,Z) \leq H(X,Y) + H(X,Z) + H(Y,Z)$$

Proof. Using the chain rule and conditioning dropping, we have

$$
\begin{aligned}
H(X,Y) &= H(X) + H(Y|X),\\
H(X,Z) &= H(X) \qquad\qquad + H(Z|X),\\
\text{and}\quad H(Y,Z) &= \qquad\qquad H(Y) + H(Z|Y).
\end{aligned}
$$

Adding up, and applying conditioning dropping, we see that the sum of the three right-hand side expressions is at least

$$2H(X) + 2H(Y|X) + 2H(Z|X,Y) = 2H(X,Y,Z),$$

with the final equality due to the chain rule. □

Here is the general form of Shearer's inequality (Chung, Graham, Frankl, and Shearer 1986).

Theorem 5.5.17 (Shearer's entropy inequality)
Let $A_1,\dots,A_s \subseteq [n]$ where each $i \in [n]$ appears in at least k sets A_j's. Let $X_1,\dots,X_n$ be jointly distributed discrete random variables. Writing $X_A := (X_i)_{i\in A}$, we have

$$kH(X_1,\dots,X_n) \leq \sum_{j\in[s]} H(X_{A_j}).$$

Exercise 5.5.18. Prove Theorem 5.5.17 by generalizing the proof of Theorem 5.5.16.

Shearer's entropy inequality is related to the generalized Hölder inequality from Section 5.3. It is a significant generalization of the projection inequality discussed in Remark 5.3.6. See Friedgut (2004) for more on these connections.

The next exercise asks you to prove a strengthening of the projection inequalities (Remark 5.3.6 and Exercise 5.3.9) by mimicking the entropy proof of Shearer's entropy inequality. The result is due to Bollobás and Thomason (1995), though their original proof does not use the entropy method.

Exercise 5.5.19 (Box theorem). For each $I \subseteq [d]$, write $\pi\colon \mathbb{R}^d \to \mathbb{R}^I$ to denote the projection obtained by omitting coordinates outside I. Show that, for every compact body $K \subseteq \mathbb{R}^d$, there exists a box $B = [a_1, b_1] \times \cdots \times [a_d, b_d] \subseteq \mathbb{R}^d$ such that $|B| = |K|$ and $|\pi_I(B)| \le |\pi_I(K)|$ for every $I \subseteq [d]$ (here $|\cdot|$ denotes volume).

Use this result to give another proof of the projection inequality from Exercise 5.3.9.

Hint: First prove it for K being a union of grid boxes. Then extend it to general K via compactness.

Let us use the entropy method to give another proof of Theorem 5.3.15, restated as follows.

Theorem 5.5.20 (The maximum number of H-colorings in a regular graph)
For every n-vertex d-regular bipartite graph F and any graph G (allowing looped vertices on G)

$$\hom(F, G) \le \hom(K_{d,d}, G)^{n/(2d)}.$$

The following proof is based on (with some further simplifications) the entropy proofs of Galvin and Tetali (2004), which was in turn based on the proof by Kahn (2001) for independent sets.

Proof. Let us first illustrate the proof for F being the following graph:

x_1 y_1
x_2 y_2
x_3 y_3

Choose $\Phi \in \mathrm{Hom}(F, G)$ uniformly at random among all homomorphisms from F to G. Let $X_1, X_2, X_3, Y_1, Y_2, Y_3 \in V(G)$ be the respective images of the vertices of G. We have

$$\begin{aligned}
&2 \log_2 \hom(F, G) \\
&\quad = 2H(X_1, X_2, X_3, Y_1, Y_2, Y_3) \\
&\quad = 2H(X_1, X_2, X_3) + 2H(Y_1, Y_2, Y_3 | X_1, X_2, X_3) && \text{[chain rule]} \\
&\quad \le H(X_1, X_2) + H(X_1, X_3) + H(X_2, X_3) \\
&\qquad\quad + 2H(Y_1 | X_1, X_2, X_3) + 2H(Y_2 | X_1, X_2, X_3) + 2H(Y_3 | X_1, X_2, X_3) && \text{[Shearer]} \\
&\quad = H(X_1, X_2) + H(X_1, X_3) + H(X_2, X_3) \\
&\qquad\quad + 2H(Y_1 | X_1, X_2) + 2H(Y_2 | X_1, X_3) + 2H(Y_3 | X_2, X_3) && \text{[cond. indep.].}
\end{aligned}$$

In the final step, we use the fact that X_3 and Y_1 are conditionally independent given X_1 and X_2 (why?), along with two other analogous statements. A more general statement is that if $S \subseteq V(F)$, then the restrictions to the different connected components of $F - S$ are conditionally independent given $(X_s)_{s \in S}$.

To complete the proof, it remains to show

$$\begin{aligned} &H(X_1, X_2) + 2H(Y_1|X_1, X_2) \le \log_2 \hom(K_{2,2}, G), \\ &H(X_1, X_3) + 2H(Y_2|X_1, X_3) \le \log_2 \hom(K_{2,2}, G), \\ \text{and} \quad &H(X_2, X_3) + 2H(Y_3|X_2, X_3) \le \log_2 \hom(K_{2,2}, G). \end{aligned}$$

They are analogous, so let us just show the first inequality. Let Y_1' be a conditionally independent copy of Y_1 given (X_1, X_2). Then (X_1, X_2, Y_1, Y_1') is the image of a homomorphism from $K_{2,2}$ to G (though not necessarily chosen uniformly).

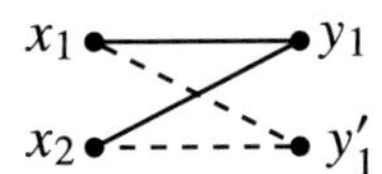

Thus, we have

$$\begin{aligned} H(X_1, X_2) + 2H(Y_1|X_1, X_2) &= H(X_1, X_2) + H(Y_1, Y_1'|X_1, X_2) \\ &= H(X_1, X_2, Y_1, Y_1') && \text{[chain rule]} \\ &\le \log_2 \hom(K_{2,2}, G) && \text{[uniform bound].} \end{aligned}$$

This concludes the proof for $F = C_6$.

Now let F be an arbitrary bipartite graph with vertex bipartition $V = A \cup B$. Let $\Phi \in \mathrm{Hom}(F, G)$ be chosen uniformly at random. For each $v \in V$, let $X_v = \Phi(v)$. For each $S \subseteq V$, write $X_S := (X_v)_{v \in S}$. We have

$$\begin{aligned} d \log_2 \hom(F, G) = dH(\Phi) &= dH(X_A) + dH(X_B|X_A) && \text{[chain rule]} \\ &\le \sum_{b \in B} H(X_{N(b)}) + d \sum_{b \in B} H(X_b|X_A) && \text{[Shearer]} \\ &= \sum_{b \in B} H(X_{N(b)}) + d \sum_{b \in B} H(X_b|X_{N(b)}). && \text{[cond. indep.]} \end{aligned}$$

For each $b \in B$, let $X_b^{(1)}, \ldots, X_b^{(d)}$ be conditionally independent copies of X_b given $X_{N(b)}$. We have

$$\begin{aligned} H(X_{N(b)}) + dH(X_b|X_{N(b)}) &= H(X_{N(b)}) + H(X_b^{(1)}, \ldots, X_b^{(d)}|X_{N(b)}) \\ &= H(X_b^{(1)}, \ldots, X_b^{(d)}, X_{N(b)}) && \text{[chain rule]} \\ &\le \log_2 \hom(K_{d,d}, G). && \text{[uniform bound]} \end{aligned}$$

Summing over all $b \in B$, and using the previous equality, we obtain

$$d \log_2 \hom(F, G) \le \frac{n}{2} \log_2 \hom(K_{d,d}, G). \qquad \square$$

Exercise 5.5.21. Prove that the following graph is Sidorenko.

Exercise 5.5.22 (△ vs. ∧ in a directed graph). Let V be a finite set, $E \subseteq V \times V$, and

$$\triangle = \left|\left\{(x,y,z) \in V^3 : (x,y),(y,z),(z,x) \in E\right\}\right|$$

(i.e., cyclic triangles; note the direction of edges) and

$$\wedge = \left|\left\{(x,y,z) \in V^3 : (x,y),(x,z) \in E\right\}\right|.$$

Prove that $\triangle \le \wedge$.

Further Reading

The book *Large Networks and Graph Limits* by Lovász (2012) contains an excellent treatment of graph homomorphism inequalities in Section 2.1 and Chapter 16.

The survey *Flag Algebras: An Interim Report* by Razborov (2013) contains a survey of results obtained using the flag algebra method.

For combinatorial applications of the entropy method, see the following surveys:

- *Entropy and Counting* by Radhakrishnan (2003), and
- *Three Tutorial Lectures on Entropy and Counting* by Galvin (2014).

Chapter Summary

- Many problems in extremal graph theory can be phrased in terms of graph homomorphism inequalities.
  - Homomorphism density inequalities are undecidable in general.
  - Many open problems remain, such as **Sidorenko's conjecture**, which says that if F is bipartite, then $t(F,G) \ge t(K_2,G)^{e(F)}$ for all graphs G.
- The set of all possible **(edge, triangle) density pairs** is known.
  - For a given edge density, the triangle density is maximized by a clique.
  - For a given edge density, the triangle density is minimized by a certain multipartite graph. (We did not prove this result in full and only established the convex hull in Section 5.4.)
- **Cauchy–Schwarz** and **Hölder** inequalities are versatile tools.
  - Simple applications of Cauchy–Schwarz inequalities can often be recognized by "reflection symmetries" in a graph that can be "folded in half."
  - **Flag algebra** leads to computerized searches of Cauchy–Schwarz proofs of subgraph density inequalities.
  - **Generalized Hölder inequality** tells us that, as an example,

    $$\int_{x,y,z} f(x,y)g(x,z)h(y,z) \le \|f\|_2 \|g\|_2 \|h\|_2 .$$

    It can be proved by repeated applications of Hölder's inequality, once for each variable. The inequality is related to **Shearer's entropy inequality**, an example of which says that for joint random variables X,Y,Z,

    $$2H(X,Y,Z) \le H(X,Y) + H(X,Z) + H(Y,Z).$$

- The **Lagrangian method** relaxes an optimization problem on graphs to one about vertex-weighted graphs, and then argues by shifting weights between vertices. We used the method to prove
  - Turán's theorem (again);
  - A linear inequality between clique densities in G is true for all G if and only if it holds whenever G is a clique.
- The **entropy method** can be used to establish various cases of Sidorenko's conjecture, including for trees, as well as for a bipartite graph with one vertex complete to the other side.

6

Forbidding 3-Term Arithmetic Progressions

Chapter Highlights

- Fourier analytic proof of Roth's theorem
- Finite field model in additive combinatorics: $\mathbb{F}_p^n$ as a model for the integers
- Basics of discrete Fourier analysis
- Density increment argument in the proof of Roth's theorem
- The polynomial method proof of Roth's theorem in $\mathbb{F}_3^n$
- Arithmetic analogue of the regularity lemma, and application to Roth's theorem with popular difference

In this chapter, we study Roth's theorem, which says that every 3-AP-free subset of $[N]$ has size $o(N)$.

Previously, in Section 2.4, we gave a proof of Roth's theorem using the graph regularity lemma. The main goal of this chapter is to give a Fourier analytic proof of Roth's theorem. This is also Roth's original proof (1953).

We begin by proving Roth's theorem in the **finite field model**. That is, we first prove an analogue of Roth's theorem in $\mathbb{F}_3^n$. The finite field vector space serves as a fruitful playground for many additive combinatorics problems. Techniques such as Fourier analysis are often simpler to carry out in the finite field model. After we develop the techniques in the finite field model, we then prove Roth's theorem in the integers. It can be a good idea to first try out ideas in the finite field model before bringing them to the integers, as there may be additional technical difficulties in the integers.

Later in Section 6.5, we will see a completely different proof of Roth's theorem in $\mathbb{F}_3^n$ using the **polynomial method**, which gives significantly better quantitative bounds. This proof surprised many people at the time of its discovery. However, unlike Fourier analysis, this polynomial method technique only applies to the finite field setting, and it is unknown how to apply it to the integers.

There is a parallel between the Fourier analytic method in this chapter and the graph regularity method from Chapter 2. In Section 6.6, we develop an arithmetic regularity lemma and use it in Section 6.7 to prove a strengthening of Roth's theorem showing **popular common differences**.

6.1 Fourier Analysis in Finite Field Vector Spaces

We review some basic facts about Fourier analysis in $\mathbb{F}_p^n$ for a prime p. Everything here can be extended to arbitrary abelian groups. As we saw in Section 3.3, eigenvalues of Cayley graphs on an abelian group and the Fourier transform are intimately related.

Throughout this section, we fix a prime p and let

$$\omega = \exp(2\pi i/p).$$

Definition 6.1.1 (Fourier transform in $\mathbb{F}_p^n$)

The Fourier transform of $f: \mathbb{F}_p^n \to \mathbb{C}$ is a function $\widehat{f}: \mathbb{F}_p^n \to \mathbb{C}$ defined by setting, for each $r \in \mathbb{F}_p^n$,

$$\widehat{f}(r) := \mathbb{E}_{x \in \mathbb{F}_p^n} f(x)\omega^{-r\cdot x} = \frac{1}{p^n} \sum_{x \in \mathbb{F}_p^n} f(x)\omega^{-r\cdot x}$$

where $r \cdot x = r_1 x_1 + \cdots + r_n x_n$.

In particular, $\widehat{f}(0) = \mathbb{E}f$ is the average of f. This value often plays a special role compared to other values $\widehat{f}(r)$.

To simplify notation, it is generally understood that the variables being averaged or summed over are varying uniformly in the domain $\mathbb{F}_p^n$.

Let us now state several important properties of the Fourier transform. We will see that all these properties are consequences of the orthogonality of the Fourier basis.

The next result allows us to write f in terms of $\widehat{f}$.

Theorem 6.1.2 (Fourier inversion formula)

Let $f: \mathbb{F}_p^n \to \mathbb{C}$. For every $x \in \mathbb{F}_p^n$,

$$f(x) = \sum_{r \in \mathbb{F}_p^n} \widehat{f}(r)\omega^{r\cdot x}.$$

The next result tells us that the Fourier transform preserves inner products.

Theorem 6.1.3 (Parseval / Plancherel)

Given $f, g: \mathbb{F}_p^n \to \mathbb{C}$, we have

$$\mathbb{E}_{x \in \mathbb{F}_p^n} f(x)\overline{g(x)} = \sum_{r \in \mathbb{F}_p^n} \widehat{f}(r)\overline{\widehat{g}(r)}.$$

In particular, as a special case ($f = g$),

$$\mathbb{E}_{x \in \mathbb{F}_p^n} |f(x)|^2 = \sum_{r \in \mathbb{F}_p^n} |\widehat{f}(r)|^2.$$

***Remark* 6.1.4** (History/naming). The names Parseval and Plancherel are often used interchangeably in practice to refer to the unitarity of the Fourier transform (i.e., the preceding theorem). Parseval derived the identity for the Fourier series of a periodic function on $\mathbb{R}$, whereas Plancherel derived it for the Fourier transform on $\mathbb{R}$.

As is nowadays the standard in additive combinatorics, we adopt the following convention for the Fourier transform in finite abelian groups:

average in physical space $\mathbb{E}f$
and sum in frequency (Fourier) space $\sum \widehat{f}$.

For example, following this convention, we define an "averaging" inner product for functions $f, g \colon \mathbb{F}_p^n \to \mathbb{C}$ by

$$\langle f, g \rangle := \mathbb{E}_{x \in \mathbb{F}_p^n} \overline{f(x)} g(x) \qquad \text{and} \qquad \|f\|_2 := \langle f, f \rangle^{1/2}.$$

In the frequency/Fourier domain, we define the "summing" inner product for functions $\alpha, \beta \colon \mathbb{F}_p^n \to \mathbb{C}$ by

$$\langle \alpha, \beta \rangle_{\ell^2} := \sum_{x \in \mathbb{F}_p^n} \overline{\alpha(x)} \beta(x) \qquad \text{and} \qquad \|\alpha\|_{\ell^2} := \langle \alpha, \alpha \rangle_{\ell^2}^{1/2}.$$

Writing $\gamma_r \colon \mathbb{F}_p^n \to \mathbb{C}$ for the function defined by

$$\gamma_r(x) := \omega^{r \cdot x}$$

(this is a ***character*** of the group $\mathbb{F}_p^n$), the Fourier transform can be written as

$$\widehat{f}(r) = \mathbb{E}_x \overline{\gamma_r(x)} f(x) = \langle \gamma_r, f \rangle. \tag{6.1}$$

Parseval's identity can be stated as

$$\langle f, g \rangle = \langle \widehat{f}, \widehat{g} \rangle_{\ell^2} \qquad \text{and} \qquad \|f\|_2 = \|\widehat{f}\|_{\ell^2}.$$

With these conventions, we often do not need to keep track of normalization factors.

The preceding identities can be proved via direct verification by plugging in the formula for the Fourier transform. We give a more conceptual proof below.

Proof of the Fourier inversion formula (Theorem 6.1.2). Let $\gamma_r(x) = \omega^{r \cdot x}$. Then the set of functions

$$\{\gamma_r : r \in \mathbb{F}_p^n\}$$

forms an orthonormal basis for the space of functions $\mathbb{F}_p^n \to \mathbb{C}$ with respect to the averaging inner product $\langle \cdot, \cdot \rangle$. Indeed,

$$\langle \gamma_r, \gamma_s \rangle = \mathbb{E}_x \omega^{(s-r) \cdot x} = \begin{cases} 1 & \text{if } r = s, \\ 0 & \text{if } r \neq s. \end{cases}$$

Furthermore, there are p^n functions γ_r (as r ranges over $\mathbb{F}_p^n$). So they form a basis of the p^n-dimensional vector space of all functions $f \colon \mathbb{F}_p^n \to \mathbb{C}$. We will call this basis the ***Fourier basis***.

Now, given an arbitrary $f \colon \mathbb{F}_p^n \to \mathbb{C}$, the "coordinate" of f with respect to the basis vector γ_r of the Fourier basis is $\langle \gamma_r, f \rangle = \widehat{f}(r)$ by (6.1). So,

$$f = \sum_r \widehat{f}(r) \gamma_r.$$

This is precisely the Fourier inversion formula. □

Proof of Parseval's identity (Theorem 6.1.3). Continuing from the previous proof, since the Fourier basis is orthonormal, we can evaluate $\langle f, g \rangle$ with respect to coordinates in this basis, thereby yielding

$$\langle f, g \rangle = \sum_{r \in \mathbb{F}_p^n} \overline{\langle \gamma_r, f \rangle} \langle \gamma_r, g \rangle = \sum_{r \in \mathbb{F}_p^n} \overline{\widehat{f}(r)} \widehat{g}(r).$$ □

The convolution is an important operation.

Definition 6.1.5 (Convolution)
Given $f, g\colon \mathbb{F}_p^n \to \mathbb{C}$, define $f * g\colon \mathbb{F}_p^n \to \mathbb{C}$ by

$$(f * g)(x) := \mathbb{E}_{y \in \mathbb{F}_p^n} f(y) g(x - y).$$

In other words, $(f * g)(x)$ is the average of $f(y)g(z)$ over all pairs (y, z) with $y + z = x$.

***Example* 6.1.6.** (a) If f is supported on $A \subseteq \mathbb{F}_p^n$ and g is supported on $B \subseteq \mathbb{F}_p^n$, then $f * g$ is supported on the sumset $A + B = \{a + b\colon a \in A, b \in B\}$.

(b) Let W be a subspace of $\mathbb{F}_p^n$. Let $\mu_W = (p^n/|W|)1_W$ be the indicator function on W normalized so that $\mathbb{E}\mu_W = 1$. Then for any $f\colon \mathbb{F}_p^n \to \mathbb{C}$, the function $f * \mu_W$ is obtained from f by replacing its value at x by its average value on the coset $x + W$.

The second example suggests that convolution can be thought of as smoothing a function, damping its potentially rough perturbations.

The Fourier transform conveniently converts convolutions to multiplication.

Theorem 6.1.7 (Convolution identity)
For any $f, g\colon \mathbb{F}_p^n \to \mathbb{C}$ and any $r \in \mathbb{F}_p^n$,

$$\widehat{f * g}(r) = \widehat{f}(r)\widehat{g}(r).$$

Proof. We have

$$\begin{aligned}\widehat{f * g}(r) &= \mathbb{E}_x (f * g)(x)\omega^{-r\cdot x} = \mathbb{E}_x \mathbb{E}_{y,z: y+z=x} f(y)g(z)\omega^{-r\cdot(y+z)} \\ &= \mathbb{E}_{y,z} f(y)g(z)\omega^{-r\cdot(y+z)} = \left(\mathbb{E}_y f(y)\omega^{-r\cdot y}\right)\left(\mathbb{E}_z g(z)\omega^{-r\cdot z}\right) = \widehat{f}(r)\widehat{g}(r). \qquad \square\end{aligned}$$

By repeated applications of the convolution identity, we have

$$(f_1 * \cdots * f_k)^\wedge = \widehat{f_1}\widehat{f_2}\cdots\widehat{f_k}.$$

(Here we write $f^\wedge$ for $\widehat{f}$ for typographical reasons.)

Now we introduce a quantity relevant to Roth's theorem on 3-APs.

Definition 6.1.8 (3-AP density)
Given $f, g, h\colon \mathbb{F}_p^n \to \mathbb{C}$, we write

$$\mathbf{\Lambda}(\boldsymbol{f}, \boldsymbol{g}, \boldsymbol{h}) := \mathbb{E}_{x,y} f(x)g(x + y)h(x + 2y), \tag{6.2}$$

and

$$\mathbf{\Lambda_3}(\boldsymbol{f}) := \Lambda(f, f, f). \tag{6.3}$$

Note that for any $A \subseteq \mathbb{F}_p^n$,

$$\Lambda(1_A) = p^{-2n} \left|\{(x, y)\colon x, x + y, x + 2y \in A\}\right| = \text{“3-AP density of } A\text{.”}.$$

Here we include "trivial" 3-APs (i.e., those with with $y = 0$).

The following identity, relating the Fourier transform and 3-APs, plays a central role in the Fourier analytic proof of Roth's theorem.

Proposition 6.1.9 (Fourier and 3-AP)
Let p be an odd prime. If $f, g, h \colon \mathbb{F}_p^n \to \mathbb{C}$, then

$$\Lambda(f, g, h) = \sum_r \widehat{f}(r)\widehat{g}(-2r)\widehat{h}(r).$$

We will give two proofs of this proposition. The first proof is more mechanically straightforward. It is similar to the proof of the convolution identity earlier. The second proof directly applies the convolution identity, and may be a bit more abstract/conceptual.

First proof. We expand the left-hand side using the formula for Fourier inversion.

$$\begin{aligned}
&\mathbb{E}_{x,y} f(x)g(x+y)h(x+2y)\\
&= \mathbb{E}_{x,y}\left(\sum_{r_1}\widehat{f}(r_1)\omega^{r_1\cdot x}\right)\left(\sum_{r_2}\widehat{g}(r_2)\omega^{r_2\cdot(x+y)}\right)\left(\sum_{r_3}\widehat{h}(r_3)\omega^{r_3\cdot(x+2y)}\right)\\
&= \sum_{r_1,r_2,r_3}\widehat{f}(r_1)\widehat{g}(r_2)\widehat{h}(r_3)\mathbb{E}_x\omega^{x\cdot(r_1+r_2+r_3)}\mathbb{E}_y\omega^{y\cdot(r_2+2r_3)}\\
&= \sum_{r_1,r_2,r_3}\widehat{f}(r_1)\widehat{g}(r_2)\widehat{h}(r_3)1_{r_1+r_2+r_3=0}1_{r_2+2r_3=0}\\
&= \sum_r \widehat{f}(r)\widehat{g}(-2r)\widehat{h}(r).
\end{aligned}$$

In the last step, we use that $r_1 + r_2 + r_3 = 0$ and $r_2 + 2r_3 = 0$ together imply $r_1 = r_3$. □

Second proof. Write $g_1(y) = g(-y/2)$. So, $\widehat{g_1}(r) = \widehat{g}(-2r)$. Applying the convolution identity,

$$\begin{aligned}
\mathbb{E}_{x,y} f(x)g(x+y)h(x+2y) &= \mathbb{E}_{x,y,z:x-2y+z=0} f(x)g(y)h(z)\\
&= \mathbb{E}_{x,y,z:x+y+z=0} f(x)g_1(y)h(z)\\
&= (f * g_1 * h)(0)\\
&= \sum_r \widehat{f * g_1 * h}(r) && \text{[Fourier inversion]}\\
&= \sum_r \widehat{f}(r)\widehat{g_1}(r)\widehat{h}(r) && \text{[Convolution identity]}\\
&= \sum_r \widehat{f}(r)\widehat{g}(-2r)\widehat{h}(r).
\end{aligned}$$

□

***Remark* 6.1.10.** In the following section, we will work in $\mathbb{F}_3^n$. Since $-2 = 1$ in $\mathbb{F}_3$ (and so $g_1 = g$ in the preceding), the proof looks even simpler. In particular, by Fourier inversion and the convolution identity,

$$\begin{aligned}
\Lambda_3(1_A) &= 3^{-2n}\left|\{(x,y,z) \in A^3 : x + y + z = 0\}\right|\\
&= (1_A * 1_A * 1_A)(0) = \sum_r (1_A * 1_A * 1_A)^\wedge(r) = \sum_r \widehat{1_A}(r)^3. \qquad (6.4)
\end{aligned}$$

When $A = -A$, the eigenvalues of the adjacency matrix of the Cayley graph $\mathrm{Cay}(\mathbb{F}_3^n, A)$ are $3^n \widehat{1_A}(r)$, $r \in \mathbb{F}_3^n$. (Recall from Section 3.3 that the eigenvalues of abelian Cayley graphs are given by the Fourier transforms.) The quantity $3^{2n}\Lambda_3(1_A)$ is the number of closed walks of length 3 in the Cayley graph $\mathrm{Cay}(\mathbb{F}_p^n, A)$. So, the preceding identity is saying that the number of closed walks of length 3 in $\mathrm{Cay}(\mathbb{F}_3^n, A)$ equals the third moment of the eigenvalues of the adjacency matrix, which is a general fact for every graph. (When $A \neq -A$, we can consider the directed or bipartite version of this argument.)

The following exercise generalizes the preceding identity.

Exercise 6.1.11. Let $a_1, \ldots, a_k$ be nonzero integers, none divisible by the prime p. Let $f_1, \ldots, f_k \colon \mathbb{F}_p^n \to \mathbb{C}$. Show that

$$\mathbb{E}_{x_1,\ldots,x_k \in \mathbb{F}_p^n : a_1x_1+\cdots+a_kx_k=0} f_1(x_1)\cdots f_k(x_k) = \sum_{r \in \mathbb{F}_p^n} \widehat{f_1}(a_1 r)\cdots \widehat{f_k}(a_k r).$$

6.2 Roth's Theorem in the Finite Field Model

In this section, we use Fourier analysis to prove the following finite field analogue of Roth's theorem (Meshulam 1995). Later in the chapter, we will convert this proof to the integer setting.

In an abelian group, a set A is said to be ***3-AP-free*** if A does not have three distinct elements of the form $x, x+y, x+2y$. A 3-AP-free subset of $\mathbb{F}_3^n$ is also called a ***cap set***. The ***cap set problem*** asks to determine the size of the largest cap set in $\mathbb{F}_3^n$.

Theorem 6.2.1 (Roth's theorem in $\mathbb{F}_3^n$)
Every 3-AP-free subset of $\mathbb{F}_3^n$ has size $O(3^n/n)$.

***Remark* 6.2.2** (General finite fields). We work in $\mathbb{F}_3^n$ mainly for convenience. The argument presented in this section also shows that for every odd prime p, there is some constant C_p so that every 3-AP-free subset of $\mathbb{F}_p^n$ has size $\leq C_p p^n/n$.

There are several equivalent interpretations of $x, y, z \in \mathbb{F}_3^n$ forming a 3-AP (allowing the possibility for a trivial 3-AP with $x = y = z$):

- $(x, y, z) = (x, x+d, x+2d)$ for some d;
- $x - 2y + z = 0$;
- $x + y + z = 0$;
- x, y, z are three distinct points of a line in $\mathbb{F}_3^n$ or are all equal;
- for each i, the ith coordinates of x, y, z are all distinct or all equal.

***Remark* 6.2.3** (SET card game). The card game SET comes with a deck of 81 cards. (See Figure 6.1). Each card one of three possibilities in each of the following four features:

- Number: 1, 2, 3;
- Symbol: diamond, squiggle, oval;
- Shading: solid, striped, open;
- Color: red, green, purple.

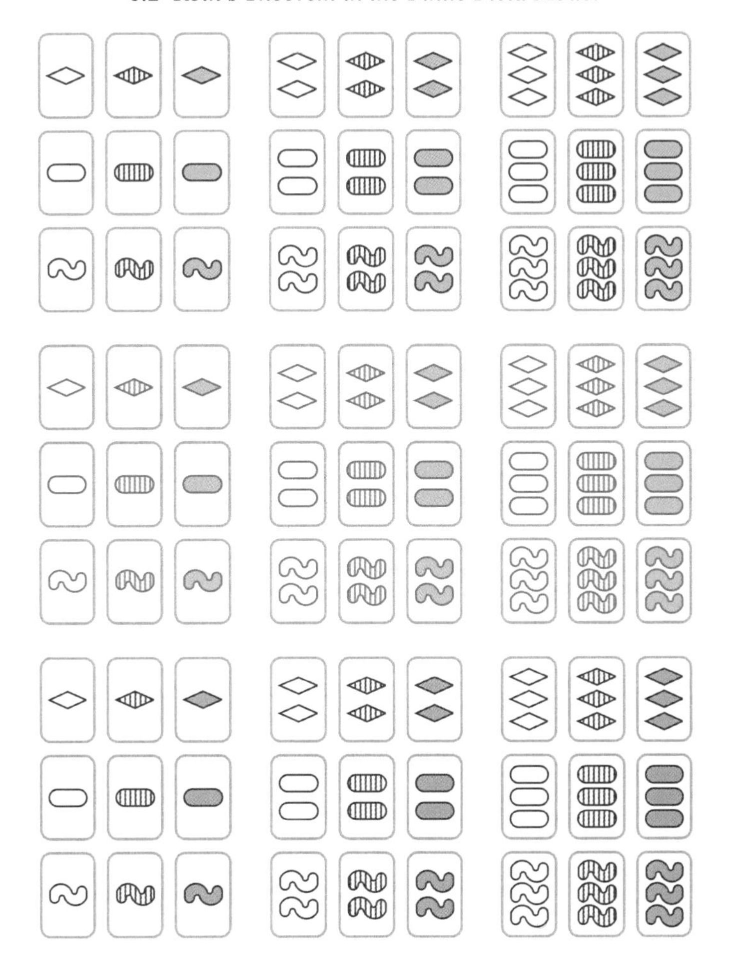

Figure 6.1 The complete deck of 81 cards in the game SET.

Each of the $3^4 = 81$ combinations appears exactly once as a card.

In this game, a combination of three cards is called a "set" if each of the four features shows up as all identical or all distinct among the three cards. For the example, the three cards shown in the following illustration form a "set": number (all distinct), symbol (all distinct), shading (all striped), color (all red).

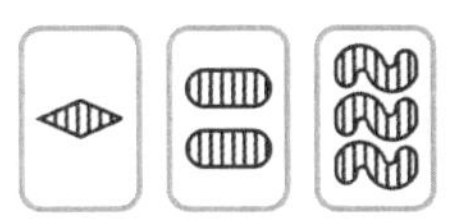

In a standard play of the game, the dealer lays down twelve cards on the table until some player finds a "set," in which case the player keeps the three cards of the set as their score, and the dealer replenishes the table by laying down more cards. If no set is found, then the dealer continues to lay down more cards until a set is found.

The cards of the game correspond to points of $\mathbb{F}_3^4$. A set is precisely a 3-AP. The cap set problem in $\mathbb{F}_3^4$ asks for the number of cards without a set. The size of the maximum cap set in $\mathbb{F}_3^4$ is 20 (Pellegrino 1970).

Here is the proof strategy of Roth's theorem in $\mathbb{F}_3^n$:
(1) A 3-AP-free set has a large Fourier coefficient.
(2) A large Fourier coefficient implies density increment on some hyperplane.
(3) Iterate.
As in the proof of the graph regularity lemma (where we refined partitions to obtain an *energy increment*), the preceding process must terminate in a bounded number of steps since the density of a subset is always between 0 and 1.

Similar to what we saw in Chapter 3 on pseudorandom graphs, a set $A \subseteq \mathbb{F}_3^n$ has pseudorandom properties if and only if all its Fourier coefficients $\widehat{1_A}(r)$, for $r \neq 0$, are small in absolute value. When A is pseudorandom in this Fourier-uniform sense, the 3-AP-density of A is similar to that of a random set with the same density. On the flip side, a large Fourier coefficient in A points to nonuniformity along the direction of the Fourier character. Then we can restrict A to some hyperplane and extract a density increment.

The following counting lemma shows that a Fourier-uniform subset of $\mathbb{F}_3^n$ has 3-AP density similar to that of a random set. It has a similar flavor as the proof that **EIG** implies $\mathbf{C_4}$ in Theorem 3.1.1. It is also related to the counting lemma for graphons (Theorem 4.5.1). Recall the 3-AP-density Λ_3 from Definition 6.1.8.

Lemma 6.2.4 (3-AP counting lemma)
Let $f\colon \mathbb{F}_3^n \to [0,1]$. Then

$$\left|\Lambda_3(f) - (\mathbb{E}f)^3\right| \le \max_{r\neq 0}|\widehat{f}(r)|\, \|f\|_2^2 .$$

Proof. By Proposition 6.1.9 (also see (6.4)),

$$\Lambda_3(f) = \sum_r \widehat{f}(r)^3 = \widehat{f}(0)^3 + \sum_{r\neq 0} \widehat{f}(r)^3.$$

Since $\mathbb{E}f = \widehat{f}(0)$, we have

$$\left|\Lambda_3(f) - (\mathbb{E}f)^3\right| \le \sum_{r\neq 0}|\widehat{f}(r)|^3 \le \max_{r\neq 0}|\widehat{f}(r)| \cdot \sum_r |\widehat{f}(r)|^2 = \max_{r\neq 0}|\widehat{f}(r)|\, \|f\|_2^2 .$$

The final step is by Parseval. □

***Remark* 6.2.5.** It would be insufficient to bound each term $|\widehat{f}(r)|^3$ by $\|\widehat{f}\|_\infty^3$. Instead, Parseval comes for the rescue. See Remark 3.1.19 for a similar issue.

Step 1. A 3-AP-free set has a large Fourier coefficient.

Lemma 6.2.6 (3-AP-free implies a large Fourier coefficient)
Let $A \subseteq \mathbb{F}_3^n$ and $\alpha = |A|/3^n$. If A is 3-AP-free and $3^n \ge 2\alpha^{-2}$, then there is $r \neq 0$ such that $|\widehat{1_A}(r)| \ge \alpha^2/2$.

Proof. Since A is 3-AP-free, $\Lambda_3(A) = |A|/3^{2n} = \alpha/3^n$, as all 3-APs are trivial (i.e., with common difference zero). By the counting lemma, Lemma 6.2.4,

$$\alpha^3 - \frac{\alpha}{3^n} = \alpha^3 - \Lambda_3(1_A) \le \max_{r \neq 0} |\widehat{1_A}(r)| \, \|1_A\|_2^2 = \max_{r \neq 0} |\widehat{1_A}(r)| \alpha.$$

By the hypothesis $3^n \ge 2\alpha^{-2}$, the left-hand side in the preceding inequality is $\ge \alpha^3/2$. So, there is some $r \neq 0$ with $|\widehat{1_A}(r)| \ge \alpha^2/2$. □

Step 2. A large Fourier coefficient implies density increment on some hyperplane.

Lemma 6.2.7 (Large Fourier coefficient implies density increment)
Let $A \subseteq \mathbb{F}_3^n$ with $\alpha = |A|/3^n$. Suppose $|\widehat{1_A}(r)| \ge \delta > 0$ for some $r \neq 0$. Then A has density at least $\alpha + \delta/2$ when restricted to some hyperplane.

Proof. We have

$$\widehat{1_A}(r) = \mathbb{E}_x 1_A(x) \omega^{-r \cdot x} = \frac{\alpha_0 + \alpha_1 \omega + \alpha_2 \omega^2}{3}$$

where $\alpha_0, \alpha_1, \alpha_2$ are densities of A on the cosets of $r^\perp$. We want to show that one of $\alpha_0, \alpha_1, \alpha_2$ is significantly larger than α. This is easy to check directly, but let us introduce a trick that we will also use later in the integer setting.

We have $\alpha = (\alpha_0 + \alpha_1 + \alpha_2)/3$. By the triangle inequality,

$$\begin{aligned} 3\delta &\le \left|\alpha_0 + \alpha_1 \omega + \alpha_2 \omega^2\right| \\ &= \left|(\alpha_0 - \alpha) + (\alpha_1 - \alpha)\omega + (\alpha_2 - \alpha)\omega^2\right| \\ &\le |\alpha_0 - \alpha| + |\alpha_1 - \alpha| + |\alpha_2 - \alpha| \\ &= \sum_{j=0}^{2} \left(|\alpha_j - \alpha| + (\alpha_j - \alpha)\right). \end{aligned}$$

Consequently, there exists j such that $|\alpha_j - \alpha| + (\alpha_j - \alpha) \ge \delta$. Note that $|t| + t$ equals $2t$ if $t > 0$ and 0 if $t \le 0$. So, $\alpha_j - \alpha \ge \delta/2$, as desired. □

Combining the previous two lemmas, here is what we have proved so far.

Lemma 6.2.8 (3-AP-free implies density increment)
Let $A \subseteq \mathbb{F}_3^n$ and $\alpha = |A|/3^n$. If A is 3-AP-free and $3^n \ge 2\alpha^{-2}$, then A has density at least $\alpha + \alpha^2/4$ when restricted to some hyperplane. □

We now view this hyperplane H as $\mathbb{F}_3^{n-1}$. (We may need to select a new origin for H if $0 \notin H$.) The restriction of A to H (i.e., $A \cap H$) is now a 3-AP-free subset of H. The density increased from α to $\alpha + \alpha^2/4$. Next we iterate this density increment.

***Remark* 6.2.9** (Translation invariance). It is important that the pattern we are forbidding (3-AP) is translation-invariant. What is wrong with the argument if instead we forbid the pattern $x + y = z$? Note that $\{x \in \mathbb{F}_3^n : x_1 = 2\}$ avoids solutions to $x + y = z$, and this set has density $1/3$.

Step 3. Iterate the density increment.

We start with a 3-AP-free $A \subseteq \mathbb{F}_3^n$. Let $V_0 := \mathbb{F}_3^n$ with density $\alpha_0 := \alpha = |A|/3^n$. Repeatedly apply Lemma 6.2.8. After i rounds, we restrict A to a codimension i affine subspace V_i (with $V_0 \supseteq V_1 \supseteq \cdots$). Let $\alpha_i = |A \cap V_i|/|V_i|$ be the density of A in V_i. As long as $2\alpha_i^{-2} \le |V_i| = 3^{n-i}$, we can apply Lemma 6.2.8 to obtain a V_{i+1} with density increment

$$\alpha_{i+1} \ge \alpha_i + \alpha_i^2/4.$$

Since $\alpha = \alpha_0 \le \alpha_1 \le \cdots \le 1$, and α_i increases by $\ge \alpha_i^2/4 \ge \alpha^2/4$ at each step, the process terminates after $m \le 4/\alpha^2$ rounds, at which point we must have $3^{n-m} < 2\alpha_m^{-2} \le 2\alpha^{-2}$ (or else we can continue via Lemma 6.2.8). So $n < m + \log_3(2\alpha^{-2}) = O(1/\alpha^2)$. Thus, $\alpha \le 1/\sqrt{n}$. This is just shy of the bound $\alpha = O(1/n)$ that we aim to prove. So let us redo the density increment analysis more carefully to analyze how quickly α_i grows.

Each round, α_i increases by at least $\alpha^2/4$. So, it takes $\le \lceil 4/\alpha \rceil$ initial rounds for α_i to double. Once $\alpha_i \ge 2\alpha$, it then increases by at least $\alpha_i^2/4$ each round afterwards, so it takes $\le \lceil 1/\alpha_i \rceil \le \lceil 1/\alpha \rceil$ additional rounds for the density to double again, and so on: the kth doubling time is at most $\lceil 4^{2-k}/\alpha \rceil$. Since the density is always at least α, the density can double at most $\log_2(1/\alpha)$ times. So, the total number of rounds is at most

$$\sum_{j \le \log_2(1/\alpha)} \left\lceil \frac{4^{2-j}}{\alpha} \right\rceil = O\left(\frac{1}{\alpha}\right).$$

Suppose the process terminates after m steps with density α_m. Then, examining the hypothesis of Lemma 6.2.8, we find that the size of the final subspace $|V_m| = 3^{n-m}$ is less than $\alpha_m^{-2} \le \alpha^{-2}$. So $n \le m + O(\log(1/\alpha)) \le O(1/\alpha)$. Thus, $\alpha = |A|/N = O(1/n)$. This completes the proof of Roth's theorem in $\mathbb{F}_3^n$ (Theorem 6.2.1).

***Remark* 6.2.10** (Quantitative bounds). Edel (2004) obtained a cap set of size $\ge 2.21^n$ for sufficiently large n. This is obtained by constructing a cap set in $\mathbb{F}_3^{480}$ of size $m = 2^{327}(2^{73} + 3^7 7^{76}) \ge 2.21^{480}$, which then implies, by a product construction, a cap set in $\mathbb{F}_3^{480k}$ of size m^k for each positive integer k.

It was an open problem of great interest whether there is an upper bound of the form c^n, with constant $c < 3$, on the size of cap sets in $\mathbb{F}_3^n$. With significant effort, the preceding Fourier analytic strategy was extended to prove an upper bound of the form $3^n/n^{1+c}$ (Bateman & Katz 2012). So it came as quite a shock to the community when a very short polynomial method proof was discovered, giving an upper bound $O(2.76^n)$ (Croot, Lev, and Pach 2017; Ellenberg and Gijswijt 2017). We will discuss this proof in Section 6.5. However, the polynomial method proof appears to be specific to the finite field model, and it is not known how to extend the strategy to the integers.

The following exercise shows why the preceding strategy does not generalize to 4-APs at least in a straightforward manner.

Exercise 6.2.11 (Fourier uniformity does not control 4-AP counts). Let

$$A = \{x \in \mathbb{F}_5^n : x \cdot x = 0\}.$$

Prove that:

(a) $|A| = (5^{-1} + o(1))5^n$ and $|\widehat{1_A}(r)| = o(1)$ for all $r \neq 0$;

(b) $|\{(x, y) \in \mathbb{F}_5^n : x, x + y, x + 2y, x + 3y \in A\}| \neq (5^{-4} + o(1))5^{2n}$.

Hint: First write 1_A as an exponential sum. Compare with the Gauss sum from Theorem 3.3.14.

Exercise 6.2.12 (Linearity testing). Show that for every prime p there is some $C_p > 0$ such that, if $f : \mathbb{F}_p^n \to \mathbb{F}_p$ satisfies

$$\mathbb{P}_{x,y\in\mathbb{F}_p^n}(f(x) + f(y) = f(x + y)) = 1 - \varepsilon,$$

then there exists some $a \in \mathbb{F}_p^n$ such that

$$\mathbb{P}_{x\in\mathbb{F}_p^n}(f(x) = a \cdot x) \geq 1 - C_p\varepsilon.$$

In the preceding $\mathbb{P}$ expressions x and y are chosen independently and uniformly at random from $\mathbb{F}_p^n$.

The following exercises introduce Gowers uniformity norms. Gowers (2001) used them to prove Szemerédi's theorem by extending the Fourier analytic proof strategy of Roth's theorem to what is now called **higher-order Fourier analysis**.

The U^2 norm in the following exercise plays a role similar to Fourier analysis.

Exercise 6.2.13 (Gowers U^2 uniformity norm). Let $f : \mathbb{F}_p^n \to \mathbb{C}$, define

$$\|f\|_{U^2} := \left(\mathbb{E}_{x,y,y'\in\mathbb{F}_p^n} f(x)\overline{f(x + y)f(x + y')}f(x + y + y')\right)^{1/4}.$$

(a) Show that the expectation here is always a nonnegative real number, so that the preceding expression is well defined. Also, show that $\|f\|_{U^2} \geq |\mathbb{E}f|$.

(b) (Gowers Cauchy–Schwarz) For $f_1, f_2, f_3, f_4 : \mathbb{F}_p^n \to \mathbb{C}$, let

$$\langle f_1, f_2, f_3, f_4\rangle = \mathbb{E}_{x,y,y'\in\mathbb{F}_p^n} f_1(x)\overline{f_2(x + y)f_3(x + y')}f_4(x + y + y').$$

Prove that

$$|\langle f_1, f_2, f_3, f_4\rangle| \leq \|f_1\|_{U^2}\, \|f_2\|_{U^2}\, \|f_3\|_{U^2}\, \|f_4\|_{U^2}.$$

(c) (Triangle inequality) Show that

$$\|f + g\|_{U^2} \leq \|f\|_{U^2} + \|g\|_{U^2}.$$

Conclude that $\|\ \|_{U^2}$ is a norm.

Hint: Note that $\langle f_1, f_2, f_3, f_4\rangle$ is multilinear. Apply (b).

(d) (Relation with Fourier) Show that

$$\|f\|_{U^2} = \|\widehat{f}\|_{\ell^4}.$$

Furthermore, deduce that if $\|f\|_\infty \leq 1$, then

$$\|\widehat{f}\|_\infty \leq \|f\|_{U^2} \leq \|\widehat{f}\|_\infty^{1/2}.$$

(The second inequality gives a so-called "inverse theorem" for the U^2 norm: if $\|f\|_{U^2} \geq \delta$, then $|\widehat{f}(r)| \geq \delta^2$ for some $r \in \mathbb{F}_p^n$. Informally, if f is not U^2-uniform, then f correlates with some exponential phase function of the form $x \mapsto \omega^{r\cdot x}$.)

The inadequacy of Fourier analysis toward understanding 4-APs is remedied by the U^3 norm, which is significantly more mysterious than the U^2 norm. Some easier properties of the U^3 norm are given in the following exercise. Understanding properties of functions with large U^3 norm (known as the inverse problem) lies at the heart of **quadratic Fourier analysis**, which we do not discuss in this book (see Further Reading). The structure of set addition, which is the topic of the next chapter, plays a central role in this theory.

Exercise 6.2.14 (Gowers U^3 uniformity norm). Let $f\colon \mathbb{F}_p^n \to \mathbb{C}$. Define

$$\|f\|_{U^3} := \Big(\mathbb{E}_{x,y_1,y_2,y_3} f(x)\overline{f(x+y_1)f(x+y_2)f(x+y_3)} \cdots \\ \cdot f(x+y_1+y_2)f(x+y_1+y_3)f(x+y_2+y_3)\overline{f(x+y_1+y_2+y_3)}\Big)^{1/8}.$$

Alternatively, for each $y \in \mathbb{F}_p^n$, define the multiplicative finite difference $\Delta_y f\colon \mathbb{F}_p^n \to \mathbb{C}$ by $\Delta_y f(x) := f(x)\overline{f(x+y)}$. We can rewrite the above expression in terms of the U^2 uniformity norm from Exercise 6.2.13 as

$$\|f\|_{U^3}^8 = \mathbb{E}_{y\in\mathbb{F}_p^n} \left\|\Delta_y f\right\|_{U^2}^4.$$

(a) (Monotonicity) Verify that the above two definitions for $\|f\|_{U^3}$ coincide and give well-defined nonnegative real numbers. Also, show that

$$\|f\|_{U^2} \le \|f\|_{U^3}.$$

(b) (Separation of norms) Let p be odd and $f\colon \mathbb{F}_p^n \to \mathbb{C}$ be defined by $f(x) = e^{2\pi i x\cdot x/p}$. Prove that $\|f\|_{U^3} = 1$ and $\|f\|_{U^2} = p^{-n/4}$.

(c) (Triangle inequality) Prove that

$$\|f+g\|_{U^3} \le \|f\|_{U^3} + \|g\|_{U^3}.$$

Conclude that $\|\ \|_{U^3}$ is a norm.

(d) (U^3 norm controls 4-APs) Let $p \ge 5$ be a prime, and $f_1, f_2, f_3, f_4\colon \mathbb{F}_p^n \to \mathbb{C}$ all taking values in the unit disk. We write

$$\Lambda(f_1,f_2,f_3,f_4) := \mathbb{E}_{x,y\in\mathbb{F}_p^n} f_1(x)f_2(x+y)f_3(x+2y)f_4(x+3y).$$

Prove that

$$|\Lambda(f_1,f_2,f_3,f_4)| \le \min_s \|f_s\|_{U^3}.$$

Furthermore, deduce that if $f,g\colon \mathbb{F}_p^n \to [0,1]$, then

$$|\Lambda(f,f,f,f) - \Lambda(g,g,g,g)| \le 4\,\|f-g\|_{U^3}.$$

Hint: Re-parametrize as in Section 2.10 and then repeatedly apply Cauchy–Schwarz.

6.3 Fourier Analysis in the Integers

Now we review the basic notions of Fourier analysis on the integers, also known as **Fourier series**. In the next section, we adapt the proof of Roth's theorem from $\mathbb{F}_3^n$ to $\mathbb{Z}$.

Here $\mathbb{R}/\mathbb{Z}$ is the set of reals mod 1. A function $f\colon \mathbb{R}/\mathbb{Z} \to \mathbb{C}$ is the same as a function $f\colon \mathbb{R} \to \mathbb{C}$ that is periodic mod 1 (i.e., $f(x+1) = f(x)$ for all $x \in \mathbb{R}$).

Definition 6.3.1 (Fourier transform in $\mathbb{Z}$)

Given a finitely supported $f\colon \mathbb{Z} \to \mathbb{C}$, define $\widehat{f}\colon \mathbb{R}/\mathbb{Z} \to \mathbb{C}$ by setting, for all $\theta \in \mathbb{R}$,

$$\boldsymbol{\widehat{f}(\theta)} := \sum_{x \in \mathbb{Z}} f(x) e(-x\theta),$$

where

$$\boldsymbol{e(t)} := \exp(2\pi i t), \qquad t \in \mathbb{R}.$$

Note that $\widehat{f}(\theta) = \widehat{f}(\theta + n)$ for all integers n. So $\widehat{f}\colon \mathbb{R}/\mathbb{Z} \to \mathbb{C}$ is well defined.

The various identities in Section 6.1 have counterparts stated in what follows. We leave the proofs as exercises for the reader.

Theorem 6.3.2 (Fourier inversion formula)

Given a finitely supported $f\colon \mathbb{Z} \to \mathbb{C}$, for any $x \in \mathbb{Z}$,

$$f(x) = \int_0^1 \widehat{f}(\theta) e(x\theta)\, d\theta.$$

Theorem 6.3.3 (Parseval / Plancherel)

Given finitely supported $f, g\colon \mathbb{Z} \to \mathbb{C}$,

$$\sum_{x \in \mathbb{Z}} \overline{f(x)} g(x) = \int_0^1 \overline{\widehat{f}(\theta)} \widehat{g}(\theta)\, d\theta.$$

In particular, as a special case ($f = g$),

$$\sum_{x \in \mathbb{Z}} |f(x)|^2 = \int_0^1 |\widehat{f}(\theta)|^2\, d\theta.$$

Note the normalization conventions: we sum in the physical space $\mathbb{Z}$ (there is no sensible way to average in $\mathbb{Z}$) and average in the frequency space $\mathbb{R}/\mathbb{Z}$.

Definition 6.3.4 (Convolution)

Given finitely supported $f, g\colon \mathbb{Z} \to \mathbb{C}$, define $f * g\colon \mathbb{Z} \to \mathbb{C}$ by

$$\boldsymbol{(f * g)(x)} := \sum_{y \in \mathbb{Z}} f(y) g(x - y).$$

Theorem 6.3.5 (Convolution identity)
Given finitely supported $f, g\colon \mathbb{Z} \to \mathbb{C}$, for any $\theta \in \mathbb{R}/\mathbb{Z}$,

$$\widehat{f * g}(\theta) = \widehat{f}(\theta)\widehat{g}(\theta).$$

Given finitely supported $f, g, h\colon \mathbb{Z} \to \mathbb{C}$, define

$$\boldsymbol{\Lambda(f, g, h)} := \sum_{x,y\in\mathbb{Z}} f(x)g(x+y)h(x+2y)$$

and

$$\boldsymbol{\Lambda_3(f)} := \Lambda(f, f, f).$$

Then for any finite set A of integers,

$$\Lambda_3(A) = |\{(x, y)\colon x, x+y, x+2y \in A\}|$$

counts the number of 3-APs in A, where each nontrivial 3-AP is counted twice, forward and backward, and each trivial 3-AP is counted once.

Proposition 6.3.6 (Fourier and 3-AP)
Given finitely supported $f, g, h\colon \mathbb{Z} \to \mathbb{C}$,

$$\Lambda(f, g, h) = \int_0^1 \widehat{f}(\theta)\widehat{g}(-2\theta)\widehat{h}(\theta)\,d\theta.$$

Exercise 6.3.7. Prove all the identities above.

Exercise 6.3.8 (Counting solutions to a single linear equation). Let $c_1, \ldots, c_k \in \mathbb{Z}$. Let $A \subseteq \mathbb{Z}$ be a finite set. Show that

$$|\{(a_1, \ldots, a_k) \in A^k\colon c_1a_1 + \cdots + c_ka_k = 0\}| = \int_0^1 \widehat{1_A}(c_1t)\widehat{1_A}(c_2t)\cdots\widehat{1_A}(c_kt)\,dt.$$

Exercise 6.3.9. Show that if a finite set A of integers contains $\beta\,|A|^2$ solutions $(a, b, c) \in A^3$ to $a + 2b = 3c$, then it contains at least $\beta^2\,|A|^3$ solutions $(a, b, c, d) \in A^4$ to $a + b = c + d$.

6.4 Roth's Theorem in the Integers

In Section 6.2 we saw a Fourier analytic proof of Roth's theorem in $\mathbb{F}_3^n$. In this section, we adapt the proof to the integers and obtain the following result. This is Roth's original proof (1953).

Theorem 6.4.1 (Roth's theorem)
Every 3-AP-free subset of $[N] = \{1, \ldots, N\}$ has size $O(N/\log\log N)$.

The proof of Roth's theorem in $\mathbb{F}_3^n$ proceeded by density increment when restricting to subspaces. An important difference between $\mathbb{F}_3^n$ and $\mathbb{Z}$ is that $\mathbb{Z}$ has no subspaces (more on this later). Instead, we will proceed in $\mathbb{Z}$ by restricting to *subprogressions*. In this section, by a ***progression*** we mean an arithmetic progression.

We have the following analoguc of Lemma 6.2.4. It says that if f and g are "Fourier-close," then they have similar 3-AP counts. We write

$$\|\widehat{f}\|_\infty := \sup_\theta |\widehat{f}(\theta)| \qquad \text{and} \qquad \|f\|_{\ell^2} := \left(\sum_{x\in\mathbb{Z}} |f(x)|^2\right)^{1/2}.$$

Proposition 6.4.2 (3-AP counting lemma)
Let $f, g\colon \mathbb{Z} \to \mathbb{C}$ be finitely supported functions. Then

$$|\Lambda_3(f) - \Lambda_3(g)| \le 3\|\widehat{f-g}\|_\infty \max\left\{\|f\|_{\ell^2}^2, \|g\|_{\ell^2}^2\right\}.$$

Proof. We have

$$\Lambda_3(f) - \Lambda_3(g) = \Lambda(f-g, f, f) + \Lambda(g, f-g, f) + \Lambda(g, g, f-g).$$

Let us bound the first term on the right-hand side. We have

$$\begin{aligned}
&|\Lambda(f-g,f,f)| \\
&= \left|\int_0^1 \widehat{(f-g)}(\theta)\widehat{f}(-2\theta)\widehat{f}(\theta)\,d\theta\right| && \text{[Prop. 6.3.6]} \\
&\le \|\widehat{f-g}\|_\infty \left|\int_0^1 \widehat{f}(-2\theta)\widehat{f}(\theta)\,d\theta\right| && \text{[Triangle ineq.]} \\
&\le \|\widehat{f-g}\|_\infty \left(\int_0^1 \left|\widehat{f}(-2\theta)\right|^2 d\theta\right)^{1/2} \left(\int_0^1 \left|\widehat{f}(\theta)\right|^2 d\theta\right)^{1/2} && \text{[Cauchy–Schwarz]} \\
&\le \|\widehat{f-g}\|_\infty \|f\|_{\ell^2}^2. && \text{[Parseval]}
\end{aligned}$$

By similar arguments, we have

$$|\Lambda(g, f-g, f)| \le \|\widehat{f-g}\|_\infty \|f\|_{\ell^2} \|g\|_{\ell^2}$$

and

$$|\Lambda(g, g, f-g)| \le \|\widehat{f-g}\|_\infty \|g\|_{\ell^2}^2.$$

Combining with the first sum gives the result. □

Now we prove Roth's theorem by following the same steps as in Section 6.2 for the finite field setting.

Step 1. A 3-AP-free set has a large Fourier coefficient.

Instead of directly studying the Fourier coefficients of 1_A (which is not a good idea since $\widehat{1_A}(\theta) \approx |A|$ is always large whenever $\theta \approx 0$), we apply a useful and standard trick and study the Fourier coefficients of the de-meaned function

$$1_A - \alpha 1_{[N]}.$$

This function has sum zero, and so its Fourier transform is zero at zero, which allows us to focus on the interesting values away from zero. Subtracting by $\alpha 1_{[N]}$ here has the same effect as considering $\widehat{1_A}(r)$ only for nonzero r in the finite field model.

Lemma 6.4.3 (3-AP-free implies a large Fourier coefficient)
Let $A \subseteq [N]$ be a 3-AP free set with $|A| = \alpha N$. If $N \geq 5\alpha^{-2}$, then there exists $\theta \in \mathbb{R}/\mathbb{Z}$ satisfying

$$\left|\sum_{x=1}^{N}(1_A - \alpha)(x)e(\theta x)\right| \geq \frac{\alpha^2}{10}N.$$

Proof. Since A is 3-AP-free, the quantity $1_A(x)1_A(x+y)1_A(x+2y)$ is nonzero only for trivial 3-APs. (Here trivial means $y = 0$.) Thus,

$$\Lambda_3(1_A) = |A| = \alpha N.$$

On the other hand, a 3-AP in $[N]$ can be counted by counting pairs of integers with the same parity to form the first and third elements of the 3-AP, yielding

$$\Lambda_3(1_{[N]}) = \lfloor N/2 \rfloor^2 + \lceil N/2 \rceil^2 \geq N^2/2.$$

Now apply the counting lemma (Proposition 6.4.2) to $f = 1_A$ and $g = \alpha 1_{[N]}$. We have $\|1_A\|_{\ell^2}^2 = |A| = \alpha N$ and $\|\alpha 1_{[N]}\|_{\ell^2}^2 = \alpha^2 N$. So,

$$\frac{\alpha^3 N^2}{2} - \alpha N \leq \alpha^3 \Lambda_3(1_{[N]}) - \Lambda_3(1_A) \leq 3\alpha N \left\|(1_A - \alpha 1_{[N]})^\wedge\right\|_\infty.$$

Thus, using $N \geq 5/\alpha^2$ in the final step in what follows,

$$\left\|(1_A - \alpha 1_{[N]})^\wedge\right\|_\infty \geq \frac{\frac{1}{2}\alpha^3 N^2 - \alpha N}{3\alpha N} = \frac{1}{6}\alpha^2 N - \frac{1}{3} \geq \frac{1}{10}\alpha^2 N.$$

Therefore, there exists some $\theta \in \mathbb{R}$ with

$$\left|\sum_{x=1}^{N}(1_A - \alpha)(x)e(\theta x)\right| = (1_A - \alpha 1_{[N]})^\wedge(\theta) \geq \frac{1}{10}\alpha^2 N. \qquad \square$$

Step 2. A large Fourier coefficient implies density increment on a subprogression.

In the finite field model, if $\widehat{1_A}(r)$ is large for some $r \in \mathbb{F}_3^n \setminus \{0\}$, then we obtain a density increment by restricting A to some coset of the hyperplane $r^\perp$.

How can we adapt this argument in the integers?

In the finite field model, we used that the Fourier character $\gamma_r(x) = \omega^{r \cdot x}$ is constant on each coset of the hyperplane $r^\perp \subseteq \mathbb{F}_3^n$. In the integer setting, we want to partition $[N]$ into subprogressions such that the character $\mathbb{Z} \to \mathbb{C}\colon x \mapsto e(x\theta)$ is roughly constant on each subprogression. As a simple example, assume that θ is a rational a/b for some fairly small b. Then, $x \mapsto e(x\theta)$ is constant on arithmetic progressions with common difference b. Thus, we could partition $[N]$ into arithmetic progressions with common difference b. This is useful as long as b is not too large. On the other hand, if b is too large, or if θ is irrational, then we would want to approximate θ as a rational number with small denominator.

We write

$$\|\theta\|_{\mathbb{R}/\mathbb{Z}} := \text{distance from } \theta \text{ to the nearest integer}.$$

Lemma 6.4.4 (Dirichlet's lemma)

Let $\theta \in \mathbb{R}$ and $0 < \delta < 1$. Then there exists a positive integer $d \leq 1/\delta$ such that $\|d\theta\|_{\mathbb{R}/\mathbb{Z}} \leq \delta$.

Proof. Let $m = \lfloor 1/\delta \rfloor$. By the pigeonhole principle, among the $m + 1$ numbers $0, \theta, \cdots, m\theta$, we can find $0 \leq i < j \leq m$ such that the fractional parts of $i\theta$ and $j\theta$ differ by at most δ. Set $d = |i - j|$. Then, $\|d\theta\|_{\mathbb{R}/\mathbb{Z}} \leq \delta$, as desired. □

Given θ, we now partition $[N]$ into subprogressions with roughly constant $e(x\theta)$ inside each progression. The constants appearing in rest of this argument are mostly unimportant.

Lemma 6.4.5 (Partition into progression level sets)

Let $0 < \eta < 1$ and $\theta \in \mathbb{R}$. Suppose $N \geq (4\pi/\eta)^6$. Then one can partition $[N]$ into subprogressions P_i, each with length

$$N^{1/3} \leq |P_i| \leq 2N^{1/3},$$

such that

$$\sup_{x,y \in P_i} |e(x\theta) - e(y\theta)| < \eta, \quad \text{for each } i.$$

Proof. By Lemma 6.4.4, there is a positive integer $d < \sqrt{N}$ such that $\|d\theta\|_{\mathbb{R}/\mathbb{Z}} \leq 1/\sqrt{N}$. Partition $[N]$ greedily into progressions with common difference d of lengths between $N^{1/3}$ and $2N^{1/3}$. Then, for two elements x, y within the same progression P_i, we have

$$|e(x\theta) - e(y\theta)| \leq |P_i|\,|e(d\theta) - 1| \leq 2N^{1/3} \cdot 2\pi \cdot N^{-1/2} \leq \eta.$$

Here we use the inequality $|e(d\theta) - 1| \leq 2\pi \, \|d\theta\|_{\mathbb{R}/\mathbb{Z}}$ from the fact that the length of a chord on a circle is at most the length of the corresponding arc. □

We can now apply this lemma to obtain a density increment.

Lemma 6.4.6 (3-AP-free implies density increment)

Let $A \subseteq [N]$ be 3-AP-free, with $|A| = \alpha N$ and $N \geq (16/\alpha)^{12}$. Then there exists a subprogression $P \subseteq [N]$ with $|P| \geq N^{1/3}$ and $|A \cap P| \geq (\alpha + \alpha^2/40)\,|P|$.

Proof. By Lemma 6.4.3, there exists θ satisfying

$$\left|\sum_{x=1}^{N}(1_A - \alpha)(x)e(x\theta)\right| \geq \frac{\alpha^2}{10}N.$$

Next, apply Lemma 6.4.5 with $\eta = \alpha^2/20$ (the hypothesis $N \geq (4\pi/\eta)^6$ is satisfied since $(16/\alpha)^{12} \geq (80\pi/\alpha^2)^6 = (4\pi/\eta)^6$) to obtain a partition $P_1, \ldots, P_k$ of $[N]$ satisfying $N^{1/3} \leq |P_i| \leq 2N^{1/3}$ and

$$|e(x\theta) - e(y\theta)| \leq \frac{\alpha^2}{20} \quad \text{for all } i \text{ and } x, y \in P_i.$$

So on each P_i,

$$\left|\sum_{x \in P_i}(1_A - \alpha)(x)e(x\theta)\right| \leq \left|\sum_{x \in P_i}(1_A - \alpha)(x)\right| + \frac{\alpha^2}{20}|P_i|.$$

Thus,

$$\begin{aligned}
\frac{\alpha^2}{10}N &\le \left|\sum_{x=1}^{N}(1_A-\alpha)(x)e(x\theta)\right| \\
&\le \sum_{i=1}^{k}\left|\sum_{x\in P_i}(1_A-\alpha)(x)e(x\theta)\right| \\
&\le \sum_{i=1}^{k}\left(\left|\sum_{x\in P_i}(1_A-\alpha)(x)\right| + \frac{\alpha^2}{20}|P_i|\right) \\
&= \sum_{i=1}^{k}\left|\sum_{x\in P_i}(1_A-\alpha)(x)\right| + \frac{\alpha^2}{20}N.
\end{aligned}$$

Thus,

$$\frac{\alpha^2}{20}N \le \sum_{i=1}^{k}\left|\sum_{x\in P_i}(1_A-\alpha)(x)\right|$$

and hence

$$\frac{\alpha^2}{20}\sum_{i=1}^{k}|P_i| \le \sum_{i=1}^{k}\big||A\cap P_i| - \alpha|P_i|\big|.$$

We want to show that there exists some P_i such that A has a density increment when restricted to P_i. The following trick is convenient. Note that

$$\begin{aligned}
\frac{\alpha^2}{20}\sum_{i=1}^{k}|P_i| &\le \sum_{i=1}^{k}\big||A\cap P_i| - \alpha|P_i|\big| \\
&= \sum_{i=1}^{k}\left(\big||A\cap P_i| - \alpha|P_i|\big| + (|A\cap P_i| - \alpha|P_i|)\right),
\end{aligned}$$

as the newly added terms in the final step sum to zero. Thus, there exists an i such that

$$\frac{\alpha^2}{20}|P_i| \le \big||A\cap P_i| - \alpha|P_i|\big| + (|A\cap P_i| - \alpha|P_i|).$$

Since $|t| + t$ is $2t$ for $t > 0$ and 0 for $t \le 0$, we deduce

$$\frac{\alpha^2}{20}|P_i| \le 2(|A\cap P_i| - \alpha|P_i|),$$

which yields

$$|A\cap P_i| \ge \left(\alpha + \frac{\alpha^2}{40}\right)|P_i|. \qquad \square$$

By translation and rescaling, we can identify P with $[N']$ with $N' = |P|$. Then $A\cap P$ becomes a subset $A' \subseteq [N']$. Note that A' is 3-AP-free. (Here we are invoking the important fact that 3-APs are translation and dilation invariant.) We can now iterate the argument. (Think about where the argument goes wrong for patterns such as $\{x, y, x+y\}$ and $\{x, x+y, x+y^2\}$.)

Step 3. Iterate the density increment.

This step is nearly identical to the proof in the finite field model. Start with $\alpha_0 = \alpha$ and $N_0 = N$. After i iterations, we arrive at a subprogression of length N_i where A has density α_i. As long as $N_i \geq (16/\alpha_i)^{12}$, we can apply apply Lemma 6.4.6 to pass down to a subprogression with

$$N_{i+1} \geq N_i^{1/3} \quad \text{and} \quad \alpha_{i+1} \geq \alpha_i + \alpha_i^2/40.$$

We double α_i from α_0 after $\leq \lceil 40/\alpha \rceil$ iterations. Once the density reaches at least 2α, the next doubling takes $\leq \lceil 20/\alpha \rceil$ iterations, and so on. In general, the kth doubling requires $\leq \lceil 40 \cdot 2^{-k}/\alpha \rceil$ iterations. There are at most $\log_2(1/\alpha)$ doublings since the density is always at most 1. Summing up, the total number of iterations is

$$m \leq \sum_{i=1}^{\log_2(1/\alpha)} \left\lceil 40 \cdot 2^{-k}/\alpha \right\rceil = O(1/\alpha).$$

When the process terminates, by Lemma 6.4.6,

$$N^{1/3^m} \leq N_m < (16/\alpha_i)^{12} \leq (16/\alpha)^{12}.$$

Rearranging gives

$$N \leq (16/\alpha)^{12 \cdot 3^m} \leq (16/\alpha)^{e^{O(1/\alpha)}}.$$

Therefore,

$$\frac{|A|}{N} = \alpha = O\left(\frac{1}{\log \log N}\right).$$

This completes the proof of Roth's theorem (Theorem 6.4.1). □

We saw that the proofs in $\mathbb{F}_3^n$ and $\mathbb{Z}$ have largely the same set of ideas, but the proof in $\mathbb{Z}$ is somewhat more technically involved. The finite field model is often a good sandbox to try out Fourier analytic ideas.

***Remark* 6.4.7** (Bohr sets). Let us compare the results in $\mathbb{F}_3^n$ and $[N]$. Write $N = 3^n$ for the size of the ambient space in both cases, for comparison. We obtained an upper bound of $O(N/\log N)$ for 3-AP-free sets in $\mathbb{F}_3^n$ and $O(N/\log \log N)$ in $[N] \subseteq \mathbb{Z}$. Where does the difference in quantitative bounds stem from?

In the density increment step for $\mathbb{F}_3^n$, at each step we pass down to a subset that has size a constant factor (namely 1/3) of the original one. However, in $[N]$, each iteration gives us a subprogression that has size equal to the cube root of the previous subprogression. The extra log for Roth's theorem in the integers comes from this rapid reduction in the sizes of the subprogressions.

Can we do better? Perhaps by passing down to subsets of $[N]$ that look more like subspaces? Indeed, this is possible. Bourgain (1999) used ***Bohr sets*** to prove an improved bound of $N/(\log N)^{1/2+o(1)}$ on Roth's theorem. Given $\theta_1, \ldots, \theta_k$, and some $\varepsilon > 0$, a Bohr set has the form

$$\left\{x \in [N] \colon \|x\theta_j\|_{\mathbb{R}/\mathbb{Z}} \leq \varepsilon \text{ for each } j = 1, \ldots, k\right\}.$$

To see why this is analogous to subspaces, note that we can define a subspace of $\mathbb{F}_3^n$ as a set of the following form,

$$\left\{x \in \mathbb{F}_3^n : r_j \cdot x = 0 \text{ for each } j = 1, \ldots, k\right\},$$

where $r_1, \ldots, r_k \in \mathbb{F}_3^n \setminus \{0\}$. Bohr sets are used widely in additive combinatorics, and in nearly all subsequent work on Roth's theorem in the integers, including the proof of the current best bound $N/(\log N)^{1+c}$ for some constant $c > 0$ (Bloom and Sisask 2020).

We will see Bohr sets again in the proof of Freiman's theorem in Chapter 7.

The next exercise is analogous to Exercise 6.2.11, which was in $\mathbb{F}_5^n$.

Exercise 6.4.8* (Fourier uniformity does not control 4-AP counts). Fix $0 < \alpha < 1$. Let N be a prime. Let

$$A = \left\{x \in [N] : x^2 \bmod N < \alpha N\right\}.$$

Viewing $A \subseteq \mathbb{Z}/N\mathbb{Z}$, prove that, as $N \to \infty$ with fixed α,

(a) $|A| = (\alpha + o(1))N$ and $\max_{r \neq 0} |\widehat{1_A}(r)| = o(1)$;

(b) $|(x, y) \in \mathbb{Z}/N\mathbb{Z} : x, x + y, x + 2y, x + 3y \in A| \neq (\alpha^4 + o(1))N^2$.

6.5 Polynomial Method

An important breakthrough of Croot, Lev, and Pach (2017) showed how to apply the **polynomial method** to Roth-type problems in the finite field model. Their method quickly found many applications. Less than a week after the Croot, Lev, and Pach paper was made public, Ellenberg and Gijswijt (2017) adapted their argument to prove the following bound on the cap set problem. The discovery came as quite a shock to the community, especially as the proof is so short.

Theorem 6.5.1 (Roth's theorem in $\mathbb{F}_3^n$: power-saving upper bound)
Every 3-AP-free subset of $\mathbb{F}_3^n$ has size $O(2.76^n)$.

The presentation of the following proof is due to Tao (2016).

Recall from linear algebra the usual ***rank*** of a matrix. Here we can view an $|A| \times |A|$ matrix over the field $\mathbb{F}$ as a function $F : A \times A \to \mathbb{F}$. A function F is said to have rank 1 if $F(x, y) = f(x)g(y)$ for some nonzero functions $f, g : A \to \mathbb{F}$. More generally, the rank of F is the minimum k so that F can be written as a sum of k rank 1 functions.

More generally, for other notions of rank, we can first define the set of rank 1 functions, and then define the rank of F to be the minimum k so that F can be written as a sum of k rank 1 functions.

Whereas a function $A \times A \to \mathbb{F}$ corresponds to a matrix, a function $A \times A \times A \to \mathbb{F}$ corresponds to a 3-tensor. There is a notion of ***tensor rank***, where the rank 1 functions are those of the form $F(x, y, z) = f(x)g(y)h(z)$. This is a standard and important notion (which comes with a lot of mystery), but it is not the one that we shall use.

Definition 6.5.2 (Slice rank)
A function $F: A \times A \times A \to \mathbb{F}$ is said to have ***slice rank 1*** if it can be written as

$$f(x)g(y,z), \quad f(y)g(x,z), \quad \text{or} \quad f(z)g(x,y),$$

for some nonzero functions $f: A \to \mathbb{F}$ and $g: A \times A \to \mathbb{F}$.

The ***slice rank*** of a function $F: A \times A \times A \to \mathbb{F}$ is the minimum k so that F can be written as a sum of k slice rank 1 functions.

Here is an easy fact about the slice rank.

Lemma 6.5.3 (Trivial upper bound for slice rank)
Every function $F: A \times A \times A \to \mathbb{F}$ has slice rank at most $|A|$.

Proof. Let F_a be the restriction of F to the "slice" $\{(x,y,z) \in A \times A \times A : x = a\}$; that is,

$$F_a(x,y,z) = \begin{cases} F(x,y,z) & \text{if } x = a, \\ 0 & \text{if } x \neq a. \end{cases}$$

F_a

Then F_a has slice rank ≤ 1 since $F_a(x,y,z) = \delta_a(x)F(a,y,z)$, where δ_a denotes the function taking value 1 at a and 0 elsewhere. Thus, $F = \sum_{a \in A} F_a$ has slice rank at most $|A|$. □

For the next lemma, we need the following fact from linear algebra.

Lemma 6.5.4 (Vector with large support)
Every k-dimensional subspace of an n-dimensional vector space (over any field) contains a point with at least k nonzero coordinates.

Proof. Form a $k \times n$ matrix M whose rows form a basis of this k-dimensional subspace W. Then M has rank k. So it has some invertible $k \times k$ submatrix with columns $S \subseteq [n]$ with $|S| = k$. Then, for every $z \in \mathbb{F}^S$, there is some linear combination of the rows whose coordinates on S are identical to those of z. In particular, there is some vector in the k-dimensional subspace W whose S-coordinates are all nonzero. □

A diagonal matrix with nonzero diagonal entries has full rank. We show that a similar statement holds true for the slice rank.

Lemma 6.5.5 (Slice rank of a diagonal)
Suppose $F: A \times A \times A \to \mathbb{F}$ satisfies $F(x,y,z) \neq 0$ if and only if $x = y = z$. Then F has slice rank $|A|$.

Proof. From Lemma 6.5.3, we already know that the slice rank of F is $\leq |A|$. It remains to prove that the slice rank of F is $\geq |A|$.

Suppose $F(x,y,z)$ can be written as a sum of functions of the form

$$f(x)g(y,z), \quad f(y)g(x,z), \quad \text{and} \quad f(z)g(x,y),$$

with m_1 summands of the first type, m_2 of the second type, and m_3 of the third type. By Lemma 6.5.4, there is some function $h: A \to \mathbb{F}$ that is orthogonal to all the f's from the third type of summands (i.e., $\sum_{x\in A} f(x)h(x) = 0$), and such that $|\operatorname{supp} h| \geq |A| - m_3$. Let

$$G(x,y) = \sum_{z\in A} F(x,y,z)h(z).$$

Only summands of the first two types remain. Each summand of the first type turns into a rank 1 function (in the matrix sense of the rank),

$$(x,y) \mapsto \sum_z f(x)g(y,z)h(z) = f(x)\widetilde{g}(y),$$

for some new function $\widetilde{g}: A \to \mathbb{F}$. Similarly with functions of the second type. So, G (viewed as an $|A| \times |A|$ matrix) has rank $\leq m_1 + m_2$. On the other hand,

$$G(x,y) = \begin{cases} F(x,x,x)h(x) & \text{if } x = y, \\ 0 & \text{if } x \neq y. \end{cases}$$

This G has rank $|\operatorname{supp} h| \geq |A| - m_3$. Combining, we get

$$|A| - m_3 \leq \operatorname{rank} G \leq m_1 + m_2.$$

So, $m_1 + m_2 + m_3 \geq |A|$. This shows that the slice rank of F is $\geq |A|$. □

Now we prove an upper bound on the slice rank by invoking the magical powers of polynomials.

Lemma 6.5.6 (Upper bound on the slice rank of $1_{x+y+z=0}$)
Define $F: A \times A \times A \to \mathbb{F}_3$ by

$$F(x,y,z) = \begin{cases} 1 & \text{if } x + y + z = 0, \\ 0 & \text{otherwise.} \end{cases}$$

Then the slice rank of F is at most

$$3 \sum_{\substack{a,b,c\geq 0 \\ a+b+c=n \\ b+2c\leq 2n/3}} \frac{n!}{a!b!c!}.$$

Proof. In $\mathbb{F}_3$, one has

$$1 - x^2 = \begin{cases} 1 & \text{if } x = 0, \\ 0 & \text{if } x \neq 0. \end{cases}$$

So, writing $x = (x_1, \ldots, x_n)$, $y = (y_1, \ldots, y_n)$, and $z = (z_1, \ldots, z_n)$, we have

$$F(x,y,z) = \prod_{i=1}^{n} (1 - (x_i + y_i + z_i)^2). \tag{6.5}$$

If we expand the right-hand side, we obtain a polynomial in $3n$ variables with degree $2n$. This is a sum of monomials, each of the form

$$x_1^{i_1} \cdots x_n^{i_n} y_1^{j_1} \cdots y_n^{j_n} z_1^{k_1} \cdots z_n^{k_n},$$

where $i_1, i_2, \ldots, i_n, j_1, \ldots, j_n, k_1, \ldots, k_n \in \{0, 1, 2\}$. For each term, by the pigeonhole principle, at least one of $i_1 + \cdots + i_n, j_1 + \cdots + j_n, k_1 + \cdots + k_n$ is at most $2n/3$. So we can split these summands into three sets:

$$\begin{aligned}\prod_{i=1}^{n}(1 - (x_i + y_i + z_i)^2) = &\sum_{i_1+\cdots+i_n \le \frac{2n}{3}} x_1^{i_1} \cdots x_n^{i_n} f_{i_1,\ldots,i_n}(y, z) \\ &+ \sum_{j_1+\cdots+j_n \le \frac{2n}{3}} y_1^{j_1} \cdots y_n^{j_n} g_{j_1,\ldots,j_n}(x, z) \\ &+ \sum_{k_1+\cdots+k_n \le \frac{2n}{3}} z_1^{k_1} \cdots z_n^{k_n} h_{k_1,\ldots,k_n}(x, y).\end{aligned}$$

Each summand has slice rank at most 1. The number of summands in the first sum is precisely the number of triples of nonnegative integers a, b, c with $a + b + c = n$ and $b + 2c \le 2n/3$. (a, b, c correspond to the numbers of i_*'s that are equal to $0, 1, 2$ respectively.) The lemma then follows. □

Here is a standard estimate. The proof is similar to that of the Chernoff bound.

Lemma 6.5.7 (A trinomial coefficient estimate)
For every positive integer n,

$$\sum_{\substack{a,b,c \ge 0 \\ a+b+c=n \\ b+2c \le 2n/3}} \frac{n!}{a!b!c!} \le 2.76^n.$$

Proof. Let $x \in [0, 1]$. The sum equals to the coefficients of all the monomials x^k with $k \le 2n/3$ in the expansion of $(1 + x + x^2)^n$. By deleting contributions x^k with $k > 2n/3$ and using $x^{2n/3} \le x^k$ whenever $k \le 2n/3$, we have

$$\sum_{\substack{a,b,c \ge 0 \\ a+b+c=n \\ b+2c \le 2n/3}} \frac{n!}{a!b!c!} \le \frac{(1 + x + x^2)^n}{x^{2n/3}}.$$

Setting $x = 0.6$ shows that the left-hand side sum is $\le (2.76)^n$. □

***Remark* 6.5.8.** Taking the optimal value $x = (\sqrt{33} - 1)/8 = 0.59307\ldots$ in the final step, we obtain $\le (2.75510\ldots)^n$. This is the true exponential asymptotics of the sum in Lemma 6.5.7 (for example, see Sanov's theorem from large deviation theory). We have no idea how close this is to the optimal bound for the cap set problem. However, quite surprisingly, such bound is tight for a variant of the cap sets known as the tricolored sum-free sets (Blasiak et al. 2017; Kleinberg et al. 2018).

Proof of Theorem 6.5.1. Let $A \subseteq \mathbb{F}_3^n$ be 3-AP-free. Define $F\colon A \times A \times A \to \mathbb{F}_3$ by

$$F(x, y, z) = \begin{cases} 1 & \text{if } x + y + z = 0, \\ 0 & \text{otherwise.} \end{cases}$$

Since A is 3-AP-free, one has $F(x, y, z) = 1$ if and only if $x = y = z \in A$. By Lemma 6.5.5, F has slice rank $|A|$. On the other hand, by Lemmas 6.5.6 and 6.5.7, F has slice rank $\leq 3(2.76)^n$. So, $|A| \leq 3(2.76)^n$. □

It is straightforward to extend the preceding proof from $\mathbb{F}_3$ to any other fixed $\mathbb{F}_p$, resulting in the following:

Theorem 6.5.9 (Roth's theorem in the finite field model)
For every odd prime p, there is some $c_p < p$ so that every 3-AP-free subset of $\mathbb{F}_p^n$ has size at most $3c_p^n$.

It remains an intriguing open problem to extend the techniques to other settings.

Open Problem 6.5.10 (Szemerédi's theorem in the finite field model)
Is there a constant $c < 5$ such that every 4-AP-free subset of $\mathbb{F}_5^n$ has size $O(c^n)$?

Open Problem 6.5.11 (Corner-free theorem in the finite field model)
Is there a constant $c < 2$ such that every corner-free subset of $\mathbb{F}_2^n \times \mathbb{F}_2^n$ has size $O(c^{2n})$? Here a corner is a configuration of the form $\{(x, y), (x + d, y), (x, y + d)\}$.

Finally, the proof technique in this section seems specific to the finite field model. It is an intriguing open problem to apply the polynomial method for Roth's theorem in the integers. Due to the Behrend example (Section 2.5), we cannot expect power-saving bounds in the integers.

Exercise 6.5.12 (Tricolor sum-free set). Let $a_1, \ldots, a_m, b_1, \ldots, b_m, c_1, \ldots, c_m \in \mathbb{F}_2^n$. Suppose that the equation $a_i + b_j + c_k = 0$ holds if and only if $i = j = k$. Show that there is some constant $c > 0$ such that $m \leq (2 - c)^n$ for all sufficiently large n.

Exercise 6.5.13 (Sunflower-free set). Three sets A, B, C form a ***sunflower*** if $A \cap B = B \cap C = A \cap C = A \cap B \cap C$. Prove that there exists some constant $c > 0$ such that if $\mathcal{F}$ is a collection of subsets of $[n]$ without a sunflower, then $|\mathcal{F}| \leq (2 - c)^n$ provided that n is sufficiently large.

6.6 Arithmetic Regularity

Here we develop an arithmetic analogue of Szemerédi's graph regularity lemma from Chapter 2. Just as the graph regularity method has powerful applications, so too does the arithmetic regularity lemma as well as the general strategy behind it.

First, we need a notion of what it means for a subset of $\mathbb{F}_p^n$ to be uniform, in a sense analogous to ε-regular pairs from the graph regularity lemma. We also saw the following notion in the Fourier analytic proof of Roth's theorem.

Definition 6.6.1 (Fourier uniformity)
We say that $A \subseteq \mathbb{F}_p^n$ is ***ε-uniform*** if $|\widehat{1_A}(r)| \leq \varepsilon$ for all $r \in \mathbb{F}_p^n \setminus \{0\}$.

The following exercises explain how Fourier uniformity is analogous to the discrepancy-type condition for ε-regular pairs in the graph regularity lemma.

Exercise 6.6.2 (Uniformity vs. discrepancy). Let $A \subseteq \mathbb{F}_p^n$ with $|A| = \alpha p^n$. We say that A satisfies **HyperplaneDISC**(η) if for every hyperplane W of $\mathbb{F}_p^n$,

$$\left| \frac{|A \cap W|}{|W|} - \alpha \right| \leq \eta.$$

(a) Prove that if A satisfies **HyperplaneDISC**(ε), then A is ε-uniform.
(b) Prove that if A is ε-uniform, then it satisfies **HyperplaneDISC**($(p-1)\varepsilon$).

Definition 6.6.3 (Fourier uniformity on affine subspaces)
For an affine subspace W of $\mathbb{F}_p^n$ (i.e., the coset of a subspace), we say that A is ***ε-uniform on W*** if $A \cap W$ is ε-uniform when viewed as a subset of W.

Here is an arithmetic analogue of Szemerédi's graph regularity lemma that we saw in Chapter 2. It is due to Green (2005a).

Theorem 6.6.4 (Arithmetic regularity lemma)
For every $\varepsilon > 0$ and prime p, there exists M so that for every $A \subseteq \mathbb{F}_p^n$, there is some subspace W of $\mathbb{F}_p^n$ with codimension at most M such that A is ε-uniform on all but at most ε-fraction of cosets of W.

The proof is very similar to the proof of the graph regularity lemma in Chapter 2. Each subspace W induces a partition of the whole space $\mathbb{F}_p^n$ into W-cosets, and we keep track of the energy (mean-squared density) of the partition. We show that if the conclusion of Theorem 6.6.4 does not hold for the current W, then we can replace W by a smaller subspace so that the energy increases significantly. Since the energy is always bounded between 0 and 1, there are at most a bounded number of iterations.

Definition 6.6.5 (Energy)
Given $A \subseteq \mathbb{F}_p^n$, and W a subspace of $\mathbb{F}_p^n$, we define the ***energy*** of W with respect to a fixed A to be

$$\boldsymbol{q_A(W)} := \mathbb{E}_{x \in \mathbb{F}_p^n} \left[\frac{|A \cap (W+x)|^2}{|W|^2} \right].$$

Given a subspace W of $\mathbb{F}_p^n$. Define $\mu_W : \mathbb{F}_p^n \to \mathbb{R}$ by

$$\boldsymbol{\mu_W} := \frac{p^n}{|W|} 1_W.$$

(One can regard μ_W as the uniform probability distribution on W; it is normalized so that $\mathbb{E}\mu_W = 1$.) Then,

$$(1_A * \mu_W)(x) = \frac{|A \cap (W+x)|}{|W|} \quad \text{for every } x \in \mathbb{F}_p^n.$$

We have (check!)

$$\widehat{\mu_W}(r) = \begin{cases} 1 & \text{if } r \in W^\perp, \\ 0 & \text{if } r \notin W^\perp. \end{cases}$$

So, by the convolution identity (Theorem 6.1.7):

$$\widehat{1_A * \mu_W}(r) = \widehat{1_A}(r)\widehat{\mu_W}(r) = \begin{cases} \widehat{1_A}(r) & \text{if } r \in W^\perp, \\ 0 & \text{if } r \notin W^\perp. \end{cases} \tag{6.6}$$

To summarize, convolving by μ_W averages 1_A along cosets of W in the physical space, and filters $W^\perp$ in the Fourier space.

Energy interacts nicely with the Fourier transform. By Parseval's identity (Theorem 6.1.3), we have

$$q_A(W) = \|1_A * \mu_W\|_2^2 = \sum_{r \in \mathbb{F}_p^n} |\widehat{1_A * \mu_W}(r)|^2 = \sum_{r \in W^\perp} |\widehat{1_A}(r)|^2. \tag{6.7}$$

The next lemma is analogous to Lemma 2.1.12. It is an easy consequence of convexity. It also directly follows from (6.7).

Lemma 6.6.6 (Energy never decreases under refinement)
Let $A \subseteq \mathbb{F}_p^n$. For subspaces $U \le W \le \mathbb{F}_p^2$, we have $q_A(U) \ge q_A(W)$. □

The next lemma is analogous to the energy boost lemma for irregular pairs in the proof of graph regularity (Lemma 2.1.13).

Lemma 6.6.7 (Local energy increment)
If $A \subseteq \mathbb{F}_p^n$ is not ε-uniform, then there is some codimension-1 subspace W with $q_A(W) > (|A|/p^n)^2 + \varepsilon^2$.

Proof. Suppose A is not ε-uniform. Then there is some $r \ne 0$ such that $|\widehat{1_A}(r)| > \varepsilon$. Let $W = r^\perp$. Then by (6.7),

$$\begin{aligned} q_A(W) &= |\widehat{1_A}(0)|^2 + |\widehat{1_A}(r)|^2 + |\widehat{1_A}(2r)|^2 + \cdots + |\widehat{1_A}((p-1)r)|^2 \\ &\ge |\widehat{1_A}(0)|^2 + |\widehat{1_A}(r)|^2 > (|A|/p^n)^2 + \varepsilon^2. \end{aligned}$$ □

By applying the preceding lemmas locally to each W-coset, we obtain the following global increment, analogous to Lemma 2.1.14:

Lemma 6.6.8 (Global energy increment)
Let $A \subseteq \mathbb{F}_p^n$. Let W be a subspace of $\mathbb{F}_p^n$. Suppose that f is not ε-uniform on $> \varepsilon$-fraction of W-cosets. Then, there is some subspace U of W with $\operatorname{codim} U - \operatorname{codim} W \le p^{\operatorname{codim} W}$ such that

$$q_A(U) > q_A(W) + \varepsilon^3.$$

Proof. By Lemma 6.6.7, for each coset W' of W on which f is not ε-uniform, we can find some $r \in \mathbb{F}_p^n \setminus W^\perp$ so that replacing W by its intersection with $r^\perp$ increases its energy on W' by more than ε^2. In other words,

$$q_{A\cap W'}(W' \cap r^\perp) > \frac{|A \cap W'|^2}{|W'|^2} + \varepsilon^2.$$

Let R be a set of such r's, one for each W-coset on which f is not ε-uniform (allowing some r's to be chosen repeatedly).

Let $U = W \cap R^\perp$. Then $\operatorname{codim} U - \operatorname{codim} W \le |R| \le |\mathbb{F}^p/W| = p^{\operatorname{codim} W}$.

Applying the monotonicity of energy (Lemma 6.6.6) on each W-coset and using the observation in the first paragraph in this proof, we see the "local" energy of U is more than that of W by $> \varepsilon^2$ on each of the $> \varepsilon$-fraction of W-cosets on which f is not ε-uniform, and is at least as great as that of W on each of the remaining W-cosets. There the energy increases by $> \varepsilon^2$ when refining from W to U. □

Proof of the arithmetic regularity lemma (Theorem 6.6.4). Starting with $W_0 = \mathbb{F}_p^n$, we construct a sequence of subspaces $W_0 \ge W_1 \ge W_2 \ge \cdots$ where each at step, unless A is ε-uniform on all but $\le \varepsilon$-fraction of W-cosets, then we apply Lemma 6.6.8 to find $W_{i+1} \le W_i$. The energy increases by $> \varepsilon^3$ at each iteration, so there are $< \varepsilon^{-3}$ iterations. We have $\operatorname{codim} W_{i+1} \le \operatorname{codim} W_i + p^{\operatorname{codim} W_i}$ at each i, so the final $W = W_m$ has codimension at most some function of p and ε. (One can check that it is an exponential tower of p's of height $O(\varepsilon^{-3})$.) This W satisfies the desired properties. □

Remark 6.6.9 (Lower bound). Recall that Gowers (1997) showed that there exist graphs whose ε-regular partition requires at least tower($\Omega(\varepsilon^{-c})$) parts (Theorem 2.1.17). There is a similar tower-type lower bound for the arithmetic regularity lemma (Green 2005a; Hosseini, Lovett, Moshkovitz, and Shapira 2016).

Remark 6.6.10 (Abelian groups). Green (2005a) also established an arithmetic regularity lemma over arbitrary finite abelian groups. Instead of subspaces, one uses Bohr sets. (See Remark 6.4.7.)

You may wish to skip ahead to Section 6.7 to see an application of the arithmetic regularity lemma.

Arithmetic Regularity Decomposition

Now let us give another arithmetic regularity result. It has the same spirit as the preceding regularity lemma, but phrased in terms of a decomposition rather than a partition. This perspective of regularity as decompositions, popularized by Tao, allows one to adapt the ideas of regularity to more general settings where we cannot neatly partition the underlying space into easily describable pieces. It is very useful and has many applications in additive combinatorics.

Theorem 6.6.11 (Arithmetic regularity decomposition)
For every sequence $\varepsilon_0 \geq \varepsilon_1 \geq \varepsilon_2 \geq \cdots > 0$, there exists M so that every $f : \mathbb{F}_p^n \to [0,1]$ can be written as

$$f = f_{\text{str}} + f_{\text{psr}} + f_{\text{sml}}$$

where

- (structured piece) $f_{\text{str}} = f * \mu_W$ for some subspace W of codimension at most M;
- (pseudorandom piece) $\|\widehat{f_{\text{psr}}}\|_\infty \leq \varepsilon_{\operatorname{codim} W}$;
- (small piece) $\|f_{\text{sml}}\|_2 \leq \varepsilon_0$.

***Remark* 6.6.12.** It is worth comparing Theorem 6.6.11 to the strong graph regularity lemma (Theorem 2.8.3). It is important that the uniformity requirement on the pseudorandom piece depends on the codim W.

In other more advanced applications, we would like f_{str} to come from some structured class of functions. For example, in higher-order Fourier analysis, f_{str} is a nilsequence.

Proof. Let $k_0 = 0$ and $k_{i+1} = \max\{k_i, \lceil \varepsilon_{k_i}^{-2} \rceil\}$ for each $i \geq 0$. Note that $k_0 \leq k_1 \leq \cdots$.

Let us label the elements $r_1, r_2, \ldots, r_{p^n}$ of $\mathbb{F}_p^n$ so that

$$|\widehat{f}(r_1)| \geq |\widehat{f}(r_2)| \geq \cdots.$$

By Parseval (Theorem 6.1.3), we have

$$\sum_{j=1}^{p^n} |\widehat{f}(r_j)|^2 = \mathbb{E} f^2 \leq 1.$$

There is some positive integer $m \leq \lceil \varepsilon_0^{-2} \rceil$ so that

$$\sum_{k_m < j \leq k_{m+1}} |\widehat{f}(r_j)|^2 \leq \varepsilon_0^2, \tag{6.8}$$

since otherwise adding up the sum over all $m \leq \lceil \varepsilon_0^{-2} \rceil$ would contradict $\sum_r |\widehat{f}(r)|^2 \leq 1$. Also, we have

$$|\widehat{f}(r_k)| \leq \frac{1}{\sqrt{k}} \quad \text{for every } k. \tag{6.9}$$

The idea now is to split

$$f(x) = \sum_{j=1}^{p^n} \widehat{f}(r_j) \omega^{r_j \cdot x}$$

into

$$f = f_{\text{str}} + f_{\text{sml}} + f_{\text{psr}}$$

according to the sizes of the Fourier coefficients. Roughly speaking, the large spectrum will go into the structured piece f_{str}, the very small spectrum will go into pseudorandom piece f_{psr}, and the remaining middle terms will form the small piece f_{sml} (which has small L^2 norm by (6.8)).

Let $W = \{r_1, \ldots, r_{k_m}\}^\perp$ and set

$$f_{\text{str}} = f * \mu_W.$$

Then, by (6.6),

$$\widehat{f_{\text{str}}}(r) = \begin{cases} \widehat{f}(r) & \text{if } r \in W^\perp, \\ 0 & \text{if } r \in W^\perp. \end{cases}$$

Let us define f_{psr} and f_{sml} via their Fourier transform (and we can recover the functions via the inverse Fourier transform). For each $j = 1, 2, \ldots, p^n$, set

$$\widehat{f_{\text{psr}}}(r_j) = \begin{cases} \widehat{f}(r_j) & \text{if } j > k_{m+1} \text{ and } r_j \notin W^\perp, \\ 0 & \text{otherwise.} \end{cases}$$

Finally, let $f_{\text{sml}} = f - f_{\text{psr}} - f_{\text{sml}}$, so that

$$\widehat{f_{\text{sml}}}(r_j) = \begin{cases} \widehat{f}(r_j) & \text{if } k_m < j \le k_{m+1} \text{ and } r_j \notin W^\perp, \\ 0 & \text{otherwise.} \end{cases}$$

Now we check that all the conditions are satisfied.

Structured piece: We have $f_{\text{str}} = f * \mu_W$ where $\operatorname{codim} W \le k_m \le k_{\lceil \varepsilon_0^{-2} \rceil}$, which is bounded as a function of the sequence $\varepsilon_0 \ge \varepsilon_1 \ge \ldots$.

Pseudorandom piece: For every $j > k_{m+1}$, we have $|\widehat{f}(r_j)| \le 1/\sqrt{k_{m+1}}$ by (6.9), which is in turn $\le \varepsilon_{k_m} \le \varepsilon_{\operatorname{codim} W}$ by the definition of k_m. It follows that $\|\widehat{f_{\text{psr}}}\| \le \varepsilon_{\operatorname{codim} W}$.

Small piece: By (6.8),

$$\|\widehat{f_{\text{sml}}}\|_2^2 \le \sum_{k_m < j \le k_{m+1}} |\widehat{f}(r_j)|^2 \le \varepsilon_0^2. \qquad \square$$

Exercise 6.6.13. Deduce Theorem 6.6.4 from Theorem 6.6.11 by using an appropriate sequence ε_i and using the same W guaranteed by Theorem 6.6.11.

***Remark* 6.6.14 (Spectral proof of the graph regularity lemma).** The proof technique of Theorem 6.6.11 can be adapted to give an alternate proof of the graph regularity lemma (along with weak and strong variants). Instead of iteratively refining partitions and tracking energy increments as we did in Chapter 2, we can first take a spectral decomposition of the adjacency matrix A of a graph:

$$A = \sum_{i=1}^{n} \lambda_i v_i v_i^\intercal,$$

where $v_1, \ldots, v_n$ is an orthonormal system of eigenvectors with eigenvalues $\lambda_1 \ge \cdots \ge \lambda_n$. Then, as in the proof of Theorem 6.6.11, we can decompose A as

$$A = A_{\text{str}} + A_{\text{psr}} + A_{\text{sml}}$$

with

$$A_{\text{str}} = \sum_{i \le k} \lambda_i v_i v_i^\intercal, \quad A_{\text{psr}} = \sum_{i > k'} \lambda_i v_i v_i^\intercal, \quad \text{and} \quad A_{\text{sml}} = \sum_{k < i \le k'} \lambda_i v_i v_i^\intercal,$$

for some appropriately chosen k and k' similar to the proof of Theorem 6.6.11.

We have

$$\sum_{i=1}^{n} \lambda_i^2 = \operatorname{tr} A^2 \le n^2.$$

So, $\lambda_i \le n/\sqrt{i}$ for each i. We can guarantee that the spectral norm of A_{psr} is small enough as a function of k and ε. Furthermore, we can guarantee that $\operatorname{tr} A_{\mathrm{sml}}^2 = \sum_{k<i\le k'} \lambda_i^2 \le \varepsilon$.

To turn A_{str} into a vertex partition, we can use the approximate level sets of the top k eigenvectors $v_1, \dots, v_k$. Some bookkeeping calculations then show that this is a regularity partition. Intuitively, A_{psr} provides us with regular pairs. Some of these regular pairs may not stay regular after adding A_{sml}, but since A_{sml} has $\le \varepsilon$ mass (in terms of L^2 norm), it destroys at most a negligible fraction of regular pairs.

See Tao (2007a, Lemma 2.11) or Tao's blog post *The Spectral Proof of the Szemerédi Regularity Lemma* (2012) for more details of the proof.

The following exercise is the arithmetic analogue of the existence of an ε-regular vertex subset in a graph (Theorem 2.1.26 and Exercise 2.1.27).

Exercise 6.6.15 (ε-uniform subspace).

(a*) Prove that, for every $0 < \varepsilon < 1/2$ and $A \subseteq \mathbb{F}_2^n$, there exists a subspace $W \subseteq \mathbb{F}_2^n$ (note that $0 \in W$) with codimension at most $\exp(C/\varepsilon)$ such that A is ε-uniform on W. Here C is some absolute constant.

(b) Let $A = \{x \in \mathbb{F}_3^n : \text{there exists } i \text{ such that } x_1 = \cdots = x_i = 0, x_{i+1} = 1\}$. Prove that A is not c-uniform on any positive dimensional subspace of $\mathbb{F}_3^n$. Here $c > 0$ is some absolute constant.

6.7 Popular Common Difference

Roth's theorem has the following qualitative strengthening. Given $A \subseteq \mathbb{F}_3^n$ with density α, there is some "popular common difference" $y \neq 0$ so that the number of 3-APs in A with common difference y is $\ge \alpha^3 - o(1)$, which is what one expects for a random A of density α. This was proved by Green (2005a) as an application of his arithmetic regularity lemma (from the previous section).

Theorem 6.7.1 (Roth's theorem with popular common difference in $\mathbb{F}_3^n$)
For all $\varepsilon > 0$, there exists $n_0 = n_0(\varepsilon)$ such that for $n \ge n_0$ and every $A \subseteq \mathbb{F}_3^n$ with $|A| = \alpha 3^n$, there exists $y \neq 0$ such that

$$\left|\{x \in \mathbb{F}_3^n : x, x+y, x+2y \in A\}\right| \ge (\alpha^3 - \varepsilon)3^n.$$

In particular, Theorem 6.7.1 implies that every 3-AP-free subset of $\mathbb{F}_3^n$ has size $o(3^n)$.

Exercise 6.7.2. Show that it is *false* that every $A \subseteq \mathbb{F}_3^n$ with $|A| = \alpha 3^n$, the number of pairs $(x, y) \in \mathbb{F}_3^n$ with $x, x+y, x+2y \in A$ is $\ge (\alpha^3 - o(1))3^{2n}$, where $o(1) \to 0$ as $n \to 0$.

We will prove Theorem 6.7.1 via the next result, which concerns the number of 3-APs with common difference coming from some subspace of bounded codimension, which is picked via the arithmetic regularity lemma.

Theorem 6.7.3 (Roth's theorem with common difference in some subspace)
For every $\varepsilon > 0$, there exists M so that for every $A \subseteq \mathbb{F}_3^n$, there exists a subspace W with codimension at most M, so that

$$\left|\{(x, y) \in \mathbb{F}_3^n \times W \colon x, x+y, x+2y \in A\}\right| \geq (\alpha^3 - \varepsilon)3^n |W|.$$

Proof. By the arithmetic regularity lemma (Theorem 6.6.4), there is some M depending only on ε and a subspace W of $\mathbb{F}_p^n$ of codimension $\leq M$ so that A is ε-uniform on all but at most ε-fraction of W-cosets.

Let $u + W$ be a W-coset on which A is ε-uniform. Denote the density of A in $u + W$ by

$$\alpha_u = \frac{|A \cap (u + W)|}{|W|}.$$

Restricting ourselves inside $u + W$ for a moment, by the 3-AP counting lemma, Lemma 6.2.4, the number of 3-APs of A (including trivial ones) that are contained in $u + W$ is

$$|\{(x, y) \in (u + W) \times W \colon x, x+y, x+2y \in A\}| \geq (\alpha_u^3 - \varepsilon) |W|^2.$$

Since A is ε-uniform on all but at most ε-fraction of W-cosets, by varying $u + W$ over all such cosets, we find that the total number of 3-APs in A with common difference in W is

$$\left|\{(x, y) \in \mathbb{F}_3^n \times W \colon x, x+y, x+2y \in A\}\right| \geq (1 - \varepsilon)(\alpha^3 - \varepsilon)3^n |W| \geq (\alpha^3 - 2\varepsilon)3^n |W|.$$

This proves the theorem (with ε replaced by 2ε). □

Exercise 6.7.4. Give another proof of Theorem 6.7.3 using Theorem 6.6.11 (arithmetic regularity decomposition $f = f_{\mathrm{str}} + f_{\mathrm{psr}} + f_{\mathrm{sml}}$).

Proof of Theorem 6.7.1. First apply Theorem 6.7.3 to find a subspace W of codimension $\leq M = M(\varepsilon)$. Choose $n_0 = M + \log_3(1/\varepsilon)$. So $n \geq n_0$ guarantees $|W| \geq 1/\varepsilon$.

We need to exclude 3-APs with common difference zero. We have

$$\begin{aligned}(\alpha^3 - \varepsilon)3^n |W| &\leq \left|\{(x, y) \in \mathbb{F}_3^n \times W \colon x, x+y, x+2y \in A\}\right| \\ &= \left|\{(x, y) \in \mathbb{F}_3^n \times (W \setminus \{0\}) \colon x, x+y, x+2y \in A\}\right| + |A|.\end{aligned}$$

We have $|A| \leq 3^n \leq \varepsilon 3^n |W|$, so,

$$(\alpha^3 - 2\varepsilon)3^n |W| \leq \left|\{(x, y) \in \mathbb{F}_3^n \times (W \setminus \{0\}) \colon x, x+y, x+2y \in A\}\right|.$$

By averaging, there exists $y \in W \setminus \{0\}$ satisfying

$$\left|\{x \in \mathbb{F}_3^n \colon x, x+y, x+2y \in A\}\right| \geq (\alpha^3 - 2\varepsilon)3^n.$$

This proves the theorem (with ε replaced by 2ε). □

By adapting the preceding proof strategy with Bohr sets, Green (2005a) proved a Roth's theorem with popular differences in finite abelian groups of odd order, as well as in the integers.

Theorem 6.7.5 (Roth's theorem with popular difference in finite abelian groups)
For all $\varepsilon > 0$, there exists $N_0 = N_0(\varepsilon)$ such that for all finite abelian groups Γ of odd order $|\Gamma| \geq N_0$, and every $A \subseteq \Gamma$ with $|A| = \alpha |\Gamma|$, there exists $y \in \Gamma \setminus \{0\}$ such that

$$|\{x \in \Gamma \colon x, x+y, x+2y \in A\}| \geq (\alpha^3 - \varepsilon) |\Gamma| .$$

Theorem 6.7.6 (Roth's theorem with popular difference in the integers)
For all $\varepsilon > 0$, there exists $N_0 = N_0(\varepsilon)$ such that for every $N \geq N_0$, and every $A \subseteq [N]$ with $|A| = \alpha N$, there exists $y \neq 0$ such that

$$|\{x \in [N] \colon x, x+y, x+2y \in A\}| \geq (\alpha^3 - \varepsilon) N.$$

See Tao's blog post *A Proof of Roth's Theorem* (2014) for a proof of Theorem 6.7.6 using Bohr sets, following an arithmetic regularity decomposition in the spirit of Theorem 6.6.11.

Remark **6.7.7** (Bounds). The preceding proof of Theorem 6.7.1 gives $n_0 = \text{tower}(\varepsilon^{-O(1)})$. The bounds Theorems 6.7.5 and 6.7.6 are also tower-type. What is the smallest $n_0(\varepsilon)$ for which Theorem 6.7.1 holds? It turns out to be $\text{tower}(\Theta(\log(1/\varepsilon)))$, as proved by Fox and Pham (2019) over finite fields and Fox, Pham, and Zhao (2022) over the integers. Although it had been known since Gowers (1997) that tower-type bounds are necessary for the regularity lemmas themselves, Roth's theorem with popular differences is the first regularity application where a tower-type bound is shown to be indeed necessary.

Using quadratic Fourier analysis, Green and Tao (2010c) extended the popular difference result over to 4-APs.

Theorem 6.7.8 (Popular difference for 4-APs)
For all $\varepsilon > 0$, there exists $N_0 = N_0(\varepsilon)$ such that for every $N \geq N_0$ and $A \subseteq [N]$ with $|A| = \alpha N$, there exists $y \neq 0$ such that

$$|\{x \colon x, x+y, x+2y, x+3y \in A\}| \geq (\alpha^4 - \varepsilon) N.$$

Surprisingly, such a statement is false for APs of length 5 or longer. This was shown by Bergelson, Host, and Kra (2005) with an appendix by Ruzsa giving a construction that is a clever modification of the Behrend construction (Section 2.5).

Theorem 6.7.9 (Popular difference fails for 5-APs)
Let $0 < \alpha < 1/2$. For all sufficiently large N, there exists $A \subseteq [N]$ with $|A| \geq \alpha N$ such that for all $y \neq 0$,

$$|\{x \colon x, x+y, x+2y, x+3y, x+4y \in A\}| \leq \alpha^{c \log(1/\alpha)} N.$$

Here $c > 0$ is some absolute constant.

For more on results of this type, as well as for popular difference for high-dimensional patterns, see Sah, Sawhney, and Zhao (2021).

Further Reading

Green has several excellent surveys and lecture notes:

- *Finite Field Models in Additive Combinatorics* (2005c) – For many additive combinatorics problems, it is a good idea to first study them in the finite field setting (also see the follow-up by Wolf [2015]).
- *Montreal Lecture Notes on Quadratic Fourier Analysis* (2007a) – An introduction to quadratic Fourier analysis and its application to the popular common difference theorem for 4-APs in $\mathbb{F}_5^n$.
- Lecture notes from his Cambridge course *Additive Combinatorics* (2009b).

Tao's FOCS 2007 tutorial *Structure and Randomness in Combinatorics* (2007a) explains many facets of arithmetic regularity and applications.

For more on algebraic methods in combinatorics (mostly predating methods in Section 6.5), see the following books:

- *Thirty-Three Miniatures* by Matoušek (2010);
- *Linear Algebra Methods in Combinatorics* by Babai and Frankl;
- *Polynomial Methods in Combinatorics* by Guth (2016);
- *Polynomial Methods and Incidence Theory* by Sheffer (2022).

In particular, the book *Fourier Analysis* by Stein and Shakarchi (2003) is a superb undergraduate textbook on Fourier analysis. The analysis viewpoint has different emphases compared to this chapter, though many standard tools (e.g., Parseval) are common to both. It is helpful for becoming familiar with general principles of Fourier analysis, such as the relationship between smoothness and decay.

Chapter Summary

- Basic tools of discrete Fourier analysis:
  - Fourier transform,
  - Fourier inversion formula,
  - Parsevel / Plancherel identity (unitarity of the Fourier transform),
  - convolution identity (Fourier transform converts convolutions to multiplication).
- The **finite field model** (e.g., $\mathbb{F}_3^n$) offers a convenient playground for Fourier analysis in additive combinatorics. Many techniques can then be adapted to the integer setting, although often with additional technicalities.
- **Roth's theorem.** Using Fourier analysis, we proved that every 3-AP-free subset has size at most
  - $O(3^n/n)$ in $\mathbb{F}^n$, and
  - $O(N/\log\log N)$ in $[N] \subseteq \mathbb{Z}$.
- The Fourier analytic proof of Roth's theorem (both in $\mathbb{F}_3^n$ and in $\mathbb{Z}$) proceeds via a **density increment argument**:
  (1) A 3-AP-free set has a large Fourier coefficient;
  (2) A large Fourier coefficient implies density increment on some hyperplane (in $\mathbb{F}_3^n$) or subprogression (in $\mathbb{Z}$);
  (3) Iterate the density increment.
- Using the **polynomial method**, we showed that every 3-AP-free subset of $\mathbb{F}_3^n$ has size $O(2.76^n)$.
- **Arithmetic regularity lemma.** Given $A \subseteq \mathbb{F}_p^n$, we can find a bounded codimensional subspace so that A is Fourier-uniform on almost all cosets.
  - An application: **Roth's theorem with popular difference.** For every $A \subseteq \mathbb{F}_3^n$, there is some "popular 3-AP common difference" with frequency at least nearly as much as if A were random.

7

Structure of Set Addition

Chapter Highlights

- Freiman's theorem: structure of sets with small doubling
- Inequalities between sizes of sumsets: Ruzsa triangle inequality and Plünnecke's inequality
- Ruzsa covering lemma
- Freiman homomorphisms: preserving partial additive structure
- Ruzsa modeling lemma
- Structure in iterated sumsets: Bogolyubov's lemma
- Geometry of numbers: Minkowski's second theorem
- Polynomial Freiman–Ruzsa conjecture
- Additive energy and the Balog–Szemerédi–Gowers theorem

Let A and B be finite subsets of some ambient abelian group. We define their ***sumset*** to be

$$\boldsymbol{A + B} := \{a + b \colon a \in A, b \in B\}.$$

Note that we view $A + B$ as a set and do not keep track of the number of ways that each element can be written as $a + b$.

The main goal of this chapter is to understand the following question.

Question 7.0.1 (Sets with small doubling)
What can we say about A if $A + A$ is small?

We will prove Freiman's theorem, which is a deep and foundational result in additive combinatorics. Freiman's theorem tells us that, if $A + A$ is at most a constant factor larger than A, then A must be a large fraction of some generalized arithmetic progression.

Most of this chapter will be devoted toward proving Freiman's theorem. We will see ideas and tools from Fourier analysis, geometry of numbers, and additive combinatorics.

In Section 7.13, we will introduce the **additive energy** of a set, which is another way to measure the additive structure of a set. We will see the Balog–Szemerédi–Gowers theorem, which relates additive energy and doubling. This section can be read independently from the earlier parts of the chapter.

These results on the structure of set addition are not only interesting on their own, but also play a key role in Gowers' proof (2001) of Szemerédi's theorem (although we do not cover it in this book; see Further Reading at the end of the chapter). Gowers' deep and foundational work shows how these topics in additive combinatorics are all highly connected.

Definition 7.0.2 (Sumset notation)
Given a positive integer k, we define the iterated sumset

$$\boldsymbol{kA} := A + \cdots + A \quad (k \text{ times}).$$

This is different from dilating a set, which is denoted by

$$\boldsymbol{\lambda \cdot A} := \{\lambda a : a \in A\}.$$

We also consider the difference set

$$\boldsymbol{A - B} = \{a - b : a \in A, b \in B\}.$$

7.1 Sets of Small Doubling: Freiman's Theorem

How small or large can $A + A$ be, given $|A|$? This is an easy question to answer.

Proposition 7.1.1 (Easy bounds on sumset size)
Let $A \subseteq \mathbb{Z}$ be a finite set. Then

$$2|A| - 1 \leq |A + A| \leq \binom{|A| + 1}{2}.$$

Furthermore, both bounds are best possible as functions of $|A|$.

Proof. Let $n = |A|$. For the lower bound $|A + A| \geq 2n - 1$, note that if the elements of A are $a_1 < a_2 < \cdots < a_n$, then

$$a_1 + a_1 < a_1 + a_2 < \cdots < a_1 + a_n < a_2 + a_n < \cdots < a_n + a_n$$

are $2n - 1$ distinct elements of $A + A$. So $|A + A| \geq 2n - 1$. Equality is attained when A is an arithmetic progression.

The upper bound $|A + A| \leq \binom{n+1}{2}$ follows from that there are $\binom{n+1}{2}$ unordered pairs of elements of A. We have equality when there are no nontrivial solutions to $a + b = c + d$ in A, such as when A consists of powers of twos. □

Exercise 7.1.2 (Sumsets in abelian groups). Show that if A is a finite subset of an abelian group, then $|A + A| \geq |A|$, with equality if and only if A is the coset of some subgroup.

What can we say about A if $A + A$ is not too much larger than A?

Definition 7.1.3 (Doubling constant)
The ***doubling constant*** of a finite subset A in an abelian group is the ratio $|A + A|/|A|$.

One of the main results of this chapter, Freiman's theorem, addresses the following question.

Question 7.1.4 (Sets of small doubling)
What is the structure of a set with bounded doubling constant (e.g. $|A + A| \leq 100|A|$)?

We've already seen an example of such a set in $\mathbb{Z}$, namely arithmetic progressions.

***Example* 7.1.5.** If $A \subseteq \mathbb{Z}$ is a finite arithmetic progression, $|A + A| = 2|A| - 1 \le 2|A|$, so it has doubling constant at most 2.

Moreover, if we delete some elements of an arithmetic progression, it should still have small doubling. In fact, if we delete even most of the elements of an arithmetic progression but leave a constant fraction of the progression remaining, we will have small doubling.

***Example* 7.1.6.** If B is a finite arithmetic progression and $A \subseteq B$ has $|A| \ge |B|/K$, then $|A + A| \le |B + B| \le 2|B| \le 2K|A|$, so A has doubling constant at most $2K$.

Now we generalize arithmetic progressions to allow multiple dimensions. Informally, we consider affine images of d-dimensional "grids", as illustrated below.

• • • •
• • • • $\longrightarrow$ ••• ••• ••• •••
• • • •

$\mathbb{Z}^2$ $\mathbb{Z}$

Definition 7.1.7 (GAP – generalized arithmetic progression)
A ***generalized arithmetic progression (GAP)*** in an abelian group Γ is defined to be an affine map

$$\phi\colon [L_1] \times \cdots \times [L_d] \to \Gamma.$$

That is, for some $a_0, \ldots, a_d \in \Gamma$,

$$\phi(x_1, \ldots, x_d) = a_0 + a_1 x_1 + \cdots + a_d x_d.$$

This GAP has ***dimension*** d and ***volume*** $L_1 \cdots L_d$. We say that this GAP is ***proper*** if ϕ is injective.

We often abuse notation and use the term GAP to refer to the image of ϕ, viewed as a set:

$$a_0 + a_1 \cdot [L_1] + \cdots + a_d \cdot [L_d] = \{a_0 + a_1 x_1 + \cdots + a_d x_d \colon x_1 \in [L_1], \ldots, x_d \in [L_d]\}.$$

***Example* 7.1.8.** A proper GAP of dimension d has doubling constant $\le 2^d$.

***Example* 7.1.9.** Let P be a proper GAP of dimension d. Let $A \subseteq P$ with $|A| \ge |P|/K$. Then A has doubling constant $\le K2^d$.

While it is often easy to check that certain sets have small doubling, the **inverse problem** is much more difficult. We would like to characterize all sets with small doubling. The following foundational result by Freiman (1973) shows that all sets with bounded doubling must look like Example 7.1.9.

Theorem 7.1.10 (Freiman's theorem)
Let $A \subseteq \mathbb{Z}$ be a finite set satisfying $|A + A| \le K|A|$. Then A is contained in a GAP of dimension at most $d(K)$ and volume at most $f(K)|A|$, where $d(K)$ and $f(K)$ are constants depending only on K.

Freiman's theorem is a deep result. We will spend most the chapter proving it.

Remark 7.1.11 (Quantitative bounds). We will present a proof giving $d(K) = \exp(K^{O(1)})$ and $f(K) = \exp(d(K))$, due to Ruzsa (1994). Chang (2002) showed that Freiman's theorem holds with $d(K) = K^{O(1)}$ and $f(K) = \exp(d(K))$ (see Exercise 7.11.2). Schoen (2011) further improved the bounds to $d(K) = K^{1+o(1)}$ and $f(K) = \exp(K^{1+o(1)})$. Sanders (2012, 2013) showed that if we change GAPs to "convex progressions" (see Section 7.12), then an analogous theorem holds with $d(K) = K(\log(2K))^{O(1)}$ and $f(K) = \exp(d(K))$.

It is easy to see that one cannot do better than $d(K) \le K - 1$ and $f(K) = e^{O(K)}$, by considering a set without additive structure.

Also see Section 7.12 on the polynomial Freiman–Ruzsa conjecture for a variant of Freiman's theorem with much better quantitative dependencies.

Remark 7.1.12 (Making the GAP proper). The conclusion of Freiman's theorem can be strengthened to force the GAP to be proper, at the cost of potentially increasing $d(K)$ and $f(K)$. For example, it is known that every GAP of dimension d is contained in some proper GAP of dimension $\le d$ with at most $d^{O(d^3)}$ factor increase in the volume; see Tao and Vu (2006, Theorem 3.40).

Remark 7.1.13 (History). Freiman's original proof (1973) was quite complicated. Ruzsa (1994) later found a simpler proof, which guided much of the subsequent work. We follow Ruzsa's presentation here. Theorem 7.1.10 is sometimes called the ***Freiman–Ruzsa theorem***. Freiman's theorem was brought into further prominence due to the role it played in the new proof of Szemerédi's theorem by Gowers (2001).

Remark 7.1.14 (Freiman's theorem in abelian groups). Green and Ruzsa (2007) proved a generalization of Freiman's theorem in an arbitrary abelian group. A ***coset progression*** is a set of the form $P + H$ where P is a GAP and H is a subgroup of the ambient abelian group. Define the ***dimension*** of this coset progression to be the dimension of P, and its ***volume*** to be $|H| \operatorname{vol} P$. Green and Ruzsa (2007) proved the following theorem.

Theorem 7.1.15 (Freiman's theorem for general abelian groups)
Let A be a subset of an abelian group satisfying $|A + A| \le K\,|A|$. Then A is contained in a coset progression of dimension at most $d(K)$ and volume at most $f(k)\,|A|$, where $d(K)$ and $f(K)$ are constants depending only on K.

7.2 Sumset Calculus I: Ruzsa Triangle Inequality

Here are some basic and useful inequalities relating the sizes of sumsets.

Theorem 7.2.1 (Ruzsa triangle inequality)
If A, B, C are finite subsets of an abelian group, then

$$|A|\,|B - C| \le |A - B|\,|A - C|\,.$$

Proof. For each $d \in B - C$, define $\mathsf{b}(d) \in B$ and $\mathsf{c}(d) \in C$ such that $d = \mathsf{b}(d) - \mathsf{c}(d)$. In other words, we fix a specific choice of b and c for each element in $B - C$. Define

$$\begin{aligned} \phi : A \times (B - C) &\longrightarrow (A - B) \times (A - C) \\ (a, d) &\longmapsto (a - \mathsf{b}(d), a - \mathsf{c}(d)). \end{aligned}$$

Then ϕ is injective since we can recover (a,d) from $\phi(a,d) = (x,y)$ via $d = y - x$ and then $a = x + \mathsf{b}(d)$. □

***Remark* 7.2.2.** By replacing B with $-B$ and/or C with $-C$, Theorem 7.2.1 implies some additional sumset inequalities:

$$|A|\,|B+C| \le |A+B|\,|A-C|\,;$$
$$|A|\,|B+C| \le |A-B|\,|A+C|\,;$$
$$|A|\,|B-C| \le |A+B|\,|A+C|\,.$$

However, this trick cannot be used to prove the similarly looking inequality

$$|A|\,|B+C| \le |A+B|\,|A+C|\,.$$

This inequality is also true, and we will prove it in the following section.

***Remark* 7.2.3** (Why is it called a triangle inequality?)**.** If we define

$$\rho(A,B) := \log \frac{|A-B|}{\sqrt{|A|\,|B|}}$$

(called a ***Ruzsa distance***), then Theorem 7.2.1 can be rewritten as

$$\rho(B,C) \le \rho(A,B) + \rho(A,C).$$

This is why Theorem 7.2.1 is called a "triangle inequality." However, one should not take the name too seriously. The function ρ is not a metric because $\rho(A,A) \ne 0$ in general.

Exercise 7.2.4 (Iterated sumsets)**.** Let A be a finite subset of an abelian group satisfying

$$|2A - 2A| \le K\,|A|\,.$$

Prove that

$$|mA - mA| \le K^{m-1}\,|A| \quad \text{for every integer } m \ge 2.$$

In the above exercise, we had to start with the assumption that $|2A - 2A| \le K\,|A|$. In the next section, we bound the sizes of iterated sumsets starting with the weaker hypothesis $|A + A| \le K\,|A|$.

7.3 Sumset Calculus II: Plünnecke's Inequality

We prove the following result, which says that having small doubling implies small iterated sumsets, with only a polynomial factor change in the expansion ratios.

Theorem 7.3.1 (Plünnecke's inequality)
Let A be a finite subset of an abelian group satisfying

$$|A + A| \le K\,|A|\,.$$

Then for all integers $m, n \ge 0$,

$$|mA - nA| \le K^{m+n}\,|A|\,.$$

Remark* 7.3.2** (History). Plünnecke (1970) proved a version of the theorem originally using graph theoretic methods. Ruzsa (1989) gave a simpler version of Plünnecke's proof and also extended it from sums to differences. Nevertheless, Ruzsa's proof was still quite long and complex. It sets up a "commutative layered graph," and uses tools from graph theory including Menger's theorem. Theorem 7.3.1 is sometimes called the ***Plünnecke–Ruzsa inequality. See Ruzsa (2009, chapter 1) or Tao and Vu (2006, chapter 6) for an account of this proof.

In a surprising breakthrough, Petridis (2012) found a very short proof of the result, which we present here.

We will prove the following more general statement. Theorem 7.3.1 is the special case $A = B$.

Theorem 7.3.3 (Plünnecke's inequality)
Let A and B be finite subsets of an abelian group satisfying

$$|A + B| \leq K\,|A|.$$

Then for all integers $m, n \geq 0$,

$$|mB - nB| \leq K^{m+n}\,|A|.$$

The following lemma plays a key role in the proof.

Lemma 7.3.4 (Expansion ratio bounds)
Let X and B be finite subsets of an abelian group, with $|X| > 0$. Suppose

$$\frac{|Y + B|}{|Y|} \geq \frac{|X + B|}{|X|} \quad \text{for all nonempty } Y \subseteq X.$$

Then for any nonempty finite subsets C of the abelian group,

$$\frac{|X + C + B|}{|X + C|} \leq \frac{|X + B|}{|X|}.$$

***Remark* 7.3.5** (Interpretation as expansion ratios). We can interpret Lemma 7.3.4 in terms of vertex expansion ratios inside the bipartite graph between two copies of the ambient abelian group, with edges $(x, x + b)$ ranging over all $x \in \Gamma$ and $b \in B$. Every vertex subset X on the left has neighbors $X + B$ on the right and thus has *vertex expansion ratio* $|X + B|/|X|$.

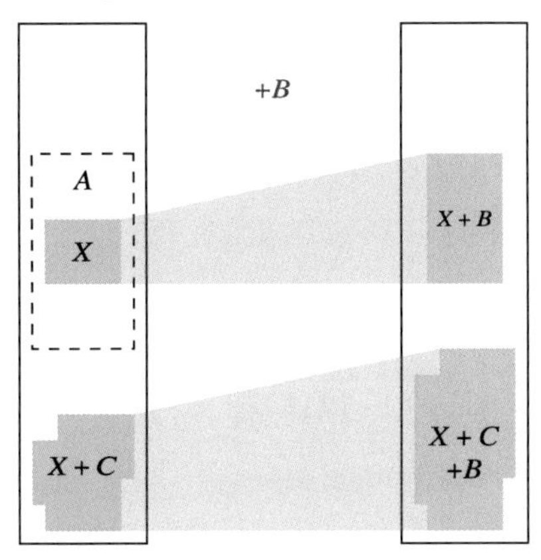

We will apply Lemma 7.3.4 by choosing X among all nonempty subsets of A with the minimum expansion ratio, so that the hypothesis of Lemma 7.3.4 is automatically satisfied. The conclusion of Lemma 7.3.4 then says that a union of translates of X has expansion ratio at most that of X.

Proof of Theorem 7.3.3 given Lemma 7.3.4. Choose X among all nonempty subsets of A with the minimum $|X + B| / |X|$ so that the hypothesis of Lemma 7.3.4 is satisfied. Also we have

$$\frac{|X + B|}{|X|} \leq \frac{|A + B|}{|A|} \leq K.$$

For every integer $n \geq 0$, applying Lemma 7.3.4 with $C = nB$, we have

$$\frac{|X + (n + 1)B|}{|X + nB|} \leq \frac{|X + B|}{|X|} \leq K.$$

So, induction on n yields, for all $n \geq 0$,

$$|X + nB| \leq K^n |X|.$$

Finally, applying the Ruzsa triangle inequality (Theorem 7.2.1), for all $m, n \geq 0$:

$$|mB - nB| \leq \frac{|X + mB| |X + nB|}{|X|} \leq K^{m+n} |X| \leq K^{m+n} |A|. \qquad \square$$

Proof of Lemma 7.3.4. We will proceed by induction on $|C|$. For the base case $|C| = 1$, note that $X + C$ is a translate of X, so $|X + C + B| = |X + B|$ and $|X + C| = |X|$.

Now for the induction step, assume that for some C,

$$\frac{|X + C + B|}{|X + C|} \leq \frac{|X + B|}{|X|}.$$

Now consider $C \cup \{c\}$ for some $c \notin C$. We wish to show that

$$\frac{|X + (C \cup \{c\}) + B|}{|X + (C \cup \{c\})|} \leq \frac{|X + B|}{|X|}.$$

By comparing the change in the left-hand side fraction, it suffices to show that

$$|(X + c + B) \setminus (X + C + B)| \leq \frac{|X + B|}{|X|} |(X + c) \setminus (X + C)|. \tag{7.1}$$

Let

$$Y = \{x \in X : x + c + B \subseteq X + C + B\} \subseteq X.$$

Then

$$|(X + c + B) \setminus (X + C + B)| \leq |X + B| - |Y + B|.$$

Furthermore, if $x \in X$ satisfies $x + c \in X + C$, then $x + c + B \subseteq X + C + B$ and hence $x \in Y$. So,

$$|(X + c) \setminus (X + C)| \geq |X| - |Y|.$$

Thus, to prove (7.1), it suffices to show

$$|X+B|-|Y+B| \le \frac{|X+B|}{|X|}(|X|-|Y|),$$

which can be rewritten as

$$|Y+B| \ge \frac{|X+B|}{|X|}|Y|,$$

which is true due to the hypothesis on X. □

Let us give a quick proof of a variant of the Ruzsa triangle inequality, mentioned in Remark 7.2.2.

Corollary 7.3.6 (Another triangle inequality)
Let A, B, and C be finite subsets of an abelian group. Then

$$|A|\,|B+C| \le |A+B|\,|A+C|.$$

Proof. Choose $X \subseteq A$ to minimize $|X+B|/|X|$. Then

$$|B+C| \le |X+B+C| \overset{\text{Lem. 7.3.4}}{\le} |X+C|\frac{|X+B|}{|X|} \le |A+C|\frac{|A+B|}{|A|}.$$ □

Exercise 7.3.7*. Show that for every sufficiently large K there is some finite set $A \subseteq \mathbb{Z}$ such that

$$|A+A| \le K|A| \qquad \text{and} \qquad |A-A| \ge K^{1.99}|A|.$$

Exercise 7.3.8* (Loomis–Whitney for sumsets)**.** Show that for every finite subsets A, B, C in an abelian group, one has

$$|A+B+C|^2 \le |A+B|\,|A+C|\,|B+C|.$$

Exercise 7.3.9* (Sumset vs. difference set)**.** Let $A \subseteq \mathbb{Z}$. Prove that

$$|A-A|^{2/3} \le |A+A| \le |A-A|^{3/2}.$$

7.4 Covering Lemma

Here is a simple and powerful tool in the study of sumsets (Ruzsa 1999).

Theorem 7.4.1 (Ruzsa covering lemma)
Let X and B be finite sets in some abelian group. If

$$|X+B| \le K|B|,$$

then there exists a subset $T \subseteq X$ with $|T| \le K$ such that

$$X \subseteq T+B-B.$$

***Remark* 7.4.2** (Geometric intuition)**.** Imagine that B is a unit ball in $\mathbb{R}^n$, and cardinality is replaced by volume. Given some region X (the shaded region in the following illustration), consider a maximal set $\mathcal{T}$ of disjoint union balls with centers in X (maximal in the sense that one cannot add an additional ball without intersecting some other ball).

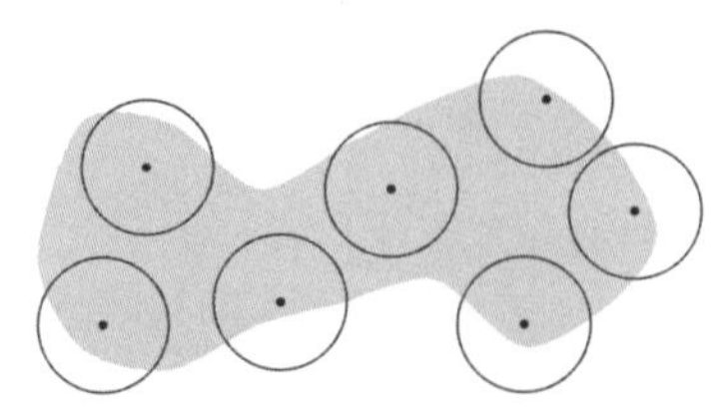

Then replacing each ball in $\mathcal{T}$ by a ball of radius 2 with the same center (i.e., replacing B by $B - B$), the resulting balls must cover the region X (which amounts to the conclusion $X \subseteq T + B - B$). For otherwise, at any uncovered point of X we could have added an additional nonoverlapping ball in the previous step.

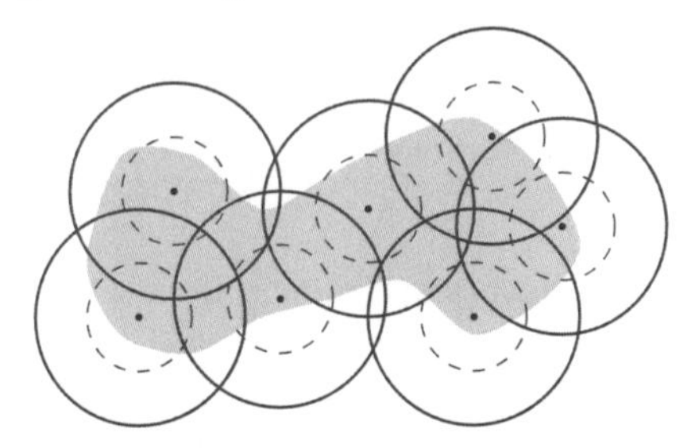

Similar arguments are important in analysis (e.g., the Vitali covering lemma).

Proof. Let $T \subseteq X$ be a maximal subset such that $t + B$ as t ranges over T are disjoint. Then

$$|T|\,|B| = |T + B| \le |X + B| \le K\,|B|\,.$$

So, $|T| \le K$.

By the maximality of T, for all $x \in X$ there exists some $t \in T$ such that $(t+B) \cap (x+B) \neq \emptyset$. In other words, there exist $t \in T$ and $b, b' \in B$ such that $t + b = x + b'$. Hence $x \in T + B - B$ for every $x \in X$. Thus, $X \subseteq T + B - B$. □

The following "more efficient" covering lemma can be used to prove a better bound in Freiman's theorem.

Exercise 7.4.3* (Chang's covering lemma)**.** Let A and B be finite sets in an abelian group satisfying

$$|A + A| \le K\,|A| \qquad \text{and} \qquad |A + B| \le K'\,|B|\,.$$

Show that there exists some set X in the abelian group so that

$$A \subseteq \Sigma X + B - B \qquad \text{and} \qquad |X| = O(K \log(KK')),$$

where ΣX denotes the set of all elements that can be written as the sum of a subset of elements of X (including zero as the sum of the empty set).

Hint: Try first finding $2K$ disjoint translates $a + B$.

7.5 Freiman's Theorem in Groups with Bounded Exponent

Let us prove a finite field model analogue of Freiman's theorem. The proof only uses the tools introduced so far, and so it is easier than Freiman's theorem in the integers.

Theorem 7.5.1 (Freiman's theorem in $\mathbb{F}_2^n$)
If $A \subseteq \mathbb{F}_2^n$ has $|A + A| \le K|A|$, then A is contained in a subspace of cardinality at most $f(K)|A|$, where $f(K)$ is a constant depending only on K.

***Remark* 7.5.2** (Quantitative bounds). We will prove Theorem 7.5.1 with $f(K) = 2^{K^4}K^2$. The exact optimal constant $f(K)$ is known for each K (Even-Zohar 2012). Asymptotically, it is $f(K) = \Theta(2^{2K}/K)$.

For a matching lower bound on $f(K)$, let $A = \{0, e_1, \dots, e_n\} \subseteq \mathbb{F}_2^n$, where e_i is the ith standard basis vector. Then $|A + A| \sim n^2/2$, and so $|A + A|/|A| \sim n/2$. However, A is not contained in a subspace of cardinality less than 2^n.

In fact, we prove a more general statement that works for any group with bounded exponent. This result and proof are due to Ruzsa (1999).

Definition 7.5.3 (Exponent of an abelian group)
The ***exponent*** of an abelian group (written additively) is the smallest positive integer r such that $rx = 0$ for all elements x of the group. If no finite r exists, we say that its exponent is infinite. (Some conventions say that the exponent is zero.)

For example, $\mathbb{F}_2^n$ has exponent 2. The cyclic group $\mathbb{Z}/N\mathbb{Z}$ has exponent N. The integer $\mathbb{Z}$ has infinite exponent.

We use $\langle A \rangle$ to refer to the subgroup of a group G generated by some subset A of G. Then the exponent of a group G is $\sup_{x \in G} |\langle x \rangle|$. When the group is a vector space (e.g., $\mathbb{F}_2^n$), $\langle A \rangle$ is the smallest subspace containing A.

Theorem 7.5.4 (Freiman's theorem in groups with bounded exponent)
Let A be a finite set in an abelian group with exponent $r < \infty$. If $|A + A| \le K|A|$, then

$$|\langle A \rangle| \le K^2 r^{K^4} |A|.$$

***Remark* 7.5.5.** This theorem is a converse of the observation that if A is a large fraction of a subgroup, then A has small doubling.

Proof. By Plünnecke's inequality (Theorem 7.3.1), we have

$$|A + (2A - A)| = |3A - A| \le K^4 |A|.$$

By the Ruzsa covering lemma (Theorem 7.4.1 applied with $X = 2A - A$ and $B = A$), there exists some $T \subseteq 2A - A$ with $|T| \le |A + (2A - A)|/|A| \le K^4$ such that

$$2A - A \subseteq T + A - A.$$

Adding A to both sides, we have

$$3A - A \subseteq T + 2A - A \subseteq 2T + A - A.$$

Iterating, for any positive integer n, we have

$$(n+1)A - A \subseteq nT + A - A \subseteq \langle T \rangle + A - A.$$

Since we are in an abelian group with bounded exponent, every element of $\langle A \rangle$ lies in nA for some n. Thus,

$$\langle A \rangle \subseteq \bigcup_{n \geq 1} (nA + A - A) \subseteq \langle T \rangle + A - A.$$

Since the exponent of the group is at most $r < \infty$,

$$|\langle T \rangle| \leq r^{|T|} \leq r^{K^4}.$$

By Plünnecke's inequality (Theorem 7.3.1),

$$|A - A| \leq K^2 |A|.$$

Thus, we have

$$|\langle A \rangle| \leq r^{K^4} K^2 |A|. \qquad \square$$

***Remark* 7.5.6.** Note the crucial use of the Ruzsa covering lemma for controlling $nA - A$. Naively bounding nA using Plünnecke's inequality is insufficient.

The preceding proof for Freiman's theorem over abelian groups of finite exponent does not immediately generalize to the integers. Indeed, in $\mathbb{Z}$, $|\langle T \rangle| = \infty$. We overcome this issue by representing subsets of $\mathbb{Z}$ inside a finite group in a way that partially preserves additive structure.

Exercise 7.5.7. Show that, for every real $K \geq 1$, there is some C_K such that for every finite set A of an abelian group with $|A + A| \leq K |A|$, one has $|nA| \leq n^{C_K} |A|$ for every positive integer n.

(If we let $f(n, K)$ denote the smallest real number so that $|A + A| \leq K |A|$ implies $|nK| \leq f(n, K) |A|$, then Plünnecke's inequality gives $f(n, K) \leq K^n$, at most a polynomial in K for a fixed n, whereas the preceding exercise gives $f(n, K) \leq n^{C_K}$, a polynomial in n for a fixed K. Does this mean that $f(n, K)$ is at most some polynomial in both n and K?)

Exercise 7.5.8* (Ball volume growth in an abelian Cayley graph). Show that there is some absolute constant C so that if S is a finite subset of an abelian group, and k is a positive integer, then

$$|2kS| \leq C^{|S|} |kS|.$$

7.6 Freiman Homomorphisms

Consider two sets of integers, depicted pictorially here as elements on the number line:

$A =$ • • • • • • • • • • • • • • • • • • • • • •

$B =$ •••• ••••• •••••• •••••••

The two sets are very similar from the point of view of additive structure. For example, the obvious bijection between A and B has the nice property that any solution to the equation

$w + x = y + z$ in one set is automatically a solution in the other. Sometimes, in additive combinatorics, it is a good idea to treat these two sets as isomorphic. Let us define this notion formally and study what it means for a map between sets to partially preserve additive structure.

Definition 7.6.1 (Freiman homomorphism)
Let A and B be subsets in two possibly different abelian groups. Let $s \geq 2$ be a positive integer. We say that $\phi\colon A \to B$ is a ***Freiman s-homomorphism*** (or ***Freiman homomorphism of order s***), if

$$\phi(a_1) + \cdots + \phi(a_s) = \phi(a_1') + \cdots + \phi(a_s')$$

whenever $a_1, \ldots, a_s, a_1', \ldots, a_s' \in A$ satisfy

$$a_1 + \cdots + a_s = a_1' + \cdots + a_s'.$$

We say that ϕ is a ***Freiman s-isomorphism*** if ϕ is a bijection, and both ϕ and ϕ^{-1} are Freiman s-homomorphisms. We say that A and B are ***Freiman s-isomorphic*** if there exists a Freiman s-isomorphism between them.

Remark **7.6.2** (Interpretation). Informally, a Freiman s-homomorphism respects s-fold sums relations. Two sets are Freiman s-isomorphic if there is a bijection between them that respects solutions to the equation $a_1 + \cdots + a_s = a_1' + \cdots + a_s'$.

Remark **7.6.3** (Composition). If ϕ_1 and ϕ_2 are both Freiman s-homomorphisms, then their composition $\phi_1 \circ \phi_2$ is also a Freiman s-homomorphism. If ϕ_1 and ϕ_2 are both Freiman s-isomorphisms, then their composition $\phi_1 \circ \phi_2$ is a Freiman s-isomorphism.

Remark **7.6.4** (Descension). Every Freiman $(s + 1)$-homomorphism is automatically a Freiman s-homomorphism (by setting $a_{s+1} = a_{s+1}'$). Likewise, every Freiman $(s + 1)$-isomorphism is automatically a Freiman s-isomorphism.

Example **7.6.5** (Freiman homomorphism).

(a) Every abelian group homomorphism is a Freiman homomorphism of every order.
(b) Let S be a set with no nontrivial solutions to $a + b = c + d$ (such a set is called a ***Sidon set***). Then every map from S to an abelian group is a Freiman 2-homomorphism.
(c) The natural embedding $\phi\colon \{0, 1\}^n \to (\mathbb{Z}/2\mathbb{Z})^n$ is the restriction of a group homomorphism from $\mathbb{Z}^n$, so it is a Freiman homomorphism of every order. This map ϕ is a bijection. However, the inverse of ϕ does not preserve some additive relations (e.g., $1 + 1 = 0 + 0 \pmod 2$). So ϕ is *not* a Freiman 2-isomorphism!
(d) Likewise, the natural embedding $\phi\colon [N] \to \mathbb{Z}/N\mathbb{Z}$ is a Freiman homomorphism of every order but not a Freiman 2-isomorphism. However, when the domain is restricted to all integers less than N/s, then ϕ becomes a Freiman s-isomorphism onto its image. (Why?)

The last example has the following easy generalization, which we will use later. The ***diameter*** of a set A is defined to be

$$\mathbf{diam}\, A := \sup_{a,b \in A} |a - b|\,.$$

Proposition 7.6.6 (Small diameter sets)
If $A \subseteq \mathbb{Z}$ has diameter $< N/s$, then A is Freiman s-isomorphic to its image mod N.

Intuitively, the idea is that there are no wraparound additive relations mod N if A has small diameter.

Proof. The mod N map $\mathbb{Z} \to \mathbb{Z}/N$ is a group homomorphism, and hence automatically a Freiman s-homomorphism. Now, if $a_1, \ldots, a_s, a'_1, \ldots, a'_s \in A$ are such that

$$(a_1 + \cdots + a_s) - (a'_1 + \cdots + a'_s) \equiv 0 \pmod{N},$$

then the left-hand side, viewed as an integer, has absolute value less than N (since $|a_i - a'_i| < N/s$ for each i). Thus the left-hand side must be 0 in $\mathbb{Z}$. So the inverse of the mod N map is a Freiman s-homomorphism over A, and thus mod N is a Freiman s-isomorphism. □

7.7 Modeling Lemma

The goal of the Ruzsa modeling lemma is to represent a set with bounded doubling inside a small cyclic group in a way that that preserves relevant additive data. This is useful since initially A may contain integers of vastly different magnitudes. On the other hand, if A is a subset of $\mathbb{Z}/N\mathbb{Z}$ with N comparable to A, then we have additional tools such as Fourier analysis (to be discussed in the following section).

As warm-up, let us first prove an easier result in the finite field model.

Proposition 7.7.1 (Modeling lemma in finite field model)
Let $A \subseteq \mathbb{F}_2^n$. Suppose $|sA - sA| \leq 2^m$ for some positive integer m. Then A is Freiman s-isomorphic to some subset of $\mathbb{F}_2^m$.

***Remark* 7.7.2.** If $|A + A| \leq K|A|$, then Plünnecke's inequality (Theorem 7.3.1) implies $|sA - sA| \leq K^{2s}|A|$. By taking m to be the smallest integer with $K^{2s}|A| \leq 2^m$, we see that the cardinality of the final vector space $\mathbb{F}_2^m$ is within a constant factor $2K^{2s}$ of $|A|$. In contrast, A initially lived in a space $\mathbb{F}_2^n$ that could potentially be much larger.

Proof. It is easy to check that the following are equivalent for a linear map $\phi\colon \mathbb{F}_2^n \to \mathbb{F}_2^m$:

(1) ϕ is a Freiman s-isomorphism when restricted to A.
(2) ϕ is injective on sA.
(3) $\phi(x) \neq 0$ for all nonzero $x \in sA - sA$.

Then let $\phi\colon \mathbb{F}_2^n \to \mathbb{F}_2^m$ be a linear map chosen uniformly at random. Each nonzero $x \in sA - sA$ violates condition (3) with probability 2^{-m}. Since there are $< 2^m$ nonzero elements in $sA - sA$ by hypothesis, (3) is satisfied with with positive probability. Therefore, the desired Freiman s-isomorphism exists. □

Starting with $A \subseteq \mathbb{Z}$ of small doubling, we will find a large fraction of A that can be modeled inside a cyclic group whose size is comparable to $|A|$. It turns out to be enough to model a large subset of A rather than all of A. We will apply the Ruzsa covering lemma later on to recover the structure of the entire set A.

Theorem 7.7.3 (Ruzsa modeling lemma)
Let $A \subseteq \mathbb{Z}$. Let $s \geq 2$ and N be positive integers. Suppose $|sA - sA| \leq N$. Then there exists $A' \subseteq A$ with $|A'| \geq |A|/s$ such that A' is Freiman s-isomorphic to a subset of $\mathbb{Z}/N\mathbb{Z}$.

Proof. Choose any prime $q > \max(sA - sA)$. For every choice of $\lambda \in [q-1]$, we define ϕ_λ as the composition of functions as follows

$$\phi = \phi_\lambda \colon \mathbb{Z} \xrightarrow{\text{mod } q} \mathbb{Z}/q\mathbb{Z} \xrightarrow{\cdot\lambda} \mathbb{Z}/q\mathbb{Z} \xrightarrow{(\text{mod } q)^{-1}} \{0, 1, \ldots, q-1\}.$$

The first map is the mod q map. The second map sends x to λx. The last map inverts the mod q map $\mathbb{Z} \to \mathbb{Z}/q\mathbb{Z}$.

If $\lambda \in [q-1]$ is chosen uniformly at random, then each nonzero integer is mapped to a uniformly random element of $[q-1]$ under ϕ_λ, and so is divisible by N with probability $\leq 1/N$. Since there are fewer than N nonzero elements in $sA - sA$, there exists a choice of λ so that

$$N \nmid \phi_\lambda(x) \qquad \text{for any nonzero } x \in sA - sA. \tag{7.2}$$

Let us fix this λ from now on and write $\phi = \phi_\lambda$.

Among the three functions whose composition defines ϕ, the first map (i.e., mod q) and the second map ($\cdot\lambda$ in $\mathbb{Z}/q\mathbb{Z}$) are group homomorphisms, and hence Freiman s-homomorphisms. The last map is not a Freiman s-homomorphism, but it becomes one when restricted to an interval of at most q/s elements (see Proposition 7.6.6). By the pigeonhole principle, we can find an interval I with

$$\operatorname{diam} I < q/s$$

such that

$$A' = \{a \in A \colon \phi(a) \in I\}$$

has $\geq |A|/s$ elements. So ϕ sends A' Freiman s-homomorphically to its image.

We further compose ϕ with the mod N map to obtain

$$\psi \colon \mathbb{Z} \xrightarrow{\phi} \{0, 1, \ldots, q-1\} \xrightarrow{\text{mod } N} \mathbb{Z}/N\mathbb{Z}.$$

We claim that ψ maps A' Freiman s-isomorphically to its image. Indeed, we saw that ψ is a Freiman s-homomorphism when restricted to A' (since both $\phi|_{A'}$ and the mod N map are). Now suppose $a_1, \ldots, a_s, a'_1, \ldots, a'_s \in A'$ satisfy

$$\psi(a_1) + \cdots + \psi(a_s) = \psi(a'_1) + \cdots + \psi(a'_s),$$

which is the same as saying that N divides

$$y := \phi(a_1) + \cdots + \phi(a_s) - \phi(a'_1) - \cdots - \phi(a'_s) \in \mathbb{Z}.$$

By swapping $(a_1, \ldots, a_s)$ with $(a'_1, \ldots, a'_s)$ if needed, we may assume that $y \geq 0$. Since $\phi(A') \subseteq I$, we have $|\phi(a_i) - \phi(a'_i)| \leq \operatorname{diam} I < q/s$ for each i, and thus

$$0 \leq y < q.$$

Let

$$x = a_1 + \cdots + a_s - a_1' - \cdots - a_s' \in sA - sA.$$

Since ϕ mod q is a group homomorphism,

$$\phi(x) \equiv \phi(a_1) + \cdots + \phi(a_s) - \phi(a_1') - \cdots - \phi(a_s') = y \pmod q.$$

Since

$$\phi(x), y \in [0, q) \cap \mathbb{Z} \quad \text{and} \quad \phi(x) \equiv y \pmod q,$$

we have $\phi(x) = y$. Since N divides $y = \phi(x)$, and by (7.2), $N \nmid \phi(x)$ for any nonzero $x \in sA - sA$, we must have $x = 0$. Thus,

$$a_1 + \cdots + a_s = a_1' + \cdots + a_s'.$$

Hence A' is a set of size $\geq |A|/s$ that is Freiman s-isomorphic via ψ to its image in $\mathbb{Z}/N\mathbb{Z}$. □

Exercise 7.7.4 (Modeling arbitrary sets of integers). Let $A \subseteq \mathbb{Z}$ with $|A| = n$.

(a) Let p be a prime. Show that there is some integer t relatively prime to p such that $\|at/p\|_{\mathbb{R}/\mathbb{Z}} \leq p^{-1/n}$ for all $a \in A$.

(b) Show that A is Freiman 2-isomorphic to a subset of $[N]$ for some $N = (4 + o(1))^n$.

(c) Show that (b) cannot be improved to $N = 2^{n-2}$.

(You may use the fact that the smallest prime larger than m has size $m + o(m)$.)

Exercise 7.7.5 (Sumset with 3-AP-free set). Let A and B be n-element subsets of the integers. Suppose A is 3-AP free. Prove that $|A + B| \geq n(\log \log n)^{1/100}$ provided that n is sufficiently large.

Hint: Ruzsa triangle inequality, Plünnecke's inequality, Ruzsa model lemma, Roth's theorem

Exercise 7.7.6 (3-AP-free subsets of arbitrary sets of integers). Prove that there is some constant $C > 0$ so that every set of n integers has a 3-AP-free subset of size $\geq ne^{-C\sqrt{\log n}}$.

7.8 Iterated Sumsets: Bogolyubov's Lemma

The goal of this section is to find a large Bohr set inside $2A - 2A$, provided that A is a relatively large subset of $\mathbb{Z}/N\mathbb{Z}$. The idea is due to Bogolyubov (1939).

Let us first explain what happens in the finite field model. Let $A \subseteq \mathbb{F}_2^n$ with $|A| \geq \alpha 2^n$. (Think of α as a constant for now.) Since A is arbitrary, we do not expect it to contain any large subspaces. But perhaps $A + A$ always does.

Question 7.8.1 (Large structure in $A + A$)
Suppose $A \subseteq \mathbb{F}_2^n$ and $|A| = \alpha 2^n$ where α is a constant independent of n. Must it be the case that $A + A$ contains a large subspace of codimension $O_\alpha(1)$?

The answer to the above question is no, as evidenced by the following example. (Niveau is French for level.)

***Example* 7.8.2** (Niveau set). Let A be the set of all points in $\mathbb{F}_2^n$ with Hamming weight (number of 1 entries) at most $(n - c\sqrt{n})/2$. Note by the central limit theorem $|A| = (\alpha + o(1))2^n$ for some constant $\alpha = \alpha(c) \in (0, 1)$. The sumset $A + A$ consists of points in the boolean cube whose Hamming weight is at most $n - c\sqrt{n}$ and thus does not contain any subspace of codimension $< c\sqrt{n}$, by Lemma 6.5.4.

It turns out that the iterated sumset $2A - 2A$ (same as $4A$ in $\mathbb{F}_2^n$) always contains a bounded codimensional subspace. The intuition is that taking sumsets "smooths" out the structure of a set, analogous to how convolutions in real analysis make functions more smooth.

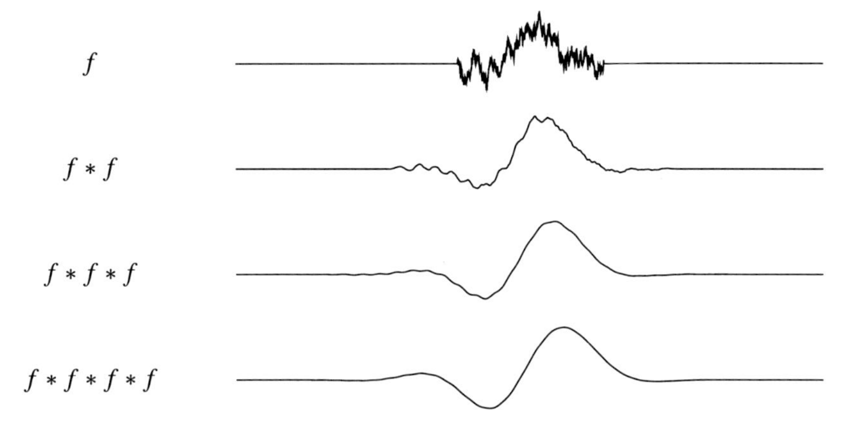

Recall some basic properties of the Fourier transform. Given $A \subseteq \mathbb{F}_p^n$ with $|A| = \alpha p^n$, we have

$$\widehat{1_A}(0) = \alpha,$$

and by Parseval's identity

$$\sum_{r \in \mathbb{F}_p^n} |\widehat{1_A}(r)|^2 = \mathbb{E}_{x \in \mathbb{F}_p^n} |1_A(x)|^2 = \alpha.$$

We write $\omega = \exp(2\pi i/p)$ in the proof below.

Theorem 7.8.3 (Bogolyubov's lemma in $\mathbb{F}_p^n$)
If $A \subseteq \mathbb{F}_p^n$ and $|A| = \alpha p^n > 0$, then $2A - 2A$ contains a subspace of codimension $< 1/\alpha^2$.

Proof. Let

$$f = 1_A * 1_A * 1_{-A} * 1_{-A},$$

which is supported on $2A - 2A$. By the convolution identity (Theorem 6.1.7), noting that $\widehat{1_{-A}}(r) = \overline{\widehat{1_A}(r)}$, we have, for every $r \in \mathbb{F}_p^n$,

$$\widehat{f}(r) = \widehat{1_A}(r)^2 \widehat{1_{-A}}(r)^2 = |\widehat{1_A}(r)|^4.$$

By the Fourier inversion formula (Theorem 6.1.2), we have

$$f(x) = \sum_{r \in \mathbb{F}_p^n} \widehat{f}(r)\omega^{r \cdot x} = \sum_{r \in \mathbb{F}_p^n} |\widehat{1_A}(r)|^4 \omega^{r \cdot x}.$$

It suffices to find a subspace where f is positive since $f(x) > 0$ implies $x \in 2A - 2A$. We will take the subspace defined by large Fourier coefficients. Let

$$R = \left\{ r \in \mathbb{F}_p^n \setminus \{0\} \colon |\widehat{1_A}(r)| > \alpha^{3/2} \right\}.$$

We can bound the size of R using Parseval's identity:

$$|R|\,\alpha^3 \le \sum_{r \in R} |\widehat{1_A}(r)|^2 < \sum_{r \in \mathbb{F}_p^n} |\widehat{1_A}(r)|^2 = \mathbb{E}_x |1_A(x)|^2 = \alpha.$$

(Skip the preceding step if R is empty.) So,

$$|R| < 1/\alpha^2.$$

If $r \notin R \cup \{0\}$, then $|\widehat{1_A}(r)| \le \alpha^{3/2}$. So, applying Parseval's identity again,

$$\begin{aligned} \sum_{r \notin R \cup \{0\}} |\widehat{1_A}(r)|^4 &\le \max_{r \notin R \cup \{0\}} |\widehat{1_A}(r)|^2 \sum_{r \notin R \cup \{0\}} |\widehat{1_A}(r)|^2 \\ &< \alpha^3 \sum_{r \in \mathbb{F}_p^n} |\widehat{1_A}(r)|^2 = \alpha^3 \mathbb{E}_x |1_A(x)|^2 = \alpha^4. \end{aligned}$$

Thus, for all $x \in R^\perp$, so that $x \cdot r = 0$ for all $r \in R$, we have

$$\begin{aligned} f(x) &= \sum_{r \in \mathbb{F}_p^n} |\widehat{1_A}(r)|^4 \operatorname{Re} \omega^{r \cdot x} \\ &\ge |\widehat{1_A}(0)|^4 + \sum_{r \in R} |\widehat{1_A}(r)|^4 - \sum_{r \notin R \cup \{0\}} |\widehat{1_A}(r)|^4 \\ &> \alpha^4 + 0 - \alpha^4 \\ &\ge 0. \end{aligned}$$

Thus, $R^\perp \subseteq \operatorname{supp}(f) = 2A - 2A$. Since $|R| < 1/\alpha^2$, we have found a subspace of codimension $< 1/\alpha^2$ contained in $2A - 2A$. □

To formulate an analogous result for a cyclic group $\mathbb{Z}/N\mathbb{Z}$, we need the notion of a Bohr set, which was mentioned earlier in the context of Roth's theorem (Remark 6.4.7).

Definition 7.8.4 (Bohr sets in $\mathbb{Z}/N\mathbb{Z}$)
Let $R \subseteq \mathbb{Z}/N\mathbb{Z}$. Define

$$\operatorname{Bohr}(R, \varepsilon) = \{x \in \mathbb{Z}/N\mathbb{Z} \colon \|rx/N\|_{\mathbb{R}/\mathbb{Z}} \le \varepsilon, \text{for all } r \in R\}$$

where $\|\cdot\|_{\mathbb{R}/\mathbb{Z}}$ denotes the distance to the nearest integer. Its ***dimension*** is $|R|$ and ***width*** is ε. (Strictly speaking, the definition of a Bohr set includes the data of R and ε and not just the set of the preceding elements.)

Bogolyubov's lemma holds over $\mathbb{Z}/N\mathbb{Z}$ after replacing subspaces by Bohr sets. Note that the dimension of a Bohr set of $\mathbb{Z}/N\mathbb{Z}$ corresponds to the codimension of a subspace in $\mathbb{F}_p^n$.

Theorem 7.8.5 (Bogolyubov's lemma in $\mathbb{Z}/N\mathbb{Z}$)
If $A \subseteq \mathbb{Z}/N\mathbb{Z}$ and $|A| = \alpha N$ then $2A - 2A$ contains some Bohr set $\operatorname{Bohr}(R, 1/4)$ with $|R| < 1/\alpha^2$.

With the right setup, the proof is essentially identical to that of Theorem 7.8.3.

Given $f\colon \mathbb{Z}/N\mathbb{Z} \to \mathbb{C}$, we define its ***Fourier transform*** to be the function $\widehat{f}\colon \mathbb{Z}/N\mathbb{Z} \to \mathbb{C}$ given by

$$\widehat{f}(r) = \mathbb{E}_{x\in\mathbb{Z}/N\mathbb{Z}} f(x)\omega^{-rx}$$

where $\omega = \exp(2\pi i/N)$. Fourier inversion, Parseval's identity, and the convolution identity all work the same way.

Proof. Let

$$f = 1_A * 1_A * 1_{-A} * 1_{-A},$$

which is supported on $2A - 2A$. By the convolution identity, for every $r \in \mathbb{Z}/N\mathbb{Z}$,

$$\widehat{f}(r) = \widehat{1_A}^2(r)\widehat{1}^2_{-A}(r) = |\widehat{1_A}(r)|^4.$$

By Fourier inversion, we have (noting that f is real-valued)

$$f(x) = \sum_{r\in\mathbb{Z}/N\mathbb{Z}} \widehat{f}(r)\omega^{rx} = \sum_{r\in\mathbb{Z}/N\mathbb{Z}} |\widehat{1_A}(r)|^4\omega^{rx}.$$

Let

$$R = \left\{r \in \mathbb{Z}/N\mathbb{Z}\setminus\{0\} : |\widehat{1_A}(r)| > \alpha^{3/2}\right\}.$$

As earlier, we can bound the size of R using Parseval's identity:

$$|R|\,\alpha^3 \le \sum_{r\in R}|\widehat{1_A}(r)|^2 < \sum_{r\in\mathbb{F}_p^n}|\widehat{1_A}(r)|^2 = \mathbb{E}_x|1_A(x)|^2 = \alpha.$$

(Skip the preceding step if R is empty.) So,

$$|R| < 1/\alpha^2.$$

We have

$$\sum_{r\notin R\cup\{0\}} |\widehat{1_A}(r)|^4 \le \alpha^3 \sum_{r\notin R\cup\{0\}} |\widehat{1_A}(r)|^2 < \alpha^4.$$

For all $x \in \operatorname{Bohr}(R, 1/4)$, every $r \in R$ satisfies $\|rx/N\|_{\mathbb{R}/\mathbb{Z}} \le 1/4$, and so $\cos(2\pi rx/N) \ge 0$. Thus, every $x \in \operatorname{Bohr}(R, 1/4)$ satisfies

$$\begin{aligned} f(x) &= \sum_{r\in\mathbb{Z}/N\mathbb{Z}} |\widehat{1_A}(r)|^4\omega^{r\cdot x} \\ &\ge |\widehat{1_A}(0)|^4 + \sum_{r\in R}|\widehat{1_A}(r)|^4 - \sum_{r\notin R\cup\{0\}} |\widehat{1_A}(r)|^4 \\ &> \alpha^4 + 0 - \alpha^4 \ge 0. \end{aligned}$$

Hence $\operatorname{Bohr}(R, 1/4) \subseteq 2A - 2A$. □

***Remark* 7.8.6** (Iterated sumsets and Goldbach conjecture)**.** The preceding proof hints at why it is easier to understand the iterated sumset kA when $k \ge 3$ than $k = 2$. (Roughly speaking, we need two iterations to just apply Parseval, and the extra room is helpful.) Exercise 7.8.7 shows that the threefold iterated sumset of every large subset of $\mathbb{F}_p^n$ contains a large affine subspace. (We do not always have a large subspace since the origin is not necessarily even in $3A$.)

A related phenomenon arises in Goldbach conjecture. Let P denote the set of primes. The still open Goldbach conjecture states that $P + P$ contains all sufficiently large even integers. On the other hand, Vinogradov (1937) showed that $P + P + P$ contains all sufficiently large odd integers (also known as the weak or ternary Goldbach problem).

Our next goal is to find a large GAP in the Bohr set produced by Bogolyubov's lemma. To do this, we need some results from the geometry of numbers.

Exercise 7.8.7 (Bogolyubov with 3-fold sums). Let $A \subseteq \mathbb{F}_p^n$ with $|A| = \alpha p^n$. Prove that $A + A + A$ contains a translate of a subspace of codimension $O(\alpha^{-3})$.

Exercise 7.8.8 (Bogolyubov with better bounds). Let $A \subseteq \mathbb{F}_p^n$ with $|A| = \alpha p^n$.

(a) Show that if $|A + A| < 0.99 \cdot 2^n$, then there is some $r \in \mathbb{F}_p^n \setminus \{0\}$ such that $|\widehat{1_A}(r)| > c\alpha^{3/2}$ for some absolute constant $c > 0$.

(b) By iterating (a), show that $A + A$ contains at least 99 percent of a subspace of codimension $O(\alpha^{-1/2})$.

(c) Deduce that $2A - 2A$ contains a subspace of codimension $O(\alpha^{-1/2})$.

7.9 Geometry of Numbers

We will need some results concerning lattices and convex bodies belonging to a topic in number theory called the geometry of numbers.

Definition 7.9.1 (Lattice)

A ***lattice*** in $\mathbb{R}^d$ is a set of the form

$$\Lambda = \mathbb{Z}v_1 \oplus \cdots \oplus \mathbb{Z}v_d = \{n_1 v_1 + \cdots + n_d v_d : n_1, \ldots, n_d \in \mathbb{Z}\}$$

where $v_1, \ldots, v_d \in \mathbb{R}^d$ are linearly independent vectors.

The ***fundamental parallelepiped*** of a lattice Λ with respect to the basis $v_1, \ldots, v_d$ is

$$\{x_1 v_1 + \cdots + x_d v_d : x_1, \ldots, x_d \in [0, 1)\}.$$

The ***determinant*** of this lattice is defined to be

$$\mathbf{det}\,\boldsymbol{\Lambda} := \left|\det\begin{pmatrix} | & \cdots & | \\ v_1 & \cdots & v_d \\ | & \cdots & | \end{pmatrix}\right|.$$

This is the absolute value of the determinant of a matrix with $v_1, \ldots, v_d$ as columns.

Given a lattice, there are many choices of a basis for the lattice. The determinant of a lattice does not depend on the choice of a basis, and equals the volume of every fundamental parallelepiped. Translations of the fundamental parallelepiped by lattice vectors tiles (i.e., partitions) the space.

An example of a lattice is illustrated here. Two different fundamental parallelepipeds are shaded.

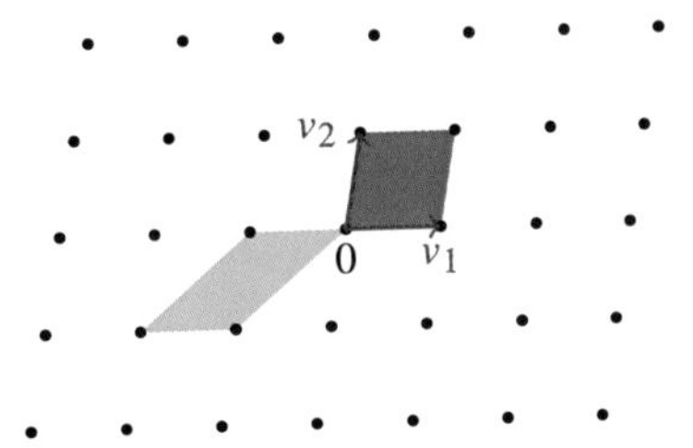

Let $\Lambda \subseteq \mathbb{R}^d$ be a lattice. Let $K \subseteq \mathbb{R}^d$ be a centrally symmetric convex body. (Here ***centrally symmetric*** means that $-x \in K$ whenever $x \in K$.) For each $\lambda \geq 0$, let $\lambda K = \{\lambda x : x \in K\}$ be the dilation of K by a factor λ.

As illustrated by the following diagram, imagine an animation where at time λ we see λK. This growing convex body initially is just the origin, and at some point it sees its first nonzero lattice point $\mathbf{b}_1$. Let us continue to grow this convex body. Later, at some point, it sees the first lattice point $\mathbf{b}_2$ in a new dimension not seen previously. And we can continue until the convex body grows big enough to contain lattice points that span all directions.

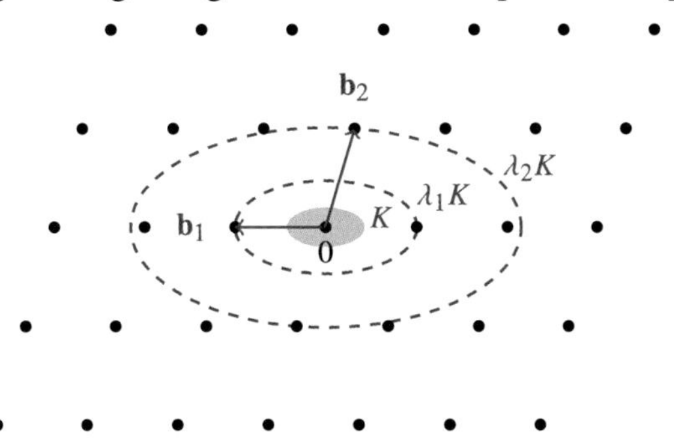

The process of dilating a convex body motivates the next definition.

Definition 7.9.2 (Successive minima)
Let Λ be a lattice in $\mathbb{R}^d$ and $K \subseteq \mathbb{R}^d$ a centrally symmetric convex body. For each $1 \leq i \leq d$, the ***ith successive minimum*** of K with respect to Λ is defined to be

$$\lambda_i = \inf\{\lambda \geq 0 : \dim(\operatorname{span}(\lambda K \cap \Lambda)) \geq i\}.$$

Equivalently, λ_i is the minimum λ such that λK contains i linearly independent lattice vectors from Λ.

A ***directional basis*** of K with respect to Λ is a basis $\mathbf{b}_1, \ldots, \mathbf{b}_d$ of $\mathbb{R}^d$ such that $\mathbf{b}_i \in \lambda_i K$ for each $i = 1, \ldots, d$.

Note that there may be more than one possible directional basis.

***Example* 7.9.3** (A directional basis does not necessarily generate the lattice)**.** Let $e_1, \ldots, e_8$ be the standard basis vectors in $\mathbb{R}^8$. Let $v = (e_1 + \cdots + e_8)/2$. Consider the lattice

$$\Lambda = \mathbb{Z}e_1 \oplus \cdots \oplus \mathbb{Z}e_7 \oplus \mathbb{Z}v = \mathbb{Z}^8 + \{0, v\}.$$

Let K be the unit ball in $\mathbb{R}^8$. Note that the directional basis of K with respect to Λ is $e_1, \ldots, e_8$, as all nonzero lattice points in Λ have length ≥ 1 (in particular, $|v| = \sqrt{2}$). This example shows that the directional basis of a convex body K is not necessarily a $\mathbb{Z}$-basis of Λ.

In the next section, we will apply the following fundamental result from the geometry of numbers (Minkowski 1896).

Theorem 7.9.4 (Minkowski's second theorem)
Let $\Lambda \in \mathbb{R}^d$ be a lattice and $K \subseteq \mathbb{R}^d$ a centrally symmetric convex body. Let $\lambda_1 \leq \cdots \leq \lambda_d$ be the successive minima of K with respect to Λ. Then

$$\lambda_1 \ldots \lambda_d \text{vol}(K) \leq 2^d \det(\Lambda).$$

***Example* 7.9.5.** Note that Minkowski's second theorem is tight when

$$K = \left[-\frac{1}{\lambda_1}, \frac{1}{\lambda_1}\right] \times \cdots \times \left[-\frac{1}{\lambda_d}, \frac{1}{\lambda_d}\right]$$

and Λ is the lattice $\mathbb{Z}^d$.

We will prove this theorem in the remainder of the section. The proof, while not long, is rather tricky. Feel free to skip the proof and jump to the next section.

Here is a simple geometric pigeonhole principle (Blichfeldt 1914).

Theorem 7.9.6 (Blichfeldt's theorem)
Let $\Lambda \subseteq \mathbb{R}^d$ be a lattice and $K \subseteq \mathbb{R}^d$ be a measurable set with $\text{vol}(K) > \det(\Lambda)$. Then there are distinct points $x, y \in K$ with $x - y \in \Lambda$.

Proof. Fix a fundamental parallelepiped P. Then $v + P$ tiles $\mathbb{R}^d$ as v ranges over Λ. Partition K by this tiling. For the portion of K lying in $v + P$, translate it by $-v$ to bring it back inside P. Then the parts of K all end up back inside P via translations by lattice vectors. Since $\text{vol}\, K > \text{vol}\, P = \det \Lambda$, some distinct pair of points $x, y \in K$ must end up at the same point of P. This then implies that $x - y \in \Lambda$. □

Here is an easy corollary (though we will not need it).

Theorem 7.9.7 (Minkowski's first theorem)
Let Λ be a lattice in $\mathbb{R}^d$ and $K \subseteq \mathbb{R}^d$ a centrally symmetric convex body. If $\text{vol}(K) > 2^d \det(\Lambda)$, then K contains a nonzero point of Λ.

Proof. We have $\text{vol}(\frac{1}{2}K) = 2^{-d}\,\text{vol}(K) > \det(\Lambda)$. By Blichfeldt's theorem there exist distinct $x, y \in \frac{1}{2}K$ such that $x - y \in \Lambda$. The point $x - y$ is the midpoint of $2x$ and $-2y$, both of which lie in K (using the fact that K is centrally symmetric) and hence $x - y$ lies in K (since K is convex). □

Note that Minkowski's first theorem is tight for $K = [-1, 1]^d$ and $\mathbb{Z}^d$.

Proof of Minkowski's second theorem (Theorem 7.9.4). The idea is to grow K until we hit a point of Λ, and then continue growing, but only in the complementary direction. However rigorously carrying out this procedure is very tricky (and easy to get wrong).

In the argument below, K is open (i.e., does not include the boundary). Fix a directional basis $\mathbf{b}_1, \ldots, \mathbf{b}_d$. For each $1 \leq j \leq d$, define map $\phi_j : K \to K$ by sending each point $x \in K$ to the center of mass of the $(j-1)$-dimensional slice of K which contains x and is parallel to $\text{span}_{\mathbb{R}}\{\mathbf{b}_1, \ldots, \mathbf{b}_{j-1}\}$. In particular, $\phi_1(x) = x$ for all $x \in K$.

Define a function $\psi\colon K \to \mathbb{R}^d$ by

$$\psi(x) = \sum_{j=1}^{d} \left(\frac{\lambda_j - \lambda_{j-1}}{2}\right) \phi_j(x),$$

where by convention we let $\lambda_0 = 0$.

For $\mathbf{x} = x_1\mathbf{b}_1 + \cdots + x_d\mathbf{b}_d \in \mathbb{R}^d$ with $x_1, \ldots, x_d \in \mathbb{R}$, we have

$$\phi_j(\mathbf{x}) = \sum_{i<j} c_{j,i}(x_j, \ldots, x_d)\mathbf{b}_i + \sum_{i \geq j} x_i \mathbf{b}_i$$

for some continuous functions $c_{j,i}$. By examining the coefficient of each $\mathbf{b}_i$, we find

$$\psi(\mathbf{x}) = \sum_{i=1}^{d} \left(\frac{\lambda_i x_i}{2} + \psi_i(x_{i+1}, \ldots, x_d)\right) \mathbf{b}_i$$

for some continuous functions $\psi_i(x_{i+1}, \ldots, x_d)$, so its Jacobian $\partial\psi(\mathbf{x})/\partial\mathbf{x}_j$ with respect to the basis $(\mathbf{b}_1, \ldots, \mathbf{b}_d)$ is upper triangular with diagonal $(\lambda_1/2, \ldots, \lambda_d/2)$. Therefore

$$\operatorname{vol}\psi(K) = \frac{\lambda_1 \cdots \lambda_d}{2^d} \operatorname{vol} K. \tag{7.3}$$

For any distinct points $\mathbf{x} = \sum x_i \mathbf{b}_i$, $\mathbf{y} = \sum y_i \mathbf{b}_i$ in K, let k be the largest index such that $x_k \neq y_k$. Then $\phi_i(\mathbf{x})$ agrees with $\phi_i(\mathbf{y})$ for all $i > k$. So,

$$\begin{aligned}\psi(\mathbf{x}) - \psi(\mathbf{y}) &= \sum_{j=1}^{d} (\lambda_j - \lambda_{j-1}) \left(\frac{\phi_j(\mathbf{x}) - \phi_j(\mathbf{y})}{2}\right) \\ &= \sum_{j=1}^{k} (\lambda_j - \lambda_{j-1}) \left(\frac{\phi_j(\mathbf{x}) - \phi_j(\mathbf{y})}{2}\right) \in \sum_{j=1}^{k} (\lambda_j - \lambda_{j-1})K = \lambda_k K.\end{aligned}$$

The $\in$ step is due to K being centrally symmetric and convex. The coefficient of $\mathbf{b}_k$ in $(\psi(\mathbf{x}) - \psi(\mathbf{y}))$ is $\lambda_k(x_k - y_k)/2 \neq 0$. So $\psi(\mathbf{x}) - \psi(\mathbf{y}) \notin \operatorname{span}_{\mathbb{R}}\{\mathbf{b}_1, \mathbf{b}_2, \ldots \mathbf{b}_{k-1}\}$. But we just saw that $\psi(\mathbf{x}) - \psi(\mathbf{y}) \in \lambda_k K$. Recall that K is open, and also $\lambda_k K \cap \Lambda$ is contained in $\operatorname{span}_{\mathbb{R}}\{\mathbf{b}_1, \mathbf{b}_2, \ldots \mathbf{b}_{k-1}\}$. Thus, $\psi(\mathbf{x}) - \psi(\mathbf{y}) \notin \Lambda$.

So, $\psi(K)$ contains no two points separated by a nonzero lattice vector. By Blichfeldt's theorem (Theorem 7.9.6), we deduce $\operatorname{vol}\psi(K) \leq \det\Lambda$. Combined with (7.3), we deduce

$$\lambda_1 \cdots \lambda_d \operatorname{vol} K \leq 2^d \operatorname{vol}\psi(K) \leq 2^d \det\Lambda. \qquad \square$$

7.10 Finding a GAP in a Bohr Set

Now we use Minkowski's second theorem to prove that a Bohr set of low dimension contains a large GAP.

Theorem 7.10.1 (Large GAP in a Bohr set)

Let N be a prime. Every Bohr set of dimension d and width $\varepsilon \in (0,1)$ in $\mathbb{Z}/N\mathbb{Z}$ contains a proper GAP with dimension at most d and volume at least $(\varepsilon/d)^d N$.

Proof. Let $R = \{r_1, \ldots, r_d\} \subseteq \mathbb{Z}/N\mathbb{Z}$. Recall that

$$\mathrm{Bohr}(R, \varepsilon) = \left\{x \in \mathbb{Z}/N\mathbb{Z} \colon \|xr/N\|_{\mathbb{R}/\mathbb{Z}} \le \varepsilon \text{ for all } r \in R\right\}.$$

Let

$$v = \left(\frac{r_1}{N}, \ldots, \frac{r_d}{N}\right).$$

Thus, for each $x = 0, 1, \ldots, N-1$, we have $x \in \mathrm{Bohr}(R, \varepsilon)$ if and only if some element of $xv + \mathbb{Z}^d$ lies in $[-\varepsilon, \varepsilon]^d$.

Let

$$\Lambda = \mathbb{Z}^d + \mathbb{Z}v \subseteq \mathbb{R}^d$$

be a lattice consisting of all points in $\mathbb{R}^d$ that are congruent mod 1 to some integer multiple of v. Note $\det(\Lambda) = 1/N$ since there are exactly N points of Λ within each translate of the unit cube. We consider the convex body $K = [-\varepsilon, \varepsilon]^d$. Let $\lambda_1, \ldots, \lambda_d$ be the successive minima of K with respect to Λ. Let $\mathbf{b}_1, \ldots, \mathbf{b}_d$ be the directional basis. We know

$$\|\mathbf{b}_j\|_\infty \le \lambda_j \varepsilon \text{ for all } j.$$

For each $1 \le j \le d$, let $L_j = \lceil 1/(\lambda_j d) \rceil$. If $0 \le l_j < L_j$, then

$$\|l_j \mathbf{b}_j\|_\infty < \frac{\varepsilon}{d}.$$

If we have integers $l_1, \ldots, l_d$ with $0 \le l_i < L_i$ for all i, then

$$\|l_1 \mathbf{b}_1 + \cdots + l_d \mathbf{b}_d\|_\infty \le \varepsilon.$$

For each $1 \le j \le d$, there is some $0 \le x_j < N$ so that $\mathbf{b}_j \in x_j v + \mathbb{Z}^d$, so its ith coordinate lies in $x_i r_i/N + \mathbb{Z}^d$. The i-th coordinate in the above L^∞ bound gives

$$\left\| \frac{(l_1 x_1 + \cdots + l_d x_d) r_i}{N} \right\|_{\mathbb{R}/\mathbb{Z}} \le \varepsilon \text{ for all } i.$$

Thus, the GAP

$$l_1 x_1 + \cdots + l_d x_d, \qquad 0 \le l_i < L_i \text{ for each } 1 \le i \le d$$

is contained in $\mathrm{Bohr}(R, \varepsilon)$. It remains to show that this GAP is large and proper. Its volume is, applying Minkowski's second theorem,

$$L_1 \cdots L_k \ge \frac{1}{\lambda_1 \cdots \lambda_d \cdot d^d} \ge \frac{\mathrm{vol}(K)}{2^d \det(\Lambda) d^d} = \frac{(2\varepsilon)^d}{2^d (1/N) d^d} = \left(\frac{\varepsilon}{d}\right)^d N.$$

Now we check that the GAP is proper. It suffices to show that if

$$l_1 x_1 + \cdots + l_d x_d \equiv l_1' x_1 + \cdots + l_d' x_d \pmod{N},$$

then we must have $l_i = l_i'$ for all i. Setting

$$\mathbf{b} = (l_1 - l_1')\mathbf{b}_1 + \cdots + (l_d - l_d')\mathbf{b}_d,$$

we have $\mathbf{b} \in \mathbb{Z}^d$. Furthermore

$$\|\mathbf{b}\|_\infty \le \sum_{i=1}^d \frac{1}{\lambda_i d} \|\mathbf{b}_i\|_\infty \le \varepsilon < 1,$$

so actually $\mathbf{b}$ must be 0. Since $b_1, \ldots, b_d$ is a basis we must have $l_i = l_i'$ for all i, as desired. □

7.11 Proof of Freiman's Theorem

We are now ready to prove Freiman's theorem by putting together all the ingredients in this chapter. Let us recall what we have proved.

- **Plünnecke's inequality** (Theorem 7.3.1): $|A + A| \le K|A|$ implies $|mA - nA| \le K^{m+n}|A|$ for all $m, n \ge 0$.
- **Ruzsa covering lemma** (Theorem 7.4.1): if $|X + B| \le K|B|$, then there exist some $T \subseteq X$ with $|T| \le K$ such that $X \subseteq T + B - B$.
- **Ruzsa modeling lemma** (Theorem 7.7.3): if $A \subseteq \mathbb{Z}$ and $|sA - sA| \le N$, then there exists $A' \subseteq A$ with $|A'| \ge |A|/s$ such that A' is Freiman s-isomorphic to a subset of $\mathbb{Z}/N\mathbb{Z}$.
- **Bogolyubov's lemma** (Theorem 7.8.5): for every $A \subseteq \mathbb{Z}/N\mathbb{Z}$ with $|A| = \alpha N$, $2A - 2A$ contains some Bohr set with dimension $< 1/\alpha^2$ and width $1/4$.
- By a geometry of numbers argument (Theorem 7.10.1), for every prime N, every Bohr set of dimension d and width $\varepsilon \in (0, 1)$ contains a proper GAP with dimension $\le d$ and volume $\ge (\varepsilon/d)^d N$.

Now we will prove Freiman's theorem. We restate it here with the bounds that we will prove.

Theorem 7.11.1 (Freiman's theorem)
Let $A \subseteq \mathbb{Z}$ be a finite set satisfying $|A + A| \le K|A|$. Then A is contained in a GAP of dimension at most $d(K)$ and volume at most $f(K)|A|$, where $d(K) \le \exp(K^C)$ and $f(K) \le \exp(\exp(K^C))$ for some absolute constant C.

Proof. By Plünnecke's theorem, we have $|8A - 8A| \le K^{16}|A|$. Let N be a prime with $K^{16}|A| \le N \le 2K^{16}|A|$ (it exists by Bertrand's postulate). By the Ruzsa modeling lemma, some $A' \subseteq A$ with $|A'| \ge |A|/8$ is Freiman 8-isomorphic to a subset B of $\mathbb{Z}/N\mathbb{Z}$.

Applying Bogolyubov's lemma on $B \subseteq \mathbb{Z}/N\mathbb{Z}$, with

$$\alpha = \frac{|B|}{N} = \frac{|A'|}{N} \ge \frac{|A|}{8N} \ge \frac{1}{16K^{16}},$$

we deduce that $2B - 2B$ contains a Bohr set with dimension $< 256K^{32}$ and width $1/4$. By Theorem 7.10.1, $2B - 2B$ contains a proper GAP with dimension $d < 256K^{32}$ and volume $\ge (4d)^{-d}N$.

Since B is Freiman 8-isomorphic to A', $2B - 2B$ is Freiman 2-isomorphic to $2A' - 2A'$. (Why?) Note GAPs are preserved by Freiman 2-isomorphisms. (Why?) Hence, the proper GAP in $2B - 2B$ is mapped to a proper GAP $Q \subseteq 2A' - 2A'$ with the same dimension ($\le d$) and volume ($\ge (4d)^{-d}N$). We have

$$|A| \le 8|A'| \le 8N \le 8(4d)^d|Q|.$$

Since $Q \subseteq 2A' - 2A' \subseteq 2A - 2A$, we have $Q + A \subseteq 3A - 2A$. By Plünnecke's inequality,

$$|Q + A| \le |3A - 2A| \le K^5|A| \le 8K^5(4d)^d|Q|.$$

By the Ruzsa covering lemma, there exists a subset X of A with $|X| \le 8K^5(4d)^d$ such that $A \subseteq X + Q - Q$. It remains to contain $X + Q - Q$ in a GAP.

By using two elements in each direction, X is contained in a GAP of dimension $|X| - 1$ and volume $\le 2^{|X|-1}$. Since Q is a proper GAP with dimension $d < 256K^{32}$ and volume $\le |2A - 2A| \le K^4 |A|$, $Q - Q$ is a GAP with dimension d and volume $\le 2^d K^4 |A|$. It follows that $A \subseteq X + Q - Q$ is contained in a GAP with

$$\text{dimension} \le |X| - 1 + d \le 8(4d)^d K^5 + d - 1 = e^{K^{O(1)}}$$

(recall $d < 256K^{32}$) and

$$\text{volume} \le 2^{|X|-1+d} K^4 |A| = e^{e^{K^{O(1)}}} |A| . \qquad \square$$

The following exercise asks to improve the quantitative bounds on Freiman's theorem.

Exercise 7.11.2 (Improved bounds on Freiman's theorem). Using a more efficient covering lemma from Exercise 7.4.3, prove Freiman's theorem with $d(K) = K^{O(1)}$ and $f(K) = \exp(K^{O(1)})$.

7.12 Polynomial Freiman–Ruzsa Conjecture

Here we explain one of the biggest open problems in additive combinatorics, known as the ***polynomial Freiman–Ruzsa conjecture (PFR)***. As mentioned in Remark 7.1.11, nearly optimal bounds $f(K) = K^{1+o(1)}$ and $d(K) = \exp(K^{1+o(1)})$ are known for Freiman's theorem. However, one can reformulate Freiman's theorem with significantly better quantitative dependencies.

PFR in the Finite Field Model

Let us first explain what happens in the finite field model $\mathbb{F}_2^n$. Theorem 7.5.1 showed that if $A \subseteq \mathbb{F}_2^n$ has $|A + A| \le K|A|$, then A is contained in a subspace of cardinality $\le f(K)|A|$. As mentioned in Remark 7.5.2, the optimal constant is known and satisfies $f(K) = \Theta(2^{2K}/K)$. An example requiring this bound is $A \subseteq \mathbb{F}^{m+n}$ defined by $A = \{e_1, \dots, e_n\} \times \mathbb{F}_2^m$ (where $e_1, \dots, e_n$ are the coordinate basis vectors of $\mathbb{F}_2^n$). Here $K = |A + A| / |A| \sim n/2$ and $|\langle A \rangle| = (2^n/n)|A|$. However, instead of trying to cover A by a single subspace, we can easily cover A by a small number of translates of a subspace with size comparable to A, namely A is covered by $\{e_1\} \times \mathbb{F}_2^m, \dots, \{e_n\} \times \mathbb{F}_2^m$, which are translates of each other and each has size $\le |A|$.

The Polynomial Freiman–Ruzsa conjecture in $\mathbb{F}_2^n$ proposes a variant of Freiman's theorem with polynomial bounds, where we are only required to cover a large fraction of A. Ruzsa (1999) attributes the conjecture to Marton.

Conjecture 7.12.1 (Polynomial Freiman–Ruzsa in $\mathbb{F}_2^n$)
If $A \subseteq \mathbb{F}_2^n$ and $|A + A| \le K|A|$, then there exists a subspace $V \subseteq \mathbb{F}_2^n$ with $|V| \le |A|$ such that A can be covered by $K^{O(1)}$ cosets of V.

The best current result says that in Conjecture 7.12.1 one can cover A by $\exp((\log K)^{O(1)})$ cosets of V (Sanders 2012). This is called a quasipolynomial bound.

This conjecture has several equivalent forms. Here we give some highlights. For more details, including proofs of equivalence, see the online note accompanying Green (2005c) titled *Notes on the Polynomial Freiman–Ruzsa Conjecture*.

For example, here is a formulation where we just need to use one subspace to cover a large fraction of A.

Conjecture 7.12.2 (Polynomial Freiman–Ruzsa in $\mathbb{F}_2^n$)
If $A \subseteq \mathbb{F}_2^n$ and $|A + A| \leq K|A|$, then there exists an affine subspace $V \subseteq \mathbb{F}_2^n$ with $|V| \leq |A|$ such that $|V \cap A| \geq K^{-O(1)}|A|$.

Proof of equivalence of Conjecture 7.12.1 and Conjecture 7.12.2. Conjecture 7.12.1 implies Conjecture 7.12.2 since by the pigeonhole principle, at least one of the cosets of V covers $\geq K^{-O(1)}$ fraction of A.

Now assume Conjecture 7.12.2. Let $A \subseteq \mathbb{F}_2^n$ with $|A + A| \leq K|A|$. Let V be as in Conjecture 7.12.2. By the Ruzsa covering lemma (Theorem 7.4.1) with $X = A$ and $B = V \cap A$ we find $T \subseteq X$ with $|T| \leq |X + B|/|X| \leq |A + A|/|A| \leq K$ such that $A \subseteq T + B - B \subseteq T + V$. The conclusion of Conjecture 7.12.1 holds. □

Here is another attractive equivalent formulation of the polynomial Freiman–Ruzsa conjecture in $\mathbb{F}_2^n$.

Conjecture 7.12.3 (Polynomial Freiman–Ruzsa in $\mathbb{F}_2^n$)
If $f : \mathbb{F}_2^n \to \mathbb{F}_2^n$ satisfies

$$\left|\{f(x, y) - f(x) - f(y) : x, y \in \mathbb{F}_2^n\}\right| \leq K,$$

then there exists a linear function $g : \mathbb{F}_2^n \to \mathbb{F}_2^n$ such that

$$\left|\{f(x) - g(x) : x \in \mathbb{F}_2^n\}\right| \leq K^{O(1)}.$$

In Conjecture 7.12.3, it is straightforward to show a bound of 2^K instead of $K^{O(1)}$, since we can extend f to a linear function based on its values at some basis.

To state our third reformulation, we need the notion of the Gowers uniformity norm. Given a finite abelian group Γ, and $f : \Gamma \to \mathbb{C}$, define the ***$U^3$ uniformity norm*** of f by

$$\|f\|_{U^3} := \Big(\mathbb{E}_{x,y_1,y_2,y_3} f(x)\overline{f(x + y_1) f(x + y_2) f(x + y_3)} \cdot \\ \cdot f(x + y_1 + y_2) f(x + y_1 + y_3) f(x + y_2 + y_3) \overline{f(x + y_1 + y_2 + y_3)}\Big)^{1/8}.$$

The U^3 norm plays a central role in Gowers' proof of Szemerédi's theorem for 4-APs. (The U^3 norm is also discussed in Exercise 6.2.14.)

If $f : \mathbb{F}_2^n \to \{-1, 1\}$ given by $f(x) = (-1)^{q(x)}$ where q is a quadratic polynomial in n variables over $\mathbb{F}_2$ (e.g., $x_1 + x_1x_2 + \cdots$), then it is not hard to check that the expression in the preceding expectation is identically 1. (It comes from taking three finite differences of q.) So, $\|f\|_{U^3} = 1$. For proving Szemerédi's theorem for 4-APs, one would like a "1% inverse result" showing that any $f : \mathbb{F}_2^n \to [-1, 1]$ satisfying $\|f\|_{U^3} \geq \delta$ must correlate with some quadratic polynomial phase function $(-1)^{q(x)}$. Such a result is known, but it remains open to find optimal quantitative bounds. The polynomial Freiman–Ruzsa conjecture in $\mathbb{F}_2^n$ is equivalent to a U^3 inverse statement with polynomial bounds (Green and Tao 2010b; Lovett 2012).

Conjecture 7.12.4 (U^3 inverse with polynomial bounds)
If $f\colon \mathbb{F}_2^n \to \mathbb{C}$ with $\|f\|_\infty \le 1$ and $\|f\|_{U_3} \ge 1/K$, then there exists a quadratic polynomial $q(x_1, \ldots, x_n)$ over $\mathbb{F}_2$ such that

$$\left|\mathbb{E}_{x\in\mathbb{F}_2^n} f(x)(-1)^{q(x)}\right| \ge K^{-O(1)}.$$

Remark **7.12.5** (Quantitative equivalence). It is known that the bounds in each of the above conjectures are equivalent to each other up to a polynomial change. This means that if one statement is true with conclusion $\le f(K)$ then all the other statements are true with conclusion $\le Cf(K)^C$ (appropriately interpreted) with some absolute constant C.

PFR in the Integers

Now we formulate the polynomial Freiman–Ruzsa conjecture in $\mathbb{Z}$ instead of $\mathbb{F}_2^n$. It is not enough to use GAPs (Lovett and Regev 2017). Instead, we need to consider convex progressions.

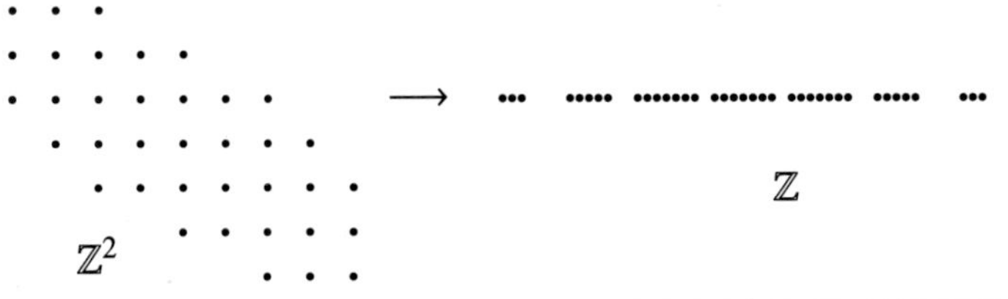

Definition 7.12.6 (Convex progression)
A ***centered convex progression*** in an abelian group Γ is defined to be an affine map

$$\phi\colon \mathbb{Z}^d \cap B \to \Gamma$$

where B is a centrally symmetric convex body. We define its ***dimension*** to be d and its ***volume*** to be $|\mathbb{Z}^d \cap B|$.

Conjecture 7.12.7 (Polynomial Freiman–Ruzsa conjecture in $\mathbb{Z}$)
If $A \subseteq \mathbb{Z}$ satisfies $|A + A| \le K\,|A|$, then one can cover A using $K^{O(1)}$ translates of some centered convex progression of dimension $O(\log K)$ and volume at most $|A|$.

More generally, one can formulate the polynomial Freiman–Ruzsa conjecture in an arbitrary abelian group.

Definition 7.12.8 (Centered convex coset progression)
In an abelian group, a ***centered convex coset progression*** is a set of the form $P + H$, where P is a centered convex progression and H is a subgroup. Its ***dimension*** is defined to be the dimension of P, and its ***volume*** is defined to be $|H|\operatorname{vol} P$.

Conjecture 7.12.9 (Polynomial Freiman–Ruzsa conjecture in abelian groups)
If A is a finite subset of an abelian group satisfying $|A + A| \le K|A|$, then one can cover A using $K^{O(1)}$ translates of some centered convex coset progression of dimension $O(\log K)$ and volume at most $|A|$.

For both Conjecture 7.12.7 and Conjecture 7.12.9, the best current result uses $\exp((\log K)^{O(1)})$ translates and dimension bound $(\log K)^{O(1)}$ (Sanders 2012, 2013).

7.13 Additive Energy and the Balog–Szemerédi–Gowers Theorem

We introduce a new way of measuring additive structure by counting the number of solutions to the equation $a + b = c + d$.

Definition 7.13.1 (Additive energy)
Let A be a finite set in an abelian group. Its ***additive energy*** is defined to be

$$\boldsymbol{E(A)} := |\{(a,b,c,d) \in A \times A \times A \times A \colon a + b = c + d\}|.$$

***Remark* 7.13.2.** The additive energy of A counts 4-cycles in the bipartite Cayley graph with generating set A. It is called an "energy" since we can write it as an L^2 quantity

$$E(A) = \sum_x r_A(x)^2$$

where

$$\boldsymbol{r_A(x)} := |\{(a,b) \in A \times A : a + b = x\}|$$

is the number of ways to write x as the sum of two elements of A.

We have the easy bound

$$2|A|^2 - |A| \le E(A) \le |A|^3.$$

The lower bound is due to trivial solutions $a + b = a + b$ and $a + b = b + a$. The lower bound is tight for sets without nontrivial solutions to $a + b = c + d$. The upper bound is due to d being determined by a, b, c when $a + b = c + d$. It is tight when A is a subgroup.

Here is the main question we explore in this section.

Question 7.13.3
What is the relationship between small doubling and large additive energy? (Both encode some notion of "lots of additive structure.")

One direction is easy.

Proposition 7.13.4 (Small doubling implies large additive energy)
Let A be a finite subset of an abelian group satisfying $|A + A| \le K|A|$. Then

$$E(A) \ge \frac{|A|^3}{K}.$$

Proof. Using $r_A(x)$ from Remark 7.13.2, by the Cauchy–Schwarz inequality

$$E(A) = \sum_{x \in A+A} r_A(x)^2 \geq \frac{1}{|A+A|}\left(\sum_{x \in A+A} r_A(x)\right)^2 = \frac{|A|^4}{|A+A|} \geq \frac{|A|^3}{K}. \qquad \square$$

The next example shows that the converse does not hold.

***Example* 7.13.5** (Large additive energy does not imply small doubling)**.** The set

$$A = [N] \cup \{2N+1, 2N+2, \ldots, 2N+2^N\}$$

is the union of a set of small doubling and a set without no additive structure. The first component has large additive energy, and so $E(A) = \Theta(N^3)$. On the other hand, the second component gives large doubling $|A+A| = \Theta(N^2)$.

However, we do have a converse if we allow passing to large subsets. Balog and Szemerédi (1994) showed that every set with large additive energy must contain a large subset with small doubling. Their proof used the regularity method, which required tower-type dependencies on the bounds. Gowers (2001) gave a new proof with much better bounds, and this result played a key role in his work on a new proof of Szemerédi's theorem. We will see Gowers' proof here. The presentation stems from Sudakov, Szemerédi, and Vu (2005).

Theorem 7.13.6 (Balog–Szemerédi–Gowers theorem)
Let A be a finite subset of an abelian group satisfying

$$E(A) \geq |A|^3 / K.$$

Then there is a subset $A' \subseteq A$ with

$$|A'| \geq K^{-O(1)} |A| \quad \text{and} \quad |A' + A'| \leq K^{O(1)} |A'|.$$

We will prove a version of the theorem allowing two different sets. Given two finite sets A and B in an abelian group, define their additive energy to be

$$\boldsymbol{E(A,B)} := |\{(a,b,a',b') \in A \times B \times A \times B \colon a + b = a' + b'\}|.$$

Then $E(A,A) = E(A)$.

Theorem 7.13.7 (Balog–Szemerédi–Gowers theorem)
Let A and B be finite subsets of the same abelian group. If $|A|, |B| \leq n$ and

$$E(A,B) \geq n^3/K,$$

then there exist subsets $A' \subseteq A$ and $B' \subseteq B$ with

$$|A'|, |B'| \geq K^{-O(1)} n \quad \text{and} \quad |A' + B'| \leq K^{O(1)} n.$$

Proof that Theorem 7.13.7 implies Theorem 7.13.6. Suppose $E(A) \geq |A|^3 / K$. Apply Theorem 7.13.7 with $B = A$ to obtain $A', B' \subseteq A$ with $|A'|, |B'| \geq K^{-O(1)} |A|$, and $|A' + B'| \leq K^{O(1)} |A|$. Then by Corollary 7.3.6, a variant of the Ruzsa triangle inequality, we have

$$|A' + A'| \le \frac{|A' + B'|^2}{|B'|} \le K^{O(1)} |A| . \qquad \square$$

We will prove Theorem 7.13.7 by setting up a graph.

Definition 7.13.8 (Restricted sumset)
Let A and B be subsets of an abelian group and let G be a bipartite graph with vertex bipartition $A \cup B$. We define the ***restricted sumset*** $A +_G B$ to be the set of sums along edges of G:

$$\boldsymbol{A +_G B} := \{a + b : (a, b) \in A \times B \text{ is an edge in } G\}.$$

Here is a graphical version of the Balog–Szemerédi–Gowers theorem.

Theorem 7.13.9 (Graph BSG)
Let A and B be finite subsets of an abelian group and let G be a bipartite graph with vertex bipartition $A \cup B$. If $|A|, |B| \le n$,

$$e(G) \ge \frac{n^2}{K} \quad \text{and} \quad |A +_G B| \le Kn,$$

then there exist subsets $A' \subseteq A$ and $B' \subseteq B$ with

$$|A'|, |B'| \ge K^{-O(1)}n \quad \text{and} \quad |A' + B'| \le K^{O(1)}n.$$

Proof that Theorem 7.13.9 implies Theorem 7.13.7. Denote the number of ways to write x as $a + b$ by

$$\boldsymbol{r_{A,B}(x)} := |\{(a, b) \in A \times B : a + b = x\}| .$$

Consider the "popular sums"

$$S = \left\{x \in A + B : r_{A,B}(x) \ge \frac{n}{2K}\right\}.$$

Build a bipartite graph G with bipartition $A \cup B$ such that $(a, b) \in A \times B$ is an edge if and only if $a + b \in S$.

We claim that G has many edges by showing that "unpopular sums" account for at most half of $E(A, B)$. Note that

$$\frac{n^3}{K} \le E(A, B) = \sum_{x \in S} r_{A,B}(x)^2 + \sum_{x \notin S} r_{A,B}(x)^2. \tag{7.4}$$

Because $r_{A,B}(x) < n/(2K)$ when $x \notin S$, we can bound the second term as

$$\sum_{x \notin S} r_{A,B}(x)^2 \le \frac{n}{2K} \sum_{x \notin S} r_{A,B}(x) \le \frac{n}{2K} |A| |B| \le \frac{n^3}{2K},$$

and setting back into (7.4) yields

$$\frac{n^3}{K} \le \sum_{x \in S} r_{A,B}(x)^2 + \frac{n^3}{2K},$$

and so

$$\sum_{x \in S} r_{A,B}(x)^2 \geq \frac{n^3}{2K}.$$

Moreover, because $r_{A,B}(x) \leq |A| \leq n$ for all x, it follows that

$$e(G) = \sum_{x \in S} r_{A,B}(x) \geq \sum_{x \in S} \frac{r_{A,B}(x)^2}{n} \geq \frac{n^2}{2K}.$$

Furthermore, $A +_G B \subseteq S$,

$$\frac{n}{2K} |A +_G B| \leq |A| |B| \leq n^2,$$

so $|A +_G B| \leq 2Kn$. Hence, we can apply Theorem 7.13.9 to find sets $A' \subseteq A$ and $B' \subseteq B$ with the desired properties. □

Proof of Graph BSG

The remainder of this section will focus on proving BSG (Theorem 7.13.9). We begin with a few lemmas.

Lemma 7.13.10 (Path of length 2 lemma)
Let $\delta, \varepsilon > 0$. Let G be a bipartite graph with vertex bipartition $A \cup B$ and at least $\delta |A| |B|$ edges. Then there is some $U \subseteq A$ with $|U| \geq \delta |A| /2$ such that at least $(1 - \varepsilon)$-fraction of the pairs $(x, y) \in U^2$ have at least $\varepsilon\delta^2 |B| /2$ neighbors common to x and y.

The proof uses the **dependent random choice** technique from Section 1.7. Instead of quoting theorems from that section, let us prove the result from scratch.

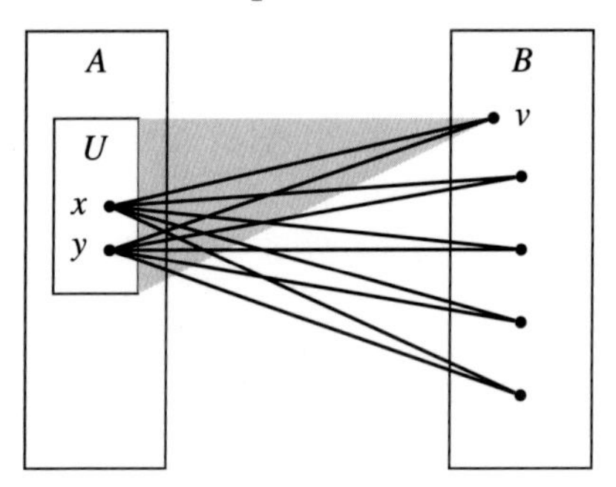

Proof. Say that a pair $(x, y) \in A^2$ is "unfriendly" if it has $< \varepsilon\delta^2 |B| /2$ common neighbors.

Choose $v \in B$ uniformly at random and let $U = N(v)$ be its neighborhood in v. We have

$$\mathbb{E} |U| = \mathbb{E} |N(v)| = \frac{e(G)}{|B|} \geq \delta |A| .$$

For each fixed pair $(x, y) \in A^2$, we have

$$\mathbb{P}(x, y \in U) = \mathbb{P}(x, y \in N(v)) = \frac{\text{codeg}(x, y)}{|B|}.$$

So if (x, y) is unfriendly, then $\mathbb{P}(x, y \in U) < \varepsilon\delta^2/2$. Let X be the number of unfriendly pairs $(x, y) \in U^2$. Then

$$\mathbb{E}X = \sum_{\substack{(x,y)\in A^2 \\ \text{unfriendly}}} \mathbb{P}(x, y \in U) < \frac{\varepsilon\delta^2}{2} |A|^2 .$$

Hence, we have

$$\mathbb{E}\left[|U|^2 - \frac{X}{\varepsilon}\right] \geq (\mathbb{E}|U|)^2 - \frac{\mathbb{E}X}{\varepsilon} > \frac{\delta^2}{2}|A|^2 .$$

So, for some $v \in B$, $U = N(v)$ satisfies

$$|U|^2 - \frac{X}{\varepsilon} \geq \frac{\delta^2}{2}|A|^2 .$$

Then this $U \subseteq A$ satisfies $|U|^2 \geq \delta^2 |A|^2/2$, and so $|U| \geq \delta|A|/2$. Moreover, we have $X \leq \varepsilon|U|^2$, so at most ε-fraction of pairs $(x, y) \in U^2$ are unfriendly. □

Lemma 7.13.11 (Path of length 3 lemma)
Let $\delta > 0$. Let G be a bipartite graph with vertex bipartition $A \cup B$ and at least $\delta|A||B|$ edges. Then there are subsets $A' \subseteq A$ and $B' \subseteq B$ with $|A'| \geq c\delta^C|A|$ and $|B'| \geq c\delta^C|B|$ such that the number of 3-edge paths joining every pair $(a, b) \in A' \times B'$ is at least $c\delta^C|A||B|$, and here $c, C > 0$ are absolute constants.

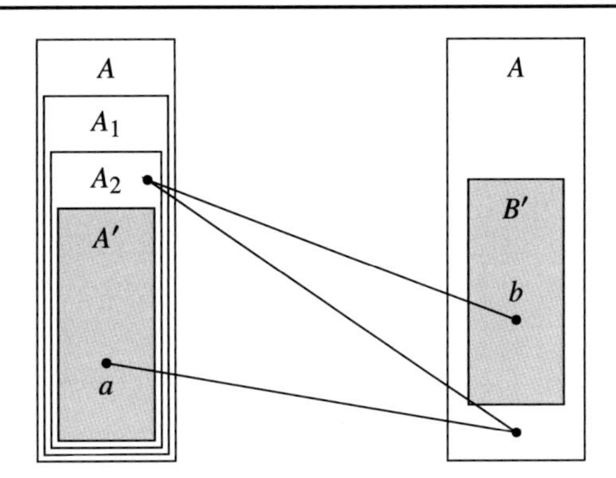

Proof. We repeatedly trim low-degree vertices.

Call vertices a pair of vertices in A "friendly" if they have $\geq \delta^3|B|/20$ common neighbors. Define

$$A_1 := \left\{a \in A\colon \deg a \geq \frac{\delta}{2}|B|\right\}.$$

Since each vertex in $A \setminus A_1$ has $< \delta|B|/2$ neighbors, $e(A \setminus A_1, B) \leq \delta|A||B|/2$. So

$$e(A_1, B) = e(A, B) - e(A \setminus A_1, B) \geq \delta|A||B| - \frac{\delta}{2}|A||B| \geq \frac{\delta}{2}|A||B| .$$

Hence $|A_1| \geq \delta|A|/2$.

Construct $A_2 \subseteq A_1$ via the path of length 2 lemma (Lemma 7.13.10) on (A_1, B) with $\varepsilon = \delta/10$. Then, $|A_2| \geq \delta|A_1|/2 \geq \delta^2|A|/4$ and $\leq (\delta/10)$-fraction pairs of vertices in A_2 are unfriendly.

Set

$$B' = \left\{ b \in B\colon \deg(b, A_2) \geq \frac{\delta}{4} |A_2| \right\}.$$

Since each vertex in $B \setminus B'$ has $< \delta |A_2| / 4$ neighbors in A_2, and so $e(A_2, B \setminus B') \leq \delta |A_2| |B| / 4$ edges. Since every vertex in A_1 has $\geq \delta |B| / 2$ neighbors in B, and $A_2 \subseteq A_1$, we have $e(A_2, B) \geq \delta |A_2| |B| / 2$. Hence

$$e(A_2, B') = e(A_2, B) - e(A_2, B \setminus B') \geq \frac{\delta}{2} |A_2| |B| - \frac{\delta}{4} |A_2| |B| \geq \frac{\delta}{4} |A_2| |B|.$$

Hence $|B'| \geq \delta |B| / 4$.

Let

$$A' = \{ a \in A_2 : a \text{ is friendly to} \geq (1 - \delta/5)\text{-fraction of } A_2 \}.$$

Since $\leq (\delta/10)$-fraction of pairs of vertices in A_2 are unfriendly, we have $|A'| \geq |A_2| / 2 \geq \delta^2 |A| / 8$.

We claim that that A' and B' satisfy the desired conclusions. Let $(a, b) \in A' \times B'$. Because b is adjacent to $\geq \delta |A_2| / 4$ vertices in A_2 and a is friendly to $\geq (1 - \delta/5) |A_2|$ vertices in A_2, there are $\geq \delta |A_2| / 20$ vertices in A_2 both friendly to a and adjacent to b. For each such $a_1 \in A_2$, there are $\geq \delta^3 |B| / 20$ vertices $b_1 \in B$ for which ab_1a_1b is a path of length 3, so the number of paths of length 3 from a to b is at least

$$\frac{\delta}{20} |A_2| \cdot \frac{\delta^3}{20} |B| \geq \frac{\delta}{20} \cdot \frac{\delta^2}{4} |A| \cdot \frac{\delta^3}{20} |B| \geq \frac{\delta^6}{1600} |A| |B|.$$

Furthermore, recall that $|A'| \geq \delta^2 |A| / 8$ and $|B'| \geq \delta |B| / 4$. □

We can use the path of length 3 lemma to prove the graph-theoretic analogue of the Balog–Szemerédi–Gowers theorem.

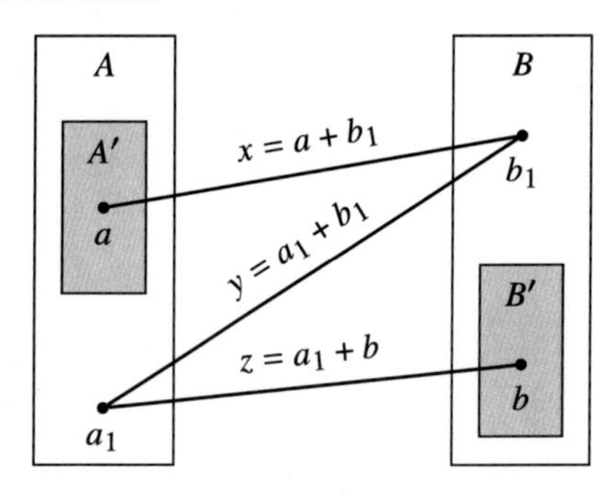

Proof of Theorem 7.13.9 (Graph BSG). Since $e(G) \geq n^2/K$, we have $|A|, |B| \geq n/K$. By the path of length 3 lemma (Lemma 7.13.11), we can find $A' \subseteq A$ and $B' \subseteq B$, each with size $\geq K^{-O(1)}n$ such that for every $(a, b) \in A' \times B'$, there are $\geq K^{-O(1)}n^2$ paths ab_1a_1b in G with $a_1 \in A$ and $b_1 \in B$. Then, with

$$x = a + b_1, \qquad y = a_1 + b_1, \qquad z = a_1 + b,$$

we have

$$a + b = x - y + z.$$

This shows that every element of $A' + B'$ can be written as $x - y + z$ for some $x, y, z \in A +_G B$ in $\geq K^{-O(1)}n^2$ ways. (For a given $(a, b) \in A' \times B'$, these choices of x, y, z are genuinely distinct; why?) Thus

$$K^{-O(1)}n^2 |A' + B'| \leq |A +_G B|^3 \leq K^3 n^3.$$

Therefore, $|A' + B'| \leq K^{O(1)}n$. □

Further Reading

See Ruzsa's lecture notes *Sumsets and Structure* (2009) for a comprehensive introduction to many topics related to set addition, including but not limited to Freiman's theorem.

Sanders' article *The Structure of Set Addition Revisited* (2013) provides a modern exposition of Freiman's theorem and his proof of the quasipolynomial Freiman–Ruzsa theorem. Lovett's article *An Exposition of Sanders' Quasi-Polynomial Freiman–Ruzsa Theorem* (2015) gives a gentle exposition of Sanders' proof in $\mathbb{F}_2^n$.

The methods discussed in this chapter play a central role in Gowers' proof of Szemerédi's theorem. The proof for 4-APs is especially worth studying, It contains many beautiful ideas and shows how these the topics in this chapter and the previous chapter are closely linked. See the original paper by Gowers (1998a) on Szemerédi's theorem for 4-APs as well as excellent lecture notes by Gowers (1998b), Green (2009b), and Soundararajan (2007).

Chapter Summary

- **Freiman's theorem.** Every $A \subseteq \mathbb{Z}$ with $|A + A| \leq K|A|$ is contained in a generalized arithmetic progression (GAP) of dimension $\leq d(K)$ and volume $\leq f(K)|A|$.
  - Informally: a set with small doubling is contained in a small GAP.
  - Up to constants, this gives a complete characterization of integer sets with bounded doubling.
- **Rusza triangle inequality.** $|A| |B - C| \leq |A - B| |A - C|$.
- **Plünnecke's inequality.** $|A + A| \leq K|A|$ implies $|mA - nA| \leq K^{m+n}|A|$.
- **Ruzsa covering lemma.** Idea: take a maximally disjoint set of translates, and their expansions must cover the entire space.
- **Freiman's theorem in groups with bounded exponent.** A set with bounded doubling is contained in a small subgroup.
- **Freiman s-homomorphisms** are maps preserving s-fold sums.
- **Ruzsa modeling lemma.** A set of integers with small doubling can be partially modeled as a large fraction of a cyclic group via a Freiman isomorphism.
- **Bogolyubov's lemma.** If A is large, then $2A - 2A$ contains a large subspace (finite field model) or GAP (cyclic group).
- A large **Bohr set** contains a large GAP. Proof uses **Minkowski's second theorem** from the **geometry of numbers**.
- **Polynomial Freiman–Ruzsa conjeture**: a central conjecture in additive combinatorics. The finite field model version has several equivalent and attractive statements, one of which says: if $A \subseteq \mathbb{F}_2^n$, and $|A + A| \leq K|A|$, then A can be covered using $K^{O(1)}$ translates of some subspace with cardinality $\leq |A|$.
- The **additive energy** $E(A)$ of a set A is the number of solutions to $a + b = c + d$ in A.
- **Balog–Szemerédi–Gowers theorem.** If $E(A) \geq |A|^3/K$, then A has a subset A' with $|A'| \geq K^{-O(1)}|A|$ and $|A' + A'| \leq K^{O(1)}|A'|$.
  - Informally: a set with large additive energy contains a large subset with small doubling.

8

Sum-Product Problem

Chapter Highlights

- The sum-product problem: show either $A + A$ or $A \cdot A$ must be large
- Erdős multiplication table problem
- Crossing number inequality: lower bound on the number of crossings in a graph drawing
- Szemerédi–Trotter theorem on point-line incidences
- Elekes' sum-product bound using incidence geometry
- Solymosi's sum-product bound via multiplicative energy

In the previous chapter we studied the ***sumset***

$$\boldsymbol{A + A} := \{a + b : a, b \in A\}.$$

Likewise we can also consider the ***product set***

$$\boldsymbol{A \cdot A = AA} := \{ab : a, b \in A\}.$$

Question 8.0.1 (Sum-product problem)
Can the sumset and the product set be simultaneously small?

Arithmetic progressions have small additive doubling, while geometric progressions have small multiplicative doubling. However, perhaps a set cannot simultaneously look like both an arithmetic and a geometric progression.

Erdős and Szemerédi (1983) conjectured that at least one of $A + A$ and AA is close to quadratic size.

Conjecture 8.0.2 (Sum-product conjecture)
For every finite subset A of $\mathbb{R}$, we have

$$\max\{|A + A|, |AA|\} \geq |A|^{2-o(1)}.$$

Here $o(1)$ is some quantity that goes to zero as $|A| \to \infty$.

Erdős and Szemerédi (1983) proved bounds of the form

$$\max\{|A + A|, |AA|\} \geq |A|^{1+c}$$

for some constant $c > 0$. In this chapter, we will give two different proofs of the above form. First, we present a proof by Elekes (1997) using incidence geometry, in particular a seminal theorem of Szemerédi and Trotter (1983) on the incidences of points and lines. Second, we present a proof by Solymosi (2009) using multiplicative energy, which gives nearly the best bound to date.

8.1 Multiplication Table Problem

Let us first explain why we need the error term $-o(1)$ in the exponent in Conjecture 8.0.2. Erdős (1955) posed the following problem.

Question 8.1.1 (Erdős multiplication table problem)
What is the size of $[N] \cdot [N]$?

This is asking for the number of distinct entries that appear in the $N \times N$ multiplication table.

1	2	3	4	5	6	7	8	9	10	⋯
2	4	6	8	10	12	14	16	18	20	⋯
3	6	9	12	15	18	21	24	27	30	⋯
4	8	12	16	20	24	28	32	36	40	⋯
5	10	15	20	25	30	35	40	45	50	⋯
6	12	18	24	30	36	42	48	54	60	⋯
7	14	21	28	35	42	49	56	63	70	⋯
8	16	24	32	40	48	56	64	72	80	⋯
9	18	27	36	45	54	63	72	81	90	⋯
10	20	30	40	50	60	70	80	90	100	⋯
⋮	⋮	⋮	⋮	⋮	⋮	⋮	⋮	⋮	⋮	⋱

After much work, we now have a satisfactory answer. A precise estimate was given by Ford (2008):

$$|[N] \cdot [N]| = \Theta\left(\frac{N^2}{(\log N)^{\delta}(\log\log N)^{3/2}}\right),$$

where $\delta = 1 - (1 + \log\log 2)/\log 2 \approx 0.086$. Here we give a short proof of some weaker estimates (Erdős 1955).

Theorem 8.1.2 (Estimates on the multiplication table problem)

$$(1 - o(1))\,\frac{N^2}{2\log N} \le |[N] \cdot [N]| = o(N^2)$$

This already shows that it is false that at least one of $A + A$ and AA has size $\ge c\,|A|^2$. So we cannot remove the $-o(1)$ term from the exponent in the sum-product conjecture.

To prove Theorem 8.1.2, we apply the following fact from number theory due to Hardy and Ramanujan (1917). A short probabilistic method proof was given by Turán (1934); also see Alon and Spencer (2016, Section 4.2).

Theorem 8.1.3 (Hardy–Ramanujan theorem)
All but $o(N)$ positive integers up to N have $(1 + o(1))\log\log N$ prime factors counted with multiplicity.

Proof of Theorem 8.1.2. First let us prove the upper bound. By the Hardy–Ramanujan theorem, all but at most $o(N^2)$ of the elements of $[N] \cdot [N]$ have $(2 + o(1)) \log\log N$ prime factors. However, by the Hardy–Ramanujan theorem again, all but $o(N^2)$ of positive integers $\leq N^2$ have $(1 + o(1)) \log\log(N^2) = (1 + o(1)) \log\log N$ prime factors, and thus cannot appear in $[N] \cdot [N]$. Hence $|[N] \cdot [N]| = o(N^2)$. (Remark: this proof gives $|[N] \cdot [N]| = O(N^2/\log\log N)$.)

Now let us prove the lower bound by giving a lower bound to the number of positive integers $\leq N^2$ of the form pm, where p is a prime in $(N^{2/3}, N]$ and $m \leq N$. Every such n has at most two such representations as pm since $n \leq N^2$ can have at most two prime factors greater than $N^{2/3}$. There are $(1 + o(1))N/\log N$ primes in $(N^{2/3}, N]$ by the prime number theorem. So the number of distinct such pm is $\geq (1/2 - o(1))N^2/\log N$. □

***Remark* 8.1.4.** The lower bound (up to a constant factor) also follows from Solymosi's sum-product estimate that we will see later in Theorem 8.3.1.

8.2 Crossing Number Inequality and Point-Line Incidences

The goal of this section is to give a proof of the following sum-product estimate, due to Elekes (1997), using incidence geometry. Recall we use $f \gtrsim g$ to mean that $f \geq cg$ for some constant $c > 0$.

Theorem 8.2.1 (Elekes' sum-product bound)
Every finite $A \subseteq \mathbb{R}$ satisfies

$$|A + A|\,|AA| \gtrsim |A|^{5/2}.$$

Corollary 8.2.2 (Elekes' sum-product bound)
Every finite $A \subseteq \mathbb{R}$ satisfies

$$\max\{|A + A|, |AA|\} \gtrsim |A|^{5/4}.$$

We introduce a basic result from geometric graph theory.

Cross Number Inequality

The ***crossing number*** $\mathrm{cr}(G)$ of a graph G is defined to be the minimum number of edge crossings in a planar drawing of G where edges are drawn with continuous curves.

The next theorem shows that every drawing of a graph with many edges necessarily has lots of edge crossings. For example, it implies that every n-vertex graph with $\Omega(n^2)$ edges has $\Omega(n^4)$ crossings; that is, a constant fraction of the edges must cross in a dense graph. This result is independently due to Ajtai, Chvátal, Newborn, and Szemerédi (1982) and Leighton (1984).

Theorem 8.2.3 (Crossing Number Inequality)
Every graph $G = (V, E)$ with $|E| \geq 4|V|$ has

$$\mathrm{cr}(G) \gtrsim \frac{|E|^3}{|V|^2}.$$

Proof. For any connected planar graph $G = (V, E)$ with at least one cycle, we have $3|F| \leq 2|E|$, with $|F|$ denoting the number of faces (including the outer face). The inequality follows from double counting using the fact that every face is adjacent to at least three edges and that every edge is adjacent to at most two faces. By Euler's formula, $|V| - |E| + |F| = 2$. Replacing $|F|$ using $3|F| \leq 2|E|$, we obtain $|E| \leq 3|V| - 6$. Therefore $|E| \leq 3|V|$ holds for every planar graph G including ones that are not connected or do not have a cycle.

If an arbitrary graph $G = (V, E)$ satisfies $|E| > 3|V|$, then any drawing of G can be made planar by deleting at most $\mathrm{cr}(G)$ edges, one for each crossing. It follows that $|E| - \mathrm{cr}(G) \geq 3|V|$. Therefore, the following inequality holds universally for all graphs $G = (V, E)$:

$$\mathrm{cr}(G) \geq |E| - 3|V|. \tag{8.1}$$

Now we apply a probabilistic method technique to "boost" the preceding inequality to denser graphs. Let $G = (V, E)$ be a graph with $|E| \geq 4|V|$. Let $p \in [0, 1]$ be some real number to be determined and let $G' = (V', E')$ be a graph obtained by independently randomly keeping each vertex of G with probability p. By (8.1), we have $\mathrm{cr}(G') \geq |E'| - 3|V'|$ for every G'. Therefore the same inequality must hold if we take the expected values of both sides:

$$\mathbb{E}\,\mathrm{cr}(G') \geq \mathbb{E}\,|E'| - 3\mathbb{E}\,|V'|.$$

We have $\mathbb{E}\,|E'| = p^2|E|$ since an edge remains in G' if and only if both of its endpoints are kept. Similarly $\mathbb{E}\,|V'| = p|V|$. By keeping the same drawing, we get the inequality $p^4\,\mathrm{cr}(G) \geq \mathbb{E}\,\mathrm{cr}(G')$. Therefore

$$\mathrm{cr}(G) \geq p^{-2}|E| - 3p^{-3}|V|.$$

Finally, set $p = 4|V|/|E| \in [0, 1]$ (here we use $|E| \geq 4|V|$) to get $\mathrm{cr}(G) \gtrsim |E|^3/|V|^2$. □

Szemerédi–Trotter Theorem on Point-Line Incidences

Given a set of points $\mathcal{P}$ and the set of lines $\mathcal{L}$, define the number of incidences to be

$$I(\mathcal{P}, \mathcal{L}) := |\{(p, \ell) \in \mathcal{P} \times \mathcal{L} : p \in \ell\}|.$$

Question 8.2.4 (Point-line incidence)
What's the maximum number of incidences between n points and m lines?

One trivial upper bound is $|\mathcal{P}|\,|\mathcal{L}|$. We can get a better bound by using the fact that every pair of points is determined by at most one line:

$$\begin{aligned}|\mathcal{P}|^2 &\geq |\{(p,p',\ell) \in \mathcal{P}\times\mathcal{P}\times\mathcal{L} : pp' \in \ell,\, p \neq p'\}| \\ &\geq \sum_{\ell\in\mathcal{L}} |\mathcal{P}\cap\ell|\,(|\mathcal{P}\cap\ell| - 1) \geq \frac{I(\mathcal{P},\mathcal{L})^2}{|\mathcal{L}|^2} - I(\mathcal{P},\mathcal{L}).\end{aligned}$$

The last inequality follows from the Cauchy–Schwarz inequality. Therefore,

$$I(\mathcal{P},\mathcal{L}) \leq |\mathcal{P}|\,|\mathcal{L}|^{1/2} + |\mathcal{L}|\,.$$

By the same argument with the roles of points and lines swapped (or by applying point-line duality),

$$I(\mathcal{P},\mathcal{L}) \leq |\mathcal{L}|\,|\mathcal{P}|^{1/2} + |\mathcal{P}|\,.$$

In particular, these inequalities tell us that n points and n lines have $O(n^{3/2})$ incidences.

The preceding bound only uses the fact that every pair of points determines at most one line. Equivalently, we are only using the fact that the bipartite point-line incidence graph is 4-cycle-free. So the $O(n^{3/2})$ bound (and the preceding proof) is the same as the $K_{2,2}$-free extremal number bound from Section 1.4. Also, the $O(n^{3/2})$ bound is tight for the finite field projective plane over $\mathbb{F}_q$ with $n = q^2 + q + 1$ points, and $n = q^2 + q + 1$ lines gives $n(q+1) \sim n^{3/2}$ incidences (this is the same construction showing that $\mathrm{ex}(n, K_{2,2}) \gtrsim n^{3/2}$ in Theorem 1.10.1).

On the other hand, in the real plane, the $n^{3/2}$ bound can be substantially improved. The following seminal result due to Szemerédi and Trotter (1983) gives a tight estimate of the number of point-line incidences in the real plane.

Theorem 8.2.5 (Szemerédi–Trotter theorem)
For any set $\mathcal{P}$ of points and $\mathcal{L}$ of lines in $\mathbb{R}^2$,

$$I(\mathcal{P},\mathcal{L}) \lesssim |\mathcal{P}|^{2/3}\,|\mathcal{L}|^{2/3} + |\mathcal{P}| + |\mathcal{L}|$$

Corollary 8.2.6
The number of point-line incidences between n points and n lines in $\mathbb{R}^2$ is $O(n^{4/3})$.

We will see a short proof using the crossing number inequality due to Székely (1997). Since the inequality is false over finite fields, any proof necessarily requires the topology of the real plane (via the application of Euler's theorem in the proof of the crossing number inequality).

***Example* 8.2.7.** The bounds in both Theorem 8.2.5 and Corollary 8.2.6 are best possible up to a constant factor. Here is an example showing that Corollary 8.2.6 is tight. Let $\mathcal{P} = [k]\times[2k^2]$ and $\mathcal{L} = \{y = mx + b \colon m \in [k], b \in [k^2]\}$. Then every line in $\mathcal{L}$ contains k points from $\mathcal{P}$. So, $I(\mathcal{P},\mathcal{L}) = k^4 = \Theta(n^{4/3})$.

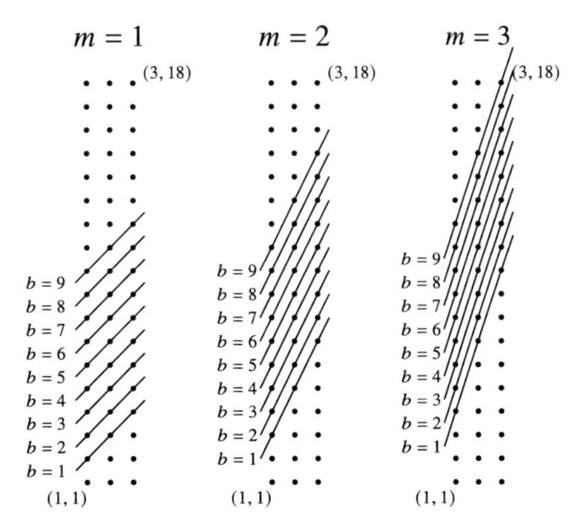

Proof of Theorem 8.2.5. We remove all lines in $\mathcal{L}$ containing at most one point in $\mathcal{P}$. These lines contribute to at most $|\mathcal{L}|$ incidences and thus do not affect the inequality we wish to prove.

Now assume that every line in $\mathcal{L}$ contains at least two points of $\mathcal{P}$. Turn every point of $\mathcal{P}$ into a vertex and each line in $\mathcal{L}$ into edges connecting consecutive points of $\mathcal{P}$ on the line. This constructs a drawing of a graph $G = (V, E)$ on the plane.

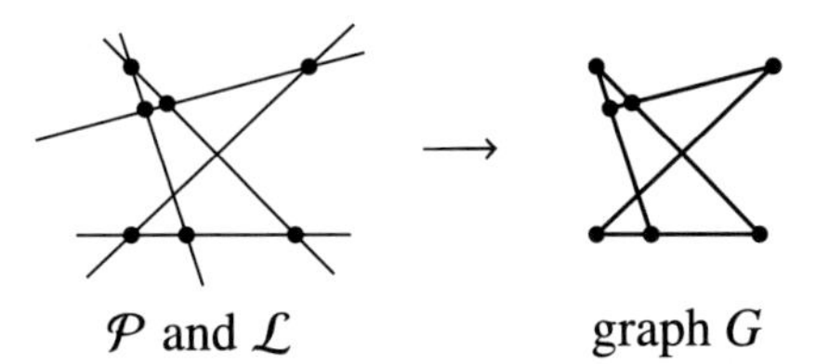

Assume that $I(\mathcal{P}, \mathcal{L}) \geq 8\,|\mathcal{P}|$ holds. (Otherwise, we are done as $I(\mathcal{P}, \mathcal{L}) \lesssim |\mathcal{P}|$.) Each line in $\mathcal{L}$ with k incidences has $k - 1 \geq k/2$ edges. So, $|E| \geq I(\mathcal{P}, \mathcal{L})/2 \geq 4\,|V|$. The crossing number inequality (Theorem 8.2.3) gives

$$\operatorname{cr}(G) \gtrsim \frac{|E|^3}{|V|^2} \gtrsim \frac{I(\mathcal{P}, \mathcal{L})^3}{|\mathcal{P}|^2}.$$

Moreover, $\operatorname{cr}(G) \leq |\mathcal{L}|^2$ since every pair of lines intersect in at most one point. Rearranging gives $I(\mathcal{P}, \mathcal{L}) \lesssim |P|^{2/3}\,|L|^{2/3}$. (Remember the linear contributions $|\mathcal{P}| + |\mathcal{L}|$ that need to be added back in due to the assumptions made earlier in the proof.) $\square$

Now we are ready to prove the sum-product estimate in Theorem 8.2.1 for $A \subseteq \mathbb{R}$:

$$|A + A|\,|AA| \gtrsim |A|^{5/2}.$$

Proof of Theorem 8.2.1. In $\mathbb{R}^2$, consider a set of points

$$\mathcal{P} = \{(x, y) \colon x \in A + A, y \in AA\}$$

and a set of lines

$$\mathcal{L} = \{y = a(x - a') : a, a' \in A\}.$$

For a line $y = a(x - a')$ in $\mathcal{L}$, $(a' + b, ab) \in \mathcal{P}$ is on the line for all $b \in A$, so each line in $\mathcal{L}$ contains $\geq |A|$ incidences. By definition of $\mathcal{P}$ and $\mathcal{L}$, we have

$$|\mathcal{P}| = |A + A|\,|AA| \qquad \text{and} \qquad |\mathcal{L}| = |A|^2.$$

By the Szemerédi–Trotter theorem (Theorem 8.2.5),

$$\begin{aligned}|A|^3 = |A|\,|\mathcal{L}| \leq I(\mathcal{P}, \mathcal{L}) &\lesssim |\mathcal{P}|^{2/3}\,|\mathcal{L}|^{2/3} + |\mathcal{P}| + |\mathcal{L}| \\ &\lesssim |A + A|^{2/3}\,|AA|^{2/3}\,|A|^{4/3}.\end{aligned}$$

The contributions from $|\mathcal{P}| + |\mathcal{L}|$ are lower order as $|\mathcal{P}| = |A + A|\,|AA| \leq |A|^4 = |\mathcal{L}|^2$ and $|\mathcal{L}| = |A|^2 \leq |A + A|^2\,|AA|^2 = |\mathcal{P}|^2$. Rearranging the preceding inequality gives

$$|A + A|\,|AA| \gtrsim |A|^{5/2}.$$

□

In Section 1.4, we proved an $O(n^{3/2})$ upper bound on the unit distance problem (Question 1.4.6) using the extremal number of $K_{2,3}$. The next exercise gives an improved bound (in fact, the best known result to date).

Exercise 8.2.8 (Unit distance bound). Using the crossing number inequality, prove that given n points in the plane, at most $O(n^{4/3})$ pairs of points are separated by exactly unit distance.

8.3 Sum-Product via Multiplicative Energy

In this chapter, we provide a different proof that gives a better sum-product estimate, due to Solymosi (2009).

Theorem 8.3.1 (Solymosi's sum-product bound)
Every finite set A of positive reals satisfies

$$|AA|\,|A + A|^2 \gtrsim \frac{|A|^4}{\log |A|}.$$

Corollary 8.3.2 (Solymosi's sum-product bound)
Every finite $A \subseteq \mathbb{R}$ satisfies

$$\max\{|A + A|, |AA|\} \geq |A|^{4/3 - o(1)}.$$

Proof of Theorem 8.3.1. We define the ***multiplicative energy*** of A to be

$$\boldsymbol{E_\times(A)} := |\{(a, b, c, d) \in A \times A \times A \times A : ab = cd\}|$$

Note that the multiplicative energy is a multiplicative version of additive energy. As with additive energy, having small multiplicative doubling implies large multiplicative energy, as seen by an application of the Cauchy–Schwarz inequality:

$$E_\times(A) = \sum_{x \in AA} \left|\{(a, b) \in A^2 : ab = x\}\right|^2 \geq \frac{|A|^4}{|AA|}.$$

Let

$$\boldsymbol{A/A} := \{a/b\colon a, b \in A\}\,.$$

Write

$$r(s) = |\{(a,b) \in A \times A\colon s = a/b\}|\,.$$

We have

$$E_\times(A) = \sum_{s \in A/A} r(s)^2.$$

By the pigeonhole principle (dyadic partitioning), there exists some nonnegative integer $k \lesssim \log|A|$ such that, setting

$$D = \{s\colon 2^k \le r(s) \le 2^{k+1}\} \quad \text{and} \quad m = |D|\,,$$

one has

$$\frac{E_\times(A)}{\log|A|} \lesssim \sum_{s \in D} r(s)^2 \le m 2^{2k+2}. \tag{8.2}$$

Let the elements of D be $s_1 < s_2 < \cdots < s_m$. For each $i \in [m]$, let ℓ_i be the line $y = s_i x$. Let ℓ_{m+1} be the vertical ray $x = \min(A)$ above ℓ_m.

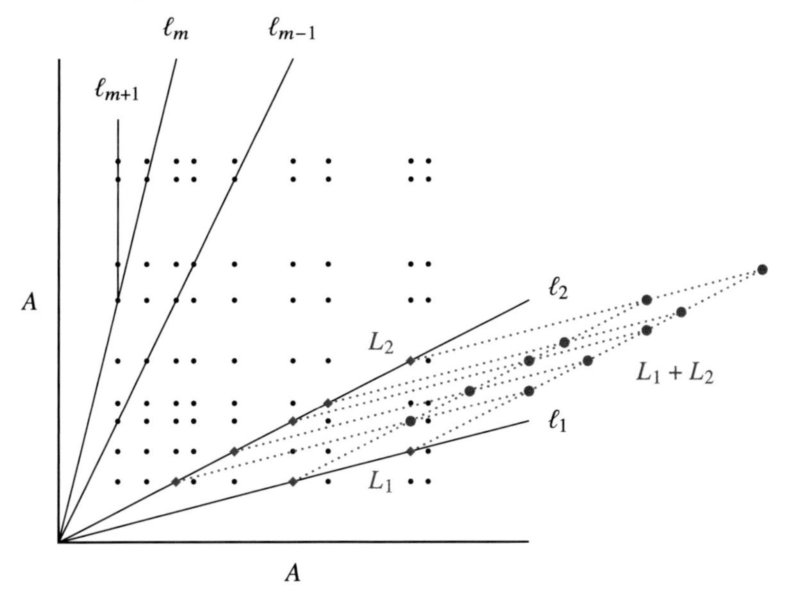

Let $L_j = (A \times A) \cap \ell_j$. Then, for each $1 \le j \le m$,

$$|L_j| = r(s_j) \ge 2^k.$$

Furthermore, $|L_{m+1}| \ge |L_m| \ge 2^k$ as well.

Since ℓ_j and ℓ_{j+1} are not parallel, we have $|L_j + L_{j+1}| = |L_j||L_{j+1}|$. Moreover, the sets $L_j + L_{j+1}$ are disjoint for different j. The sumset $A \times A + A \times A$ (here $A \times A$ is the Cartesian product) contains $L_j + L_{j+1}$ for each $1 \le j \le m$, so, using (8.2),

$$|A+A|^2 = |A \times A + A \times A| \geq \sum_{j=1}^{m} |L_j + L_{j+1}| = \sum_{j=1}^{m} |L_j||L_{j+1}| \geq m 2^{2k} \gtrsim \frac{E_\times(A)}{\log |A|}.$$

Combining with $E_\times(A) \geq |A|^4 / |AA|$, which we obtained at the beginning of the proof, we obtain

$$|A+A|^2 \, |AA| \log |A| \gtrsim |A|^4 . \qquad \square$$

***Remark* 8.3.3** (Improvements). Konyagin and Shkredov (2015) improved Solymosi's sum-product bound to $\max\{|A+A|, |AA|\} \geq |A|^{4/3+c}$ for a small constant $c > 0$. This constant c was improved in subsequent works but still remains quite small.

***Remark* 8.3.4** (Sum-product in $\mathbb{F}_p$). Bourgain, Katz, and Tao (2004), combined with a later result of Bourgain, Glibichuk, and Konyagin (2006), proved the following sum-product estimate in $\mathbb{F}_p$ with p prime:

Theorem 8.3.5 (Sum-product in prime finite fields)
For every $\varepsilon > 0$ there exists $\delta > 0$ and $c > 0$ so that every $A \subseteq \mathbb{F}_p$, with p prime, and $1 < |A| < p^{1-\varepsilon}$, satisfies

$$\max\{|A+A|, |AA|\} \geq c\,|A|^{1+\delta} .$$

The statement is false over nonprime fields, since we could take A to be a subfield. Informally, the preceding theorem says that a prime field does not have any approximate subrings.

Further Reading

Dvir's survey *Incidence Theorems and Their Applications* (2012) discusses many interesting related topics including incidence geometry and additive combinatorics together with their applications to computer science.

Guth's book *The Polynomial Method in Combinatorics* (2016) gives an in-depth discussion of incidence geometry in $\mathbb{R}^2$ and $\mathbb{R}^3$ leading to a proof of the solution of the Erdős distinct distances problem by Guth and Katz (2015).

Sheffer's book *Polynomial Methods and Incidence Theory* (2022) provides an introduction to incidence geometry and related topics.

Chapter Summary

- **Sum-product conjecture.** $\max\{|A+A|, |AA|\} \geq |A|^{2-o(1)}$ for all $A \subseteq \mathbb{R}$.
- **Elekes' bound.** $\max\{|A+A|, |AA|\} \gtrsim |A|^{5/4}$.
  - Proof uses point-line incidences.
  - **Crossing number inequality.** Every graph G with n vertices and $m \geq 4n$ edges has $\gtrsim m^3/n^2$ crossings in every drawing.
  - **Szemerédi–Trotter theorem.** m lines and n points in $\mathbb{R}^2$ form $O(m^{2/3}n^{2/3} + m + n)$ incidences.
- **Solymosi's bound:** $\max\{|A+A|, |AA|\} \gtrsim |A|^{4/3-o(1)}$.

9

Progressions in Sparse Pseudorandom Sets

Chapter Highlights

- The Green–Tao theorem: proof strategy
- A relative Szemerédi theorem and its proof: a central ingredient in the proof of the Green–Tao theorem
- Transference principle: applying Szemerédi's theorem as a black box to the sparse pseudorandom setting
- A graph theoretic approach
- Dense model theorem: modeling a sparse set by a dense set
- Sparse triangle counting lemma

In this chapter we discuss a celebrated theorem by Green and Tao (2008) that settled a folklore conjecture about primes.

Theorem 9.0.1 (Green–Tao theorem)
The primes contain arbitrarily long arithmetic progressions.

The proof of this stunning result uses sophisticated ideas from both combinatorics and number theory. As stated in the abstract of their paper:

[T]he main new ingredient of this paper . . . is a certain transference principle. This allows us to deduce from Szemerédi's theorem that any subset of a sufficiently pseudorandom set (or measure) of positive relative density contains progressions of arbitrary length.

The main goal of this chapter is to explain what the preceding paragraph means. As Green (2007b) writes (emphasis in original):

Our main advance, then, lies not in our understanding of the primes but rather in what we can say about *arithmetic progressions*.

We will abstract away ingredients related to prime numbers (see Further Reading at the end of the chapter) and instead focus on the central combinatorial result: a **relative Szemerédi theorem**. We follow the graph theoretic approach by Conlon, Fox, and Zhao (2014, 2015), which simplified both the hypotheses and the proof of the relative Szemerédi theorem.

9.1 Green–Tao Theorem

In this section, we give a high-level overview of the proof strategy of the Green–Tao theorem. Recall Szemerédi's theorem:

Theorem 9.1.1 (Szemerédi's theorem)
Fix $k \geq 3$. Every k-AP-free subset of $[N]$ has size $o(N)$.

By the prime number theorem,

$$\#\{\text{primes} \leq N\} = (1 + o(1))\frac{N}{\log N}.$$

So Szemerédi's theorem does not automatically imply the Green–Tao theorem.

***Remark* 9.1.2** (Quantitative bounds). It is possible that better quantitative bounds on Szemerédi's theorem might eventually imply the Green–Tao theorem based on the density of primes alone. For example, Erdős famously conjectured that any $A \subseteq \mathbb{N}$ with divergent harmonic series (i.e., $\sum_{a \in A} 1/a = \infty$) contains arbitrarily long arithmetic progressions (Conjecture 0.2.5). The current best quantitative bounds on Szemerédi's theorem for k-APs is $|A| \leq N(\log\log N)^{-c_k}$ (Gowers 2001), which are insufficient for the primes, although better bounds are known for $k = 3, 4$. More recently, Bloom and Sisask (2020) proved that for $k = 3$, $|A| \leq N(\log N)^{-1-c}$ for some constant $c > 0$, thereby implying the Green–Tao theorem for 3-APs via the density of primes alone.

We will be quite informal here in order to highlight some key ideas of the proof of the Green–Tao theorem. Fix $k \geq 3$. The idea is to embed the primes in a slightly larger "pseudorandom host set":

$$\{\text{primes}\} \subseteq \{\text{“almost primes”}\}.$$

Very roughly speaking, "almost primes" are numbers with no small prime divisors. The "almost primes" are much easier to analyze compared to the primes. Using analytic number theory (involving techniques related to the problem of *small gaps between primes*), one can construct "almost primes" satisfying the following properties.

Properties of the "almost primes":

(1) The primes occupy at least a positive constant fraction of the "almost primes":

$$\frac{\#\{\text{primes} \leq N\}}{\#\{\text{“almost primes”} \leq N\}} \geq \delta_k.$$

(2) The "almost primes" behave pseudorandomly with respect to certain pattern counts.

The next key ingredient plays a central role in the proof of the Green–Tao theorem, as mentioned at the beginning of this chapter. It will be nicer to work in $\mathbb{Z}/N\mathbb{Z}$ rather than $[N]$.

Relative Szemerédi Theorem (Informal) *Fix $k \geq 3$. If $S \subseteq \mathbb{Z}/N\mathbb{Z}$ satisfies certain pseudorandomness hypotheses, then every k-AP-free subset of S has size $o(|S|)$.*

Here imagine a sequence $S = S_N \subseteq \mathbb{Z}/N\mathbb{Z}$ of size $o(N)$ (or else the relative Szemerédi theorem would already follow from Szemerédi's theorem), and $|S| \geq N^{1-c_k}$ for some small constant $c_k > 0$. In the proof of the Green–Tao theorem, the set S will be the "almost primes" (so that $|S| = \Theta(N/\log N)$), subject to various other technical modifications such as the W-trick discussed in Remark 9.1.4.

The relative Szemerédi theorem and the construction of the "almost primes" together tell us that the primes contain a k-AP. It also implies the following.

Theorem 9.1.3 (Green–Tao)
Fix $k \geq 3$. If A is a k-AP-free subset of the primes, then

$$\lim_{N \to \infty} \frac{|A \cap [N]|}{|\text{Primes} \cap [N]|} = 0.$$

In other words, every subset of primes with positive *relative density* contains arbitrarily long arithmetic progressions.

Remark **9.1.4** (Residue biases in the primes and the W-trick). There are certain local biases that get in the way of pseudorandomness for primes. For example, all primes greater than 2 are odd, all primes greater than 3 are not divisible by 3, and so on. In this way, the primes look different from a subset of positive integers where each n is included with probability $1/\log n$ independently at random.

The ***W*-trick** corrects these residue class biases. Let $w = w(N)$ be a function with $w \to \infty$ slowly as $N \to \infty$. Let $W = \prod_{p \leq w} p$ be the product of primes up to w. The W-trick tells us to only consider primes that are congruent to 1 mod W. The resulting set of "W-tricked primes" $\{n : nW + 1 \text{ is prime}\}$ does not have any bias modulo a small fixed prime. The relative Szemerédi theorem should be applied to the W-tricked primes.

The goal of the rest of the chapter is to state and prove the relative Szemerédi theorem. We shall not dwell on the analytic number theoretic arguments here. See Further Reading at the end of the chapter for references. For example, Conlon, Fox, and Zhao (2014, Sections 8 and 9) gives an exposition of the construction of the "almost primes" and the proofs of its properties.

9.2 Relative Szemerédi Theorem

In this section, we formulate a relative Szemerédi theorem. For concreteness, we mostly discuss 3-APs, though everything generalizes to k-APs straightforwardly.

Recall Roth's theorem:

Theorem 9.2.1 (Roth's theorem)
Every 3-AP-free subset of $\mathbb{Z}/N\mathbb{Z}$ has size $o(N)$.

We would like to formulate a result of the following form, where $\mathbb{Z}/N\mathbb{Z}$ is replaced by a **sparse pseudorandom host set** S.

Relative Roth Theorem (Informal) *If $S \subseteq \mathbb{Z}/N\mathbb{Z}$ satisfies certain pseudorandomness conditions, then every 3-AP-free subset of S has size $o(|S|)$.*

In what sense should S behave pseudorandomly? It will be easiest to explain the pseudorandom hypothesis using a graph.

Consider the following construction of a graph G_S that we saw in Chapter 6 (in particular Sections 2.4 and 2.10).

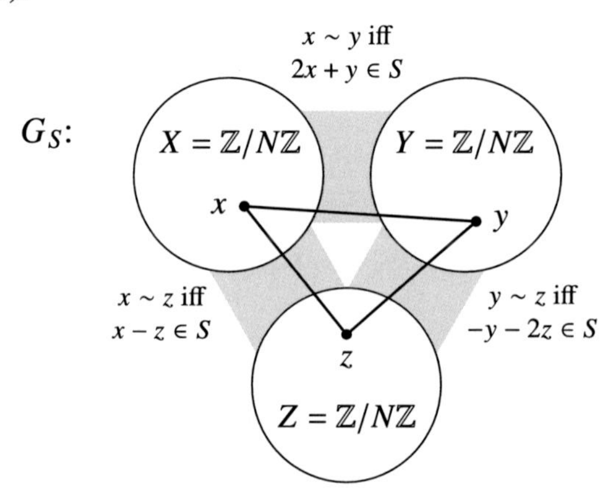

Here G_S is a tripartite graph with vertex sets X, Y, Z, each being a copy of $\mathbb{Z}/N\mathbb{Z}$. Its edges are:

- $(x, y) \in X \times Y$ whenever $2x + y \in S$;
- $(x, z) \in X \times Z$ whenever $x - z \in S$;
- $(y, z) \in Y \times Z$ whenever $-y - 2z \in S$.

This graph G_S is designed so that $(x, y, z) \in X \times Y \times Z$ is a triangle if and only if

$$2x + y, \quad x - z, \quad -y - 2z \in S.$$

Note that these three terms form a 3-AP with common difference $-x - y - z$. So the triangles in G_S precisely correspond to 3-APs in S. (It is an N-to-1 correspondence.)

The following definition is a variation of homomorphism density from Section 4.3.

Definition 9.2.2 (F-density)
Let F and G be tripartite graphs with three labeled parts. Define ***F*-density in *G***, denoted $\boldsymbol{t(F, G)}$, to be the probability that a random map $V(F) \to V(G)$ is a graph homomorphism $F \to G$, where each vertex in the first vertex part of F is sent to a uniform vertex of the first vertex part of G, and likewise with the second and third parts, all independently.

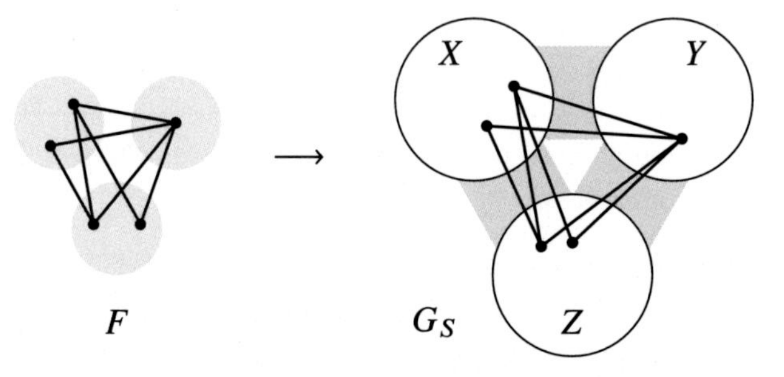

Now we define the desired pseudorandomness hypotheses on $S \subseteq \mathbb{Z}/N\mathbb{Z}$, which says that the associated graph G_S has certain subgraph counts close to random.

Definition 9.2.3 (3-linear forms condition)
We say that $S \subseteq \mathbb{Z}/N\mathbb{Z}$ satisfies the ***3-linear forms condition with tolerance*** $\boldsymbol{\varepsilon}$ if, setting $p = |S|/N$, one has

$$(1-\varepsilon)p^{e(F)} \le t(F, G_S) \le (1+\varepsilon)p^{e(F)} \qquad \text{whenever } F \subseteq K_{2,2,2}.$$

(Here $F \subseteq K_{2,2,2}$ means that is a subgraph of the labeled tripartite graph $K_{2,2,2}$; an example is illustrated in what follows.)

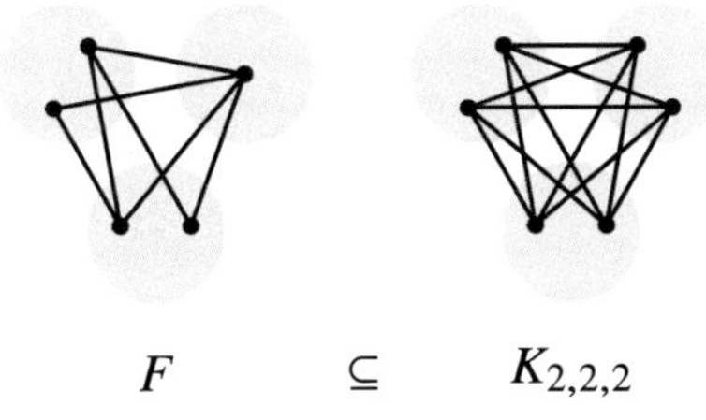

$F \quad \subseteq \quad K_{2,2,2}$

In other words, comparing the graph G_S to a random tripartite graph with the same edge density p, these two graphs have approximately the same F-density whenever $F \subseteq K_{2,2,2}$.

Alternatively, we can state the 3-linear forms condition explicitly without referring to graphs. This is done by expanding the definition of G_S. Let $x_0, x_1, y_0, y_1, z_0, z_1 \in \mathbb{Z}/N\mathbb{Z}$ be chosen independently and uniformly at random. Then $S \subseteq \mathbb{Z}/N\mathbb{Z}$ with $|S| = pN$ satisfies the 3-linear forms condition with tolerance ε if the probability that

$$\left\{\begin{matrix} -y_0 - 2z_0, & x_0 - z_0, & 2x_0 + y_0, \\ -y_1 - 2z_0, & x_1 - z_0, & 2x_1 + y_0, \\ -y_0 - 2z_1, & x_0 - z_1, & 2x_0 + y_1, \\ -y_1 - 2z_1, & x_1 - z_1, & 2x_1 + y_1 \end{matrix}\right\} \subseteq S$$

lies in the interval $(1 \pm \varepsilon)p^{12}$. And furthermore, the same holds if we erase any subset of the above 12 linear forms and also change the "12" in p^{12} to the number of linear forms remaining.

Remark **9.2.4.** This $K_{2,2,2}$ condition is reminiscent of the C_4-count condition for the quasirandom graph in Theorem 3.1.1 by Chung, Graham, and Wilson (1989). Just as $C_4 = K_{2,2}$ is a 2-blow-up of a single edge, $K_{2,2,2}$ is a 2-blow-up of a triangle.

The 3-linear forms condition can be viewed as a "second moment" condition with respect to triangles. It is needed in the proof of the sparse triangle counting lemma later.

We are now ready to state a precise formulation of the relative Roth theorem.

Theorem 9.2.5 (Relative Roth theorem)
For every $\delta > 0$, there exist $\varepsilon > 0$ and N_0 so that for all odd $N \geq N_0$, if $S \subseteq \mathbb{Z}/N\mathbb{Z}$ satisfies the 3-linear forms condition with tolerance ε, then every 3-AP-free subset of S has size less than $\delta |S|$.

To extend these definitions and results to k-APs, we set up a $(k-1)$-uniform hypergraph. We use a procedure similar to the deduction of Szemerédi's theorem from the hypergraph removal lemma in Section 2.10.

Let us illustrate it first for 4-APs. We say that $S \subseteq \mathbb{Z}/N\mathbb{Z}$ satisfies the ***4-linear forms condition with tolerance* ε** if, given random $w_0, w_1, x_0, x_1, y_0, y_1, z_0, z_1 \in \mathbb{Z}/N\mathbb{Z}$ (independent and uniform as always), the probability that

$$\left\{\begin{array}{llll} 3w_0 + 2x_0 + y_0, & 2w_0 + x_0 - z_0, & w_0 - y_0 - 2z_0, & -x_0 - 2y_0 - 3z_0, \\ 3w_0 + 2x_0 + y_1, & 2w_0 + x_0 - z_1, & w_0 - y_0 - 2z_1, & -x_0 - 2y_0 - 3z_1, \\ 3w_0 + 2x_1 + y_0, & 2w_0 + x_1 - z_0, & w_0 - y_1 - 2z_0, & -x_0 - 2y_1 - 3z_0, \\ 3w_0 + 2x_1 + y_1, & 2w_0 + x_1 - z_1, & w_0 - y_1 - 2z_1, & -x_0 - 2y_1 - 3z_1, \\ 3w_1 + 2x_0 + y_0, & 2w_1 + x_0 - z_0, & w_1 - y_0 - 2z_0, & -x_1 - 2y_0 - 3z_0, \\ 3w_1 + 2x_0 + y_1, & 2w_1 + x_0 - z_1, & w_1 - y_0 - 2z_1, & -x_1 - 2y_0 - 3z_1, \\ 3w_1 + 2x_1 + y_0, & 2w_1 + x_1 - z_0, & w_1 - y_1 - 2z_0, & -x_1 - 2y_1 - 3z_0, \\ 3w_1 + 2x_1 + y_1, & 2w_1 + x_1 - z_1, & w_1 - y_1 - 2z_1, & -x_1 - 2y_1 - 3z_1 \end{array}\right\} \subseteq S$$

lies within the interval $(1 \pm \varepsilon)p^{32}$. And furthermore, the same is true if we erase any subset of the preceding 32 factors and replace the "32" in p^{32} by the number of linear forms remaining.

Here is the statement for k-APs. (You may wish to skip it and simply imagine how it should generalize based on the preceding examples.)

Definition 9.2.6 (k-linear forms condition)
For each $1 \leq r \leq k$, let

$$L_r(x_1, \ldots, x_k) = kx_1 + (k-1)x_2 + \cdots + x_k - r(x_1 + \cdots + x_k).$$

We say that $S \subseteq \mathbb{Z}/N\mathbb{Z}$ satisfies the ***k-linear forms condition with tolerance* ε** if for every $R \subseteq [k] \times \{0,1\}^k$, with each variable $x_{i,j} \in \mathbb{Z}/N\mathbb{Z}$ chosen independently and uniformly at random, the probability that

$$L_r(x_{1,j_1}, \ldots, x_{k,j_k}) \in S \quad \text{for all } (r, j_1, \ldots, j_k) \in R$$

lies within the interval $(1 \pm \varepsilon)p^{|R|}$.

Theorem 9.2.7 (Relative Szemerédi theorem)
For every $k \geq 3$ and $\delta > 0$, there exist $\varepsilon > 0$ and N_0 so that for all $N \geq N_0$ coprime to $(k-1)!$, if $S \subseteq \mathbb{Z}/N\mathbb{Z}$ satisfies the k-linear forms condition with tolerance ε, then every k-AP-free subset of S has size less than $\delta |S|$.

***Remark* 9.2.8** (History). The preceding formulations of relative Roth and Szemerédi theorems are due to Conlon, Fox, and Zhao (2015). The original approach by Green and Tao (2008) required in addition another technical hypothesis on S known as the "correlation condition," which is no longer needed.

***Remark* 9.2.9** (Szemerédi's theorem in a random set). Instead of a pseudorandom host set S, what happens if S is a random subset of $\mathbb{Z}/N\mathbb{Z}$ obtained by keeping each element with probability $p = p_N \to 0$ as $N \to \infty$? A second moment argument shows that, provided that p_N tends to zero sufficiently slowly, the random set S indeed satisfies the k-linear forms condition (see Exercise 9.2.11). However, this argument is rather lossy. The following sharp result was proved independently by Conlon and Gowers (2016) and Schacht (2016). In the following statement, there is no substantive difference between $[N]$ and $\mathbb{Z}/N\mathbb{Z}$.

Theorem 9.2.10 (Szemerédi's theorem in a random set)
For every $k \geq 3$ and $\delta > 0$, there is some C such that, as long as $p > CN^{-1/(k-1)}$, with probability approaching 1 as $N \to \infty$, given a random $S \subseteq [N]$ where every element is included independently with probability p, every k-AP-free subset of S has size at most $\delta\,|S|$.

The threshold $CN^{-1/(k-1)}$ is optimal up to the constant C. Indeed, the expected number of k-APs in S is $O(p^k N^2)$, which is less than half of $\mathbb{E}\,|S| = pN$ if $p < cN^{-1/(k-1)}$ for a sufficiently small constant $c > 0$. One can delete from S an element from each k-AP contained in S. So with high probability, this process deletes at most half of S, and the remaining subset of S is k-AP-free.

The **hypergraph container method** gives another proof of the preceding result, plus much more (Balogh, Morris, and Samotij 2015; Saxton and Thomason 2015). See the survey *The method of hypergraph containers* by Balogh, Morris, and Samotij (2018) for more on this topic.

Exercise 9.2.11 (Random sets and the linear forms condition). Let $S \subseteq \mathbb{Z}/N\mathbb{Z}$ be a random set where every element of $\mathbb{Z}/N\mathbb{Z}$ is included in S independently with probability p.

Prove that there is some $c > 0$ so that, for every $\varepsilon > 0$, there is some $C > 0$ so that as long as $p > CN^{-c}$ and N is large enough, with probability at least $1 - \varepsilon$, S satisfies the 3-linear forms condition with tolerance ε . What is the optimal c?

Hint: Use the second moment method; see Alon and Spencer (2016, Chapter 4).

9.3 Transference Principle

To prove the relative Szemerédi theorem, we shall assume Szemerédi's theorem and apply it as a black box to the sparse pseudorandom setting. It may be surprising that we can apply Szemerédi's theorem this way. Green and Tao developed a method known as a ***transference principle*** for bringing Szemerédi's theorem to the sparse pseudorandom setting. The idea also appeared earlier in the work of Green (2005b) establishing Roth's theorem in the primes. The transference principle is an influential idea, and it can be applied to other extremal problems in combinatorics.

Let us sketch the outline of the proof of the relative Szemerédi theorem. We are given

$$A \subseteq S \qquad \text{with } |A| \geq \delta\,|S|\,.$$

Here $S \subseteq \mathbb{Z}/N\mathbb{Z}$ is a sparse pseudorandom set satisfying the k-linear forms condition.

Step 1. Approximate A by a dense model.

We will prove a **dense model theorem** that produces a "dense model" B of A. In particular, the density of B in $\mathbb{Z}/N\mathbb{Z}$ is similar to the relative density of A in S:

$$\frac{|B|}{N} \approx \frac{|A|}{|S|} \geq \delta.$$

And furthermore, B will be close to A with respect to a "cut norm" derived from the graphon cut norm. (See Chapter 4 on graph limits.) Recall that the graphon cut norm is closely linked to ε-regularity from the regularity lemma (Chapter 2) and the discrepancy condition **DISC** from quasirandom graphs (Chapter 3).

Step 2. Count k-APs in A and B.

We will prove a **sparse counting lemma** to show that the number of k-APs in A is similar to the number of k-APs in B, after an appropriate density normalization. In other words, setting $p = |S|/N$ for the normalizing density, we will show

$$|\{k\text{-APs in } A\}| \approx p^k |\{k\text{-APs in } B\}|.$$

Szemerédi's theorem says that every subset of $[N]$ with size $\geq \delta N$ contains a k-AP (provided that N is sufficiently large compared with the constant $\delta > 0$). In fact, we can bootstrap Szemerédi's theorem to show that a subset of $[N]$ with size $\geq \delta N$ must contain lots of k-APs. The deduction uses a sampling argument and is attributed to Varnavides (1959). (This was Exercise 1.3.7 from Section 1.3 on supersaturation.)

Theorem 9.3.1 (Szemerédi's theorem, counting version)
For every $\delta > 0$, there exists $c > 0$ and N_0 such that for every $N \geq N_0$, every subset of $\mathbb{Z}/N\mathbb{Z}$ with $\geq \delta N$ elements contains $\geq cN^2$ k-APs.

Since the "dense model" B has size $\geq \delta N/2$, by the counting version of Szemerédi's theorem, B has $\gtrsim_\delta N^2$ k-APs, and hence A has $\gtrsim_\delta p^k N^2$ k-APs by the sparse counting lemma. So in particular, A cannot be k-AP-free. This finishes the proof sketch of the relative Szemerédi theorem.

Now that we have seen the above outline, it remains to formulate and prove:

- a dense model theorem, and
- a sparse counting lemma.

We will focus on explaining the 3-AP case (i.e., relative Roth theorem) in the rest of this chapter. The 3-AP setting is notationally simpler than that of k-AP. It is straightforward to generalize the 3-AP proof to k-APs following the $(k-1)$-uniform hypergraph setup discussed in the previous section.

9.4 Dense Model Theorem

In this section, Γ is any finite abelian group. We will only need the case $\Gamma = \mathbb{Z}/N\mathbb{Z}$ in subsequent sections.

Given $f\colon \Gamma \to \mathbb{R}$, we define the following "cut norm" similar to the cut norm from graph limits (Chapter 4):

$$\|f\|_{\square} := \sup_{A,B\subseteq\Gamma} \left|\mathbb{E}_{x,y\in\Gamma}[f(x+y)1_A(x)1_B(y)]\right|.$$

This is essentially the graphon cut norm applied to the function $\Gamma\times\Gamma \to \mathbb{R}$ given by $(x,y)\mapsto f(x+y)$.

As should be expected from the equivalence of **DISC** and **EIG** for quasirandom Cayley graphs (Theorem 3.5.3), having small cut norm is equivalent to being Fourier uniform.

Exercise 9.4.1. Show that for all $f\colon \Gamma \to \mathbb{R}$,

$$c\|\widehat{f}\|_\infty \le \|f\|_\square \le \|\widehat{f}\|_\infty,$$

where c is some absolute constant (not depending on Γ or f).

***Remark* 9.4.2** (Generalizations to k-APs). The preceding definition is tailored to 3-APs. For 4-APs, we should define the corresponding norm of f as

$$\sup_{A,B,C\subseteq\Gamma\times\Gamma} \left|\mathbb{E}_{x,y,z\in\Gamma}[f(x+y+z)1_A(x,y)1_B(x,z)1_C(y,z)]\right|.$$

(The more obvious guess of using $1_A(x)1_B(y)1_C(z)$ instead of the preceding turns out to be insufficient for proving the relative Szemerédi theorem. A related issue in the context of hypergraph regularity was discussed in Section 2.11.) The generalization to k-APs is straightforward. However, for $k \ge 4$, the above norm is no longer equivalent to Fourier uniformity. This is why we study $\|f\|_\square$ norm instead of $\|\widehat{f}\|_\infty$ in this section.

Informally, the main result of this section says that if a sparse set S is close to random in normalized cut norm, then every subset $A \subseteq S$ can be approximated by some dense $B \subseteq \mathbb{Z}/N\mathbb{Z}$ in normalized cut norm.

Theorem 9.4.3 (Dense model theorem)
For every $\varepsilon > 0$, there exists $\delta > 0$ such that the following holds. For every finite abelian group Γ and sets $A \subseteq S \subseteq \Gamma$ such that, setting $p = |S|/|\Gamma|$,

$$\|1_S - p\|_\square \le \delta p,$$

there exists $g\colon \Gamma \to [0,1]$ such that

$$\|1_A - pg\|_\square \le \varepsilon p.$$

***Remark* 9.4.4** (3-linear forms condition implies small cut norm). The cut norm hypothesis is weaker than the 3-linear forms condition, as can be proved by two applications of the Cauchy–Schwarz inequality. (For example, see the proof of Lemma 9.5.2 in the next section.) In short, $\|\nu - 1\|_\square^4 \le t(K_{2,2}, \nu - 1)$.

***Remark* 9.4.5** (Set instead of function). We can replace the function g by a random set $B \subseteq \Gamma$ where each $x \in \Gamma$ is included in B with probability $g(x)$. By standard concentration bounds, changing g to B induces a negligible effect on ε if Γ is large enough. It is important here that $g(x) \in [0,1]$ for all $x \in \Gamma$.

So the preceding theorem says that, given a sparse pseudorandom host set S, any subset of S can be modeled by a dense set B that is close to A with respect to the normalized cut norm.

It will be more natural to prove the preceding theorem a bit more generally where sets $A \subseteq S \subseteq \Gamma$ are replaced by functional analogs. Since these are sparse sets, we should scale indicator functions as follows:

$$f = p^{-1} 1_A \quad \text{and} \quad \nu = p^{-1} 1_S.$$

Then $f \le \nu$ pointwise. Note that f and ν take values in $[0, p^{-1}]$, unlike g, which takes values in $[0, 1]$. The normalization is such that $\mathbb{E}\nu = 1$. Here is the main result of this section.

Theorem 9.4.6 (Dense model theorem)
For every $\varepsilon > 0$, there exists $\delta > 0$ such that the following holds. For every finite abelian group Γ and functions $f, \nu \colon \Gamma \to [0, \infty)$ satisfying

$$\|\nu - 1\|_\square \le \delta$$

and

$$f \le \nu \quad \text{pointwise,}$$

there exists a function $g \colon \Gamma \to [0, 1]$ such that

$$\|f - g\|_\square \le \varepsilon.$$

The rest of this section is devoted to proving the preceding theorem. First, we reformulate the cut norm using convex geometry.

Let Φ denote the set of all functions $\Gamma \to \mathbb{R}$ that can be written as a convex combination of convolutions of the form $1_A * 1_B$ or $-1_A * 1_B$, where $A, B \subseteq \Gamma$. Equivalently,

$$\Phi = \text{ConvexHull}\left(\{1_A * 1_B : A, B \subseteq \Gamma\} \cup \{-1_A * 1_B : A, B \subseteq \Gamma\}\right).$$

Note that Φ is a centrally symmetric convex set of functions $\Gamma \to \mathbb{R}$.

Lemma 9.4.7 (Multiplicative closure)
The set Φ is closed under pointwise multiplication; that is, if $\varphi, \varphi' \in \Phi$, then $\varphi\varphi' \in \Phi$.

Proof. Given $A, B, C, D \subseteq \Gamma$, we have

$$\begin{aligned}
(1_A * 1_B)(x)(1_C * 1_D)(x) &= \mathbb{E}_{a,b,c,d:a+b=c+d=x} 1_A(a)1_B(b)1_C(c)1_D(d) \\
&= \mathbb{E}_{a,b,s:a+b=x} 1_A(a)1_B(b)1_C(a+s)1_D(b-s) \\
&= \mathbb{E}_s \mathbb{E}_{a,b:a+b=x} 1_{A\cap(C-s)}(a)1_{B\cap(D+s)}(b) \\
&= \mathbb{E}_s (1_{A\cap(C-s)} * 1_{B\cap(D+s)})(x).
\end{aligned}$$

Thus the pointwise product of $1_A * 1_B$ and $1_C * 1_D$ lies in Φ since it is an average of various functions of the form $1_S * 1_T$. Since Φ is the convex hull of functions of the form $1_A * 1_B$ and $-1_A * 1_B$, Φ is closed under pointwise multiplication. □

Given $f, g \colon \Gamma \to \mathbb{R}$, define the inner product by

$$\langle f, g \rangle := \mathbb{E}_{x\in\Gamma} f(x)g(x).$$

Since

$$\mathbb{E}_{x,y\in\Gamma} f(x+y)1_A(x)1_B(y) = \langle f, 1_A * 1_B \rangle,$$

we have

$$\|f\|_\square = \sup_{A,B\subseteq\Gamma} |\langle f, 1_A * 1_B\rangle| = \sup_{\varphi\in\Phi} \langle f, \varphi\rangle .$$

Since Φ is a centrally symmetric convex body, $\| \ \|_\square$ is indeed a norm. Its dual norm is thus given by, for any nonzero $\psi \colon \Gamma \to \mathbb{R}$,

$$\|\psi\|_\square^* = \sup_{\substack{f\colon \Gamma\to\mathbb{R} \\ \|f\|_\square \le 1}} \langle f, \psi\rangle = \sup\left\{ r \in \mathbb{R} : r^{-1}\psi \in \Phi \right\}.$$

In other words, Φ is the unit ball for $\| \ \|_\square^*$ norm. The following inequality holds for all $f, \psi \colon \Gamma \to \mathbb{R}$:

$$\langle f, \psi\rangle \le \|f\|_\square \, \|\psi\|_\square^* .$$

Lemma 9.4.8 (Submultiplicativity of the dual cut norm)
The norm $\|\cdot\|_\square^*$ is submultiplicative; that is, for all $\psi, \psi' \colon \Gamma \to \mathbb{R}$,

$$\|\psi\psi'\|_\square^* \le \|\psi\|_\square^* \, \|\psi'\|_\square^* .$$

Proof. The inequality is not affected if we multiply ψ and ψ' each by a constant. So we can assume that $\|\psi\|_\square^* = \|\psi'\|_\square^* = 1$. Then $\psi, \psi' \in \Phi$. Hence $\psi\psi' \in \Phi$ by Lemma 9.4.7. This implies that $\|\psi\psi'\|_\square' \le 1$. □

We need two classical results from analysis and convex geometry.

Theorem 9.4.9 (Weierstrass polynomial approximation theorem)
Let $a, b \in \mathbb{R}$ and $\varepsilon > 0$. Let $F \colon [a,b] \to \mathbb{R}$ be a continuous function. Then there exists a polynomial P such that $|F(t) - P(t)| \le \varepsilon$ for all $t \in [a,b]$.

Theorem 9.4.10 (Separating hyperplane theorem)
Given a closed convex set $K \subseteq \mathbb{R}^n$ and a point $p \notin K$, there exists a hyperplane separating K and p.

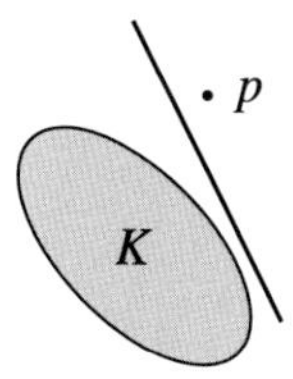

Proof idea of the dense model theorem. If no $g \colon \Gamma \to [0,1]$ satisfies $\|f - g\|_\square \le \varepsilon$, then f does not lie in the convex set containing all functions of the form $g + g'$ where $g \colon \Gamma \to [0,1]$ and $\|g'\|_\square \le \varepsilon$. The separating hyperplane theorem then gives us a function ψ so that $\langle f, \psi\rangle > 1$ and $\langle g + g', \psi\rangle \le 1$ for all such g, g'. (It helps to pretend a bit of extra slack here, say $\langle f, \psi\rangle > 1 + \varepsilon$.) Using the Weierstrass polynomial approximation theorem, choose a polynomial $P(t)$ so that $P(t) \approx \max\{0,t\}$ pointwise for all $|t| \leq \|\psi\|_\square^* = O_\varepsilon(1)$. Writing $\psi_+(x) = \max\{0,\psi(x)\}$ for the positive part of ψ, we have

$$\langle f,\psi\rangle \leq \langle f,\psi_+\rangle \leq \langle \nu,\psi_+\rangle \approx \langle \nu, P\psi\rangle = \langle \nu - 1, P\psi\rangle + \langle 1, P\psi\rangle \,.$$

We can show that $\|\psi\|_\square^* = O_\varepsilon(1)$. As P is a polynomial, by the triangle inequality and the submultiplicativity of $\|\ \|_\square^*$, we find that $\|P\psi\|_\square^* = O_\varepsilon(1)$. And so

$$\langle \nu - 1, P\psi\rangle \leq \|\nu - 1\|_\square \|P\psi\|_\square^* \leq \delta \|P\psi\|_\square^*$$

can be made arbitrarily small by making δ small. We also have $\langle 1, P\psi\rangle \approx \langle 1,\psi_+\rangle$, which is at most around 1. Together, we see that $\langle f,\psi\rangle$ is at most around 1, which would contradict $\langle f,\psi\rangle > 1$ from earlier (assuming enough slack). ∎

Proof of the dense model theorem (Theorem 9.4.6). We will show that the conclusion holds with $\delta > 0$ chosen to be sufficiently small as a function of ε. We may assume that $0 < \varepsilon < 1/2$. We will prove the existence of a function $g\colon \Gamma \to [0, 1 + \varepsilon/2]$ such that $\|f - g\|_\square \leq \varepsilon/2$. (To obtain the function $\Gamma \to [0,1]$ in the theorem, we can replace g by $\min\{g,1\}$.)

We are trying to prove that one can write f as $g + g'$ with

$$g \in K := \left\{\text{functions } \Gamma \to [0, 1 + \tfrac{\varepsilon}{2}]\right\}$$

and

$$g' \in K' := \left\{\text{functions } \Gamma \to \mathbb{R} \text{ with } \|\cdot\|_\square \leq \tfrac{\varepsilon}{2}\right\}.$$

We can view the sets K and K' as convex bodies (both containing the origin) in the space of all functions $\Gamma \to \mathbb{R}$. Our goal is to show that $f \in K + K'$.

Let us assume the contrary. By the separating hyperplane theorem applied to $f \notin K + K'$, there exists a function $\psi\colon \Gamma \to \mathbb{R}$ (which is a normal vector to the separating hyperplane) such that

(a) $\langle f,\psi\rangle > 1$, and
(b) $\langle g + g',\psi\rangle \leq 1$ for all $g \in K$ and $g' \in K'$

Taking $g = (1 + \frac{\varepsilon}{2})1_{\psi \geq 0}$ and $g' = 0$ in (b), we have

$$\langle 1,\psi_+\rangle \leq \frac{1}{1 + \varepsilon/2}. \tag{9.1}$$

Here we write ψ_+ for the function $\psi_+(x) := \max\{\psi(x), 0\}$.

On the other hand, setting $g = 0$, we have

$$1 \geq \sup_{g' \in K'} \langle g',\psi\rangle = \sup_{\|g'\|_\square \leq \varepsilon/2} \langle g',\psi\rangle = \frac{\varepsilon}{2} \|\psi\|_\square^* \,.$$

So

$$\|\psi\|_\square^* \leq \frac{2}{\varepsilon}.$$

Setting $g = 0$ and $g' = \pm\frac{\varepsilon}{2}N1_x$ for a single $x \in \Gamma$ (i.e., g' is supported on a single element of Γ), we have $\|g'\|_\square \leq \varepsilon/2$ and $1 \geq \langle g',\psi\rangle = \pm\frac{\varepsilon}{2}\psi(x)$. So $|\psi(x)| \leq 2/\varepsilon$. This holds for every $x \in \Gamma$. Thus,

$$\|\psi\|_\infty \leq \frac{2}{\varepsilon}.$$

By the Weierstrass polynomial approximation theorem, there exists some real polynomial $P(x) = p_d x^d + \cdots + p_1 x + p_0$ such that

$$|P(t) - \max\{t, 0\}| \le \frac{\varepsilon}{20} \quad \text{whenever } |t| \le \frac{2}{\varepsilon}.$$

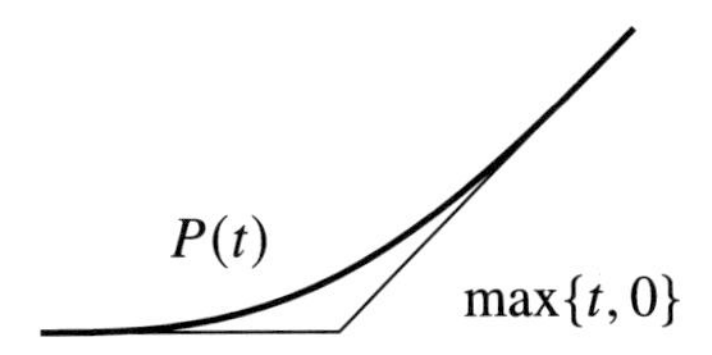

Set

$$R = \sum_{i=0}^{d} |p_i| \left(\frac{2}{\varepsilon}\right)^i,$$

which is a constant that depends only on ε. (A more careful analysis gives $R = \exp(\varepsilon^{-O(1)})$.)

Write $P\psi \colon \Gamma \to \mathbb{R}$ to mean the function given by $P\psi(x) = P(\psi(x))$. By the triangle inequality and the submultiplicativity of $\|\cdot\|_{\square}^*$ (Lemma 9.4.8),

$$\|P\psi\|_{\square}^* \le \sum_{i=0}^{d} |p_i| \, \|\psi^i\|_{\square}^* \le \sum_{i=0}^{d} |p_i| \, (\|\psi\|_{\square}^*)^i \le \sum_{i=0}^{d} |p_i| \left(\frac{2}{\varepsilon}\right)^i = R.$$

Let us choose

$$\delta = \min\left\{\frac{\varepsilon}{20R}, 1\right\}.$$

Then $\|\nu - 1\|_{\square} \le \delta$ implies that

$$|\langle \nu - 1, P\psi \rangle| \le \|\nu - 1\|_{\square} \, \|P\psi\|_{\square}^* \le \delta R \le \frac{\varepsilon}{20}. \tag{9.2}$$

Earlier we showed that $\|\psi\|_\infty \le 2/\varepsilon$, and also $|P(t) - \max\{t, 0\}| \le \varepsilon/20$ whenever $|t| \le 2/\varepsilon$. Thus,

$$\|P\psi - \psi_+\|_\infty \le \frac{\varepsilon}{20}. \tag{9.3}$$

Hence,

$$\begin{aligned}
\langle \nu, P\psi \rangle &= \langle 1, P\psi \rangle + \langle \nu - 1, P\psi \rangle \\
&\le \langle 1, P\psi \rangle + \frac{\varepsilon}{20} && \text{[by (9.2)]} \\
&\le \langle 1, \psi_+ \rangle + \frac{\varepsilon}{10} && \text{[by (9.3)]} \\
&\le \frac{1}{1 + \varepsilon/2} + \frac{\varepsilon}{10}. && \text{[by (9.1)].}
\end{aligned}$$

Also,

$$\langle \nu - 1, 1 \rangle \le \|\nu - 1\|_{\square} \le \delta.$$

Thus,

$$\|\nu\|_1 \le 1 + \|\nu - 1\|_1 \le 1 + \delta \le 2.$$

So by (9.3),

$$\langle \nu, \psi_+ - P\psi \rangle \le \|\nu\|_1 \, \|\psi_+ - P\psi\|_\infty \le 2 \cdot \frac{\varepsilon}{20} \le \frac{\varepsilon}{10}. \tag{9.4}$$

Thus, using that $0 \le f \le \nu$,

$$\begin{aligned} \langle f, \psi \rangle \le \langle f, \psi_+ \rangle &\le \langle \nu, \psi_+ \rangle \\ &\le \langle \nu, P\psi \rangle + \langle \nu, \psi_+ - P\psi \rangle \\ &\le \frac{1}{1+\varepsilon/2} + \frac{\varepsilon}{10} + \frac{\varepsilon}{10} \le 1 - \frac{\varepsilon}{10}. \end{aligned}$$

This contradicts (a) from earlier. This concludes the proof of the theorem. □

***Remark* 9.4.11 (History).** An early version of the density model theorem was used by Green and Tao (2008), where it was proved using a regularity-type energy increment argument. The above significantly simpler proof is due to Gowers (2010) and Reingold, Trevisan, Tulsiani, and Vadhan (2008) independently. Before the work of Conlon, Fox, and Zhao (2015), one needed to consider the Gowers uniformity norm rather than the simpler cut norm as we did earlier. The use of the cut norm further simplifies the proof of the corresponding dense model theorem, as noted by Zhao (2014).

Exercise 9.4.12. State and prove a dense model theorem for k-APs.

9.5 Sparse Counting Lemma

Let us prove an extension of the triangle counting lemma from Section 4.5. Here we work with a sparse graph (represented by f) that is a subgraph of a sparse pseudorandom host graph (represented by ν) satisfying a 3-linear forms condition (involving $K_{2,2,2}$ densities). The conclusion is that if f is close in cut norm to another dense graph g, then f and g have similar triangle densities (we normalize f for density).

Setup for This Section. Throughout this section, we have three finite sets X, Y, Z (which can also be probability spaces) representing the vertex sets of a tripartite graph. The following functions represent edge-weighted tripartite graphs:

$$f, g, \nu \colon (X \times Y) \cup (X \times Z) \cup (Y \times Z) \to \mathbb{R}.$$

- ν represents the normalized edge-indicator function of a possibly sparse pseudorandom host graph (arising from $S \subseteq \mathbb{Z}/N\mathbb{Z}$ in the statement of the relative Roth theorem).
- f represents the normalized edge-indicator function of a relatively dense subset $A \subseteq S$.
- g represents the dense model of f.

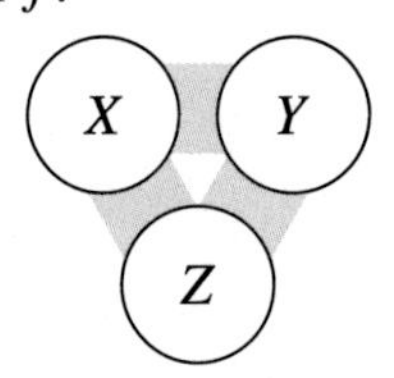

For any tripartite graph F, we write $t(F, f)$ for the F-density in f (and likewise with g and ν). Some examples:

$$t(K_3, f) = \mathbb{E}_{x,y,z} f(x,y) f(x,z) f(y,z) \quad \text{and}$$

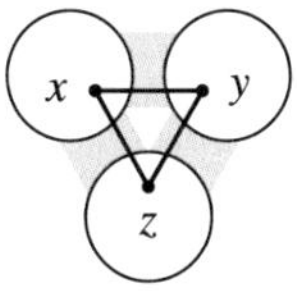

$$t(K_{2,1,1}, F) = \mathbb{E}_{x,x',y,z} f(x,y) f(x',y) f(x,z) f(x',z) f(y,z).$$

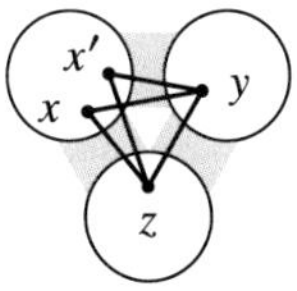

We maintain the convention that x, x' range uniformly over X, and so on.

The functions f, g, ν are assumed to satisfy the following conditions:

- $0 \le f \le \nu$ pointwise;
- $0 \le g \le 1$ pointwise;
- The **3-linear forms condition**:

$$|t(F,\nu) - 1| \le \varepsilon \quad \text{whenever } F \subseteq K_{2,2,2};$$

- When restricted to each of $X \times Y$, $X \times Z$, and $Y \times Z$, we have

$$\|f - g\|_\square \le \varepsilon.$$

For example, when restricted to $X \times Y$, the left-hand-side quantity denotes

$$\sup_{A \subseteq X, B \subseteq Y} \left| \mathbb{E}_{x,y} (f - g)(x,y) 1_A(x) 1_B(y) \right|.$$

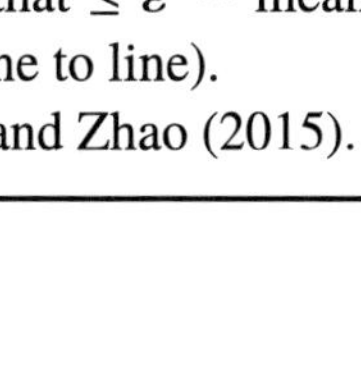

Throughout we assume that $\varepsilon > 0$ is sufficiently small, so that $\le \varepsilon^{\Omega(1)}$ means $\le C\varepsilon^c$ for some absolute constants $c, C > 0$ (which could change from line to line).

Here is the main result of this section, due to Conlon, Fox, and Zhao (2015).

Theorem 9.5.1 (Sparse triangle counting lemma)
Assume the setup at the beginning of this section. Then

$$|t(K_3, f) - t(K_3, g)| \le \varepsilon^{\Omega(1)}.$$

You should now pause and review the proof of the "dense" triangle counting lemma from Proposition 4.5.4, which says that if in addition we assume $0 \le f \le 1$ (that is, assuming $\nu = 1$ identically), then

$$|t(K_3, f) - t(K_3, g)| \le 3 \|f - g\|_\square \le 3\varepsilon.$$

Roughly speaking, the proof of the dense triangle counting lemma proceeds by replacing f by g one edge at a time, each time incurring at most an $\|f - g\|_\square$ loss.

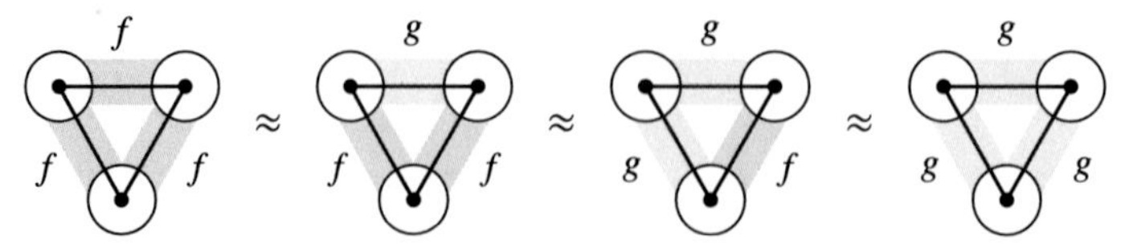

Having $\nu = 1$ should be thought of as the "dense" case. Indeed, $\nu = 1$ correponds to $S = \mathbb{Z}/N\mathbb{Z}$ rather than having a sparse pseudorandom set S. In general, starting with a general "sparse" ν, our strategy is to reduce the problem to another triangle counting problem where ν is replaced by 1 on one of the edges of the triangle.

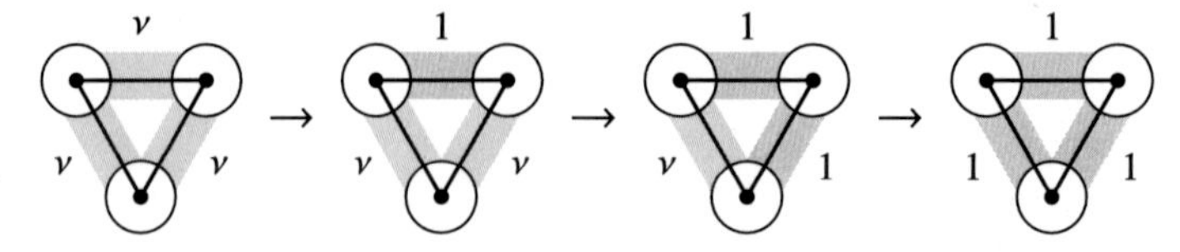

This **densification** strategy reduces a sparse triangle counting problem to a progressively easier triangle counting problem where some of the sparse bipartite graphs among X, Y, Z become dense.

Let **Sparsity(ν)** be the number of elements of $\{X \times Y, X \times Z, Y \times Z\}$ on which ν differs from 1. We will prove the statement:

SparseTCL(k): the sparse triangle counting lemma is true whenever Sparsity$(\nu) \leq k$. (The hidden constants may depend on k.)

We already proved the base case **SparseTCL**(0), which is the "dense" case corresponding to $\nu = 1$, as discussed earlier. We will prove the implications

$$\textbf{SparseTCL}(0) \implies \textbf{SparseTCL}(1) \implies \textbf{SparseTCL}(2) \implies \textbf{SparseTCL}(3).$$

We phrase our argument as an induction (a slightly unusual induction setup, as $0 \leq k \leq 3$). For the induction step, it suffices to prove the conclusion of the sparse triangle counting lemmas under the following hypothesis.

Induction Hypothesis: SparseTCL($k-1$) holds with k = Sparsity(ν), and ν is not identically 1 on $X \times Y$.

The next lemma shows that ν is close to 1 in a strong sense, provided that ν satisfies the 3-linear forms condition.

Lemma 9.5.2 (Strong linear forms)
Assume the setup at the beginning of this section. We have

$$\left|\mathbb{E}_{x,y,z,z'}(\nu(x,y) - 1)f(x,z)f(x,z')f(y,z)f(y,z')\right| \leq \varepsilon^{\Omega(1)}.$$

The same statement holds if any subset of the four f factors are replaced by g.

Proof. The proof uses two applications of the Cauchy–Schwarz inequality. Let us write down the proof in the case when none of the four f's are replaced by g's. The other cases are similar (basically apply $g \leq 1$ instead of $f \leq \nu$ wherever appropriate).

Here is a figure illustrating the first application of the Cauchy–Schwarz inequality.

Here are the inequalities written out:

$$
\begin{aligned}
&\left|\mathbb{E}_{x,y,z,z'}(\nu(x,y)-1)f(x,z)f(x,z')f(y,z)f(y,z')\right|^2 \\
&= \left|\mathbb{E}_{y,z,z'}\mathbb{E}_x\left[(\nu(x,y)-1)f(x,z)f(x,z')\right]f(y,z)f(y,z')\right|^2 \\
&\le \left(\mathbb{E}_{y,z,z'}\left(\mathbb{E}_x(\nu(x,y)-1)f(x,z)f(x,z')\right)^2 f(y,z)f(y,z')\right)\mathbb{E}_{y,z,z'}f(y,z)f(y,z') \\
&\le \left(\mathbb{E}_{y,z,z'}\left(\mathbb{E}_x(\nu(x,y)-1)f(x,z)f(x,z')\right)^2 \nu(y,z)\nu(y,z')\right)\mathbb{E}_{y,z,z'}\nu(y,z)\nu(y,z').
\end{aligned}
$$

Note that we are able to apply $f \le \nu$ in the preceding final step due to the nonnegativity of the square, which arose from the Cauchy–Schwarz inequality. We could not have applied $f \le \nu$ at the very beginning.

The second factor above is at most $1+\varepsilon$ due to the 3-linear forms condition. It remains to show that the first factor is $\le \varepsilon^{\Omega(1)}$. The first factor expands to

$$
\mathbb{E}_{x,x',y,z,z'}(\nu(x,y)-1)(\nu(x',y)-1)f(x,z)f(x,z')f(x',z)f(x',z')\nu(y,z)\nu(y,z').
$$

We can upper bound the above quantity as illustrated here, using a second application of the Cauchy–Schwarz inequality.

On the right-hand side, the first factor is $\le \varepsilon^{\Omega(1)}$ by the 3-linear forms condition. Indeed, $|t(F,\nu)-1| \le \varepsilon$ for any $F \subseteq K_{2,2,2}$. If we expand all the $\nu - 1$ in the first factor, then it becomes an alternating sum of various $t(F,\nu) \in [1-\varepsilon, 1+\varepsilon]$ with $F \subseteq K_{2,2,2}$, with the main contribution 1 from each term canceling each other out. The second factor is $\le 1+\varepsilon$ again by the 3-linear forms condition.

Putting everything together, this completes the proof of the lemma. □

Define $\nu_\wedge, f_\wedge, g_\wedge \colon X \times Y \to [0,\infty)$ by

$$
\begin{aligned}
\nu_\wedge(x,y) &:= \mathbb{E}_z \nu(x,z)\nu(y,z), \\
f_\wedge(x,y) &:= \mathbb{E}_z f(x,z)f(y,z), \\
g_\wedge(x,y) &:= \mathbb{E}_z g(x,z)g(y,z).
\end{aligned}
$$

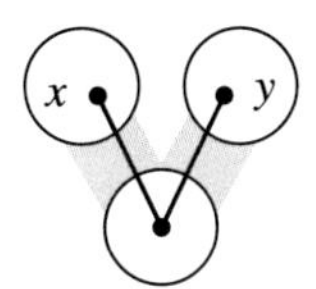

They represent codegrees. Even though ν and f are possibly unbounded, the new weighted graphs $\nu_\wedge$ and $f_\wedge$ behave like dense graphs because the sparseness is somehow smoothed out. (This is a key observation.) On a first reading of the proof, you may wish to pretend that $\nu_\wedge$ and $f_\wedge$ are uniformly bounded above by 1. (In reality, we need to control the negligible bit of ν exceeding 1.)

We have

$$t(K_3, f) = \langle f, f_\wedge \rangle,$$
$$\text{and} \quad t(K_3, g) = \langle g, g_\wedge \rangle$$

So

$$\begin{aligned} t(K_3, f) - t(K_3, g) &= \langle f, f_\wedge \rangle - \langle g, g_\wedge \rangle \\ &= \langle f, f_\wedge - g_\wedge \rangle + \langle f - g, g_\wedge \rangle. \end{aligned}$$

We have

$$|\langle f - g, g_\wedge \rangle| \le \|f - g\|_\square \le \varepsilon.$$

by the same argument as in the dense triangle counting lemma (Proposition 4.5.4), as $0 \le g \le 1$. So, it remains to show $|\langle f, f_\wedge - g_\wedge \rangle| \le \varepsilon^{\Omega(1)}$.

By the Cauchy–Schwarz inequality, we have

$$\langle f, f_\wedge - g_\wedge \rangle^2 = \mathbb{E}[f(f_\wedge - g_\wedge)]^2 \le \mathbb{E}[f(f_\wedge - g_\wedge)^2]\, \mathbb{E}f \le \mathbb{E}[\nu(f_\wedge - g_\wedge)^2]\, \mathbb{E}\nu.$$

The second factor is $\mathbb{E}\nu \le 1 + \varepsilon$ by the 3-linear forms condition. So, it remains to show that

$$\mathbb{E}[\nu(f_\wedge - g_\wedge)^2] = \langle \nu, (f_\wedge - g_\wedge)^2 \rangle \le \varepsilon^{\Omega(1)}.$$

By Lemma 9.5.2

$$\left|\langle \nu - 1, (f_\wedge - g_\wedge)^2 \rangle\right| \le \varepsilon^{\Omega(1)}.$$

(To see this inequality, first expand $(f_\wedge - g_\wedge)^2$ and then apply Lemma 9.5.2 term by term.) Thus,

$$\mathbb{E}[\nu(f_\wedge - g_\wedge)^2] \le \mathbb{E}[(f_\wedge - g_\wedge)^2] + \varepsilon^{\Omega(1)}.$$

Thus, to prove the induction step (as stated earlier) for the sparse triangle counting lemma, it remains to prove the following.

Lemma 9.5.3 (Densified triangle counting)

Assuming the setup at the beginning of the section as well as the induction hypothesis, we have

$$\mathbb{E}[(f_\wedge - g_\wedge)^2] \le \varepsilon^{\Omega(1)}. \tag{9.5}$$

Let us first sketch the idea of the proof of Lemma 9.5.3. Expanding, we have

$$\text{LHS of (9.5)} = \langle f_\wedge, f_\wedge\rangle - \langle f_\wedge, g_\wedge\rangle - \langle g_\wedge, f_\wedge\rangle + \langle g_\wedge, g_\wedge\rangle. \tag{9.6}$$

Each term represents some 4-cycle density.

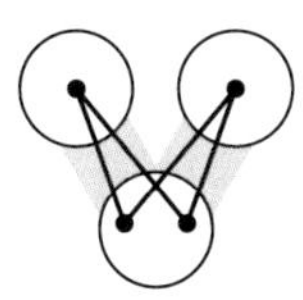

So, it suffices to show that each of the preceding four terms differs from $\langle g_\wedge, g_\wedge\rangle$ by $\le \varepsilon^{\Omega(1)}$. We are trying to show that $\langle f_\wedge, f_\wedge\rangle \approx \langle g_\wedge, g_\wedge\rangle$. Expanding the second factor in each $\langle\cdot,\cdot\rangle$, we are trying to show that

$$\begin{aligned} &\mathbb{E}_{x,y,z} f_\wedge(x,y) f(x,z) f(y,z) \\ &\approx \mathbb{E}_{x,y,z} g_\wedge(x,y) g(x,z) g(y,z). \end{aligned}$$

However, this is just another instance of the sparse triangle counting lemma! And importantly, this instance is easier than the one we started with. Indeed, we have $\|f_\wedge - g_\wedge\|_\square \le \varepsilon^{\Omega(1)}$. (This can be proved by invoking the induction hypothesis.) Furthermore, the first factor $f_\wedge(x,y)$ now behaves more like a bounded function (corresponding to a dense graph rather than a sparse graph). Let us pretend for a second that $f_\wedge \le 1$, ignoring the negligible part of $f_\wedge$ exceeding 1. Then we have reduced the original problem to a new instance of the triangle counting lemma, except that now $f \le \nu$ on $X \times Y$ has been replaced by $f_\wedge \le 1$. (This is the key point where **densification** occurs.) Lemma 9.5.3 then follows from the induction hypothesis, as we have reduced the sparsity of the pseudorandom host graph.

Coming back to the proof, as discussed earlier, while $f_\wedge$ is not necessarily ≤ 1, it is almost so. We need to handle the error term arising from replacing $f_\wedge$ by its capped version $\overline{f_\wedge}\colon X \times Y \to [0,1]$ defined by

$$\overline{f_\wedge} = \min\{f_\wedge, 1\} \quad \text{pointwise.}$$

We have

$$0 \le f_\wedge - \overline{f_\wedge} = \max\{f_\wedge - 1, 0\} \le \max\{\nu_\wedge - 1, 0\} \le |\nu_\wedge - 1|. \tag{9.7}$$

Also,

$$(\mathbb{E}|\nu_\wedge - 1|)^2 \le \mathbb{E}[(\nu_\wedge - 1)^2] = \mathbb{E}\nu_\wedge^2 - 2\mathbb{E}\nu_\wedge + 1 \le 3\varepsilon \tag{9.8}$$

by the 3-linear forms condition, since $\mathbb{E}\nu_\wedge^2$ and $\mathbb{E}\nu_\wedge$ are both within ε of 1. So,

$$\begin{aligned} \left|\langle f_\wedge, f_\wedge\rangle - \langle \overline{f_\wedge}, f_\wedge\rangle\right| = \left|\langle f_\wedge - \overline{f_\wedge}, f_\wedge\rangle\right| &\le \mathbb{E}\,|\nu_\wedge - 1|\,\nu_\wedge \\ &= \mathbb{E}\,|\nu_\wedge - 1|\,(\nu_\wedge - 1) + \mathbb{E}\,|\nu_\wedge - 1| \\ &\le \mathbb{E}[(\nu_\wedge - 1)^2] + \mathbb{E}\,|\nu_\wedge - 1| \\ &\le \varepsilon^{\Omega(1)}. \qquad \text{[by (9.8)]} \end{aligned} \tag{9.9}$$

Lemma 9.5.4 (Cut norm between codegrees)
With the same assumptions as Lemma 9.5.3,

$$\|\overline{f_\wedge} - g_\wedge\|_\square \le \varepsilon^{\Omega(1)}.$$

Proof. Indeed, for any $A \subseteq X$ and $B \subseteq Y$, we have

$$\langle \overline{f_\wedge} - g_\wedge, 1_{A\times B}\rangle = \langle \overline{f_\wedge} - f_\wedge, 1_{A\times B}\rangle + \langle f_\wedge - g_\wedge, 1_{A\times B}\rangle.$$

By (9.7) followed by (9.8)

$$\langle \overline{f_\wedge} - f_\wedge, 1_{A\times B}\rangle \le \mathbb{E}|\overline{f_\wedge} - f_\wedge| \le \mathbb{E}|\nu_\wedge - 1| \le \varepsilon^{\Omega(1)}.$$

So it remains to show that

$$|\langle f_\wedge - g_\wedge, 1_{A\times B}\rangle| \le \varepsilon^{\Omega(1)}.$$

This is true since

$$\langle f_\wedge, 1_{A\times B}\rangle = \mathbb{E}_{x,y,z} 1_{A\times B}(x,y) f(x,z) f(y,z)$$
$$\text{and } \langle g_\wedge, 1_{A\times B}\rangle = \mathbb{E}_{x,y,z} 1_{A\times B}(x,y) g(x,z) g(y,z)$$

satisfy the hypothesis of the sparse counting lemma with f, g, ν on $X \times Y$ replaced by $1_{A\times B}, 1_{A\times B}, 1$, thereby decreasing the sparsity of ν by 1, and hence we can apply the induction hypothesis. □

Proof of Lemma 9.5.3. We need to show that each of the four terms on the right-hand side of (9.6) is within $\varepsilon^{\Omega(1)}$ of $\langle g_\wedge, g_\wedge\rangle$. Let us show that

$$|\langle f_\wedge, f_\wedge\rangle - \langle g_\wedge, g_\wedge\rangle| \le \varepsilon^{\Omega(1)}.$$

By (9.9), $\langle f_\wedge, f_\wedge\rangle$ differs from $\langle \overline{f_\wedge}, f_\wedge\rangle$ by $\le \varepsilon^{\Omega(1)}$, and thus it suffices to show that

$$\langle \overline{f_\wedge}, f_\wedge\rangle = \mathbb{E}_{x,y,z} \overline{f_\wedge}(x,y) f(x,z) f(y,z)$$

and

$$\langle g_\wedge, g_\wedge\rangle = \mathbb{E}_{x,y,z} g_\wedge(x,y) g(x,z) g(y,z)$$

differ by $\le \varepsilon^{\Omega(1)}$. To show this, we apply the induction hypothesis to the setting where f, g, ν on $X \times Y$ are replaced by $\overline{f_\wedge}, g, 1$ (recall from Lemma 9.5.4 that $\|\overline{f_\wedge} - g\|_\square \le \varepsilon^{\Omega(1)}$), which reduces the sparsity of ν by 1. So the induction hypothesis implies

$$\left|\langle \overline{f_\wedge}, f_\wedge\rangle - \langle g_\wedge, g_\wedge\rangle\right| \le \varepsilon^{\Omega(1)}.$$

Thus $|\langle f_\wedge, f_\wedge\rangle - \langle g_\wedge, g_\wedge\rangle| \le \varepsilon^{\Omega(1)}$. Likewise, the other terms on the right-hand side of (9.9) are within $\varepsilon^{\Omega(1)}$ of $\langle g_\wedge, g_\wedge\rangle$ (Exercise!). The conclusion $\mathbb{E}[(f_\wedge - g_\wedge)^2] \le \varepsilon^{\Omega(1)}$ then follows. □

Exercise 9.5.5. State and prove a generalization of the sparse counting lemma to count an arbitrary but fixed subgraph (replacing the triangle). How about hypergraphs?

9.6 Proof of the Relative Roth Theorem

Now we combine the dense model theorem and the sparse triangle counting lemma to prove the relative Roth theorem:

Theorem 9.2.5 (restated). For every $\delta > 0$, there exist $\varepsilon > 0$ and N_0 so that for all $N \geq N_0$, if $S \subseteq \mathbb{Z}/N\mathbb{Z}$ satisfies the 3-linear forms condition with tolerance ε, then every 3-AP-free subset of S has size less than $\delta\,|S|$.

Recall that with $x_0, x_1, y_0, y_1, z_0, z_1 \in \mathbb{Z}/N\mathbb{Z}$ chosen independently and uniformly at random, the set $S \subseteq \mathbb{Z}/N\mathbb{Z}$ with $|S| = pN$ satisfies the ***3-linear forms condition with tolerance ε*** if the probability that

$$\left\{\begin{matrix} -y_0 - 2z_0, & x_0 - z_0, & 2x_0 + y_0, \\ -y_1 - 2z_0, & x_1 - z_0, & 2x_1 + y_0, \\ -y_0 - 2z_1, & x_0 - z_1, & 2x_0 + y_1, \\ -y_1 - 2z_1, & x_1 - z_1, & 2x_1 + y_1 \end{matrix}\right\} \subseteq S$$

lies in the interval $(1 \pm \varepsilon)p^{12}$. And furthermore, the same holds if we erase any subset of the preceding 12 linear forms and also change the "12" in p^{12} to the number of linear forms remaining.

The proof follows the strategy outlined in Section 9.3 on the transference principle. We need a counting version of Roth's theorem. As in Chapter 6, we define, for $f: \mathbb{Z}/N\mathbb{Z} \to \mathbb{R}$, its 3-AP density by

$$\Lambda_3(f) := \mathbb{E}_{x,d \in \mathbb{Z}/N\mathbb{Z}} f(x)f(x+d)f(x+2d).$$

Theorem 9.6.1 (Roth's theorem, functional/counting version)
For every $\delta > 0$, there exists $c = c(\delta) > 0$ such that every $f: \mathbb{Z}/N\mathbb{Z} \to [0,1]$ with $\mathbb{E}f \geq \delta$,

$$\Lambda_3(f) \geq c.$$

Exercise 9.6.2. Deduce the above version of Roth's theorem from the existence version (namely that every 3-AP-free subset of $[N]$ has size $o(N)$.)

Proof of the relative Roth theorem (Theorem 9.2.5). Let $p = |S|\,/N$. Define

$$\nu: \mathbb{Z}/N\mathbb{Z} \to [0, \infty) \qquad \text{by} \qquad \nu = p^{-1} 1_S.$$

Let $X = Y = Z = \mathbb{Z}/N\mathbb{Z}$. Consider the associated edge-weighted tripartite graph

$$\nu': (X \times Y) \cup (X \times Z) \cup (Y \times Z) \to [0, \infty)$$

defined by, for $x \in X$, $y \in Y$, and $z \in Z$,

$$\nu'(x,y) = \nu(2x+y), \qquad \nu'(x,z) = \nu(x-z), \qquad \nu'(y,z) = \nu(-y-2z).$$

Since ν satisfies the 3-linear forms condition (as a function on $\mathbb{Z}/N\mathbb{Z}$), ν' also satisfies the 3-linear forms condition in the sense of Section 9.5. Likewise,

$$\|\nu - 1\|_\square = \|\nu' - 1\|_\square$$

where $\|\nu - 1\|_\square$ on the left-hand side is in the sense of Section 9.4 and $\|\nu' - 1\|_\square$ is defined as in Section 9.5 where ν' is restricted to $X \times Y$. (The same would be true had we restricted to $X \times Z$ or $Y \times Z$). Indeed,

$$\|\nu - 1\|_\square = \sup_{A \subseteq X, B \subseteq Y} \mathbb{E}(\nu(x+y) - 1)1_A(x)1_B(y)$$

whereas

$$\begin{aligned}\|\nu' - 1\|_\square &= \sup_{A \subseteq X, B \subseteq Y} \mathbb{E}(\nu'(x,y) - 1)1_A(x)1_B(y) \\ &= \sup_{A \subseteq X, B \subseteq Y} \mathbb{E}(\nu(2x+y) - 1)1_A(x)1_B(y),\end{aligned}$$

and these two expressions are equal to each other after a change of variables $x \leftrightarrow 2x$ (which is a bijection as N is odd).

By Lemma 9.5.2 (or simply two applications of the Cauchy–Schwarz inequality followed by the 3-linear forms condition), we obtain

$$\|\nu - 1\|_\square \le \varepsilon^{\Omega(1)}.$$

Now suppose $A \subseteq S$ and $|A| \ge \delta N$. Define $f: \mathbb{Z}/N\mathbb{Z} \to [0, \infty)$ by

$$f = p^{-1} 1_A$$

so that $0 \le f \le \nu$ pointwise. Then by the dense model theorem (Theorem 9.4.6), there exists a function $g: \mathbb{Z}/N\mathbb{Z} \to [0,1]$ such that

$$\|f - g\|_\square \le \eta,$$

where $\eta = \eta(\varepsilon)$ is some quantity that tends to zero as $\varepsilon \to 0$.

Define the associated edge-weighted tripatite graphs

$$f', g' : (X \times Y) \cup (X \times Z) \cup (Y \times Z) \to [0, \infty)$$

where, for $x \in X$, $y \in Y$, and $z \in Z$,

$$\begin{aligned} f'(x,y) &= f(2x+y), & f'(x,z) &= f(x-z), & f'(y,z) &= f(-y-2z), \\ g'(x,y) &= g(2x+y), & g'(x,z) &= g(x-z), & g'(y,z) &= g(-y-2z). \end{aligned}$$

Note that g' takes values in $[0,1]$. Then

$$\|f' - g'\|_\square = \|f - g\|_\square \le \eta$$

when $f' - g'$ is interpreted as restricted to $X \times Y$ (and the same for $X \times Z$ or $Y \times Z$). Thus, by the sparse triangle counting lemma (Theorem 9.5.1), we have

$$|t(K_3, f') - t(K_3, g')| \le \eta^{\Omega(1)}.$$

Note that

$$\begin{aligned} t(K_3, f') &= \mathbb{E}_{x,y,z} f'(x,y) f'(x,z) f'(y,z) \\ &= \mathbb{E}_{x,y,z \in \mathbb{Z}/N\mathbb{Z}} f(2x+y) f(x-z) f(-y-2z) \\ &= \mathbb{E}_{x,d \in \mathbb{Z}/N\mathbb{Z}} f(x) f(x+d) f(x+2d) \\ &= \Lambda_3(f). \end{aligned}$$

Likewise, $t(K_3, g') = \Lambda_3(g)$. And so,

$$|\Lambda_3(f) - \Lambda_3(g)| \le \eta^{\Omega(1)}. \tag{9.10}$$

We have

$$\mathbb{E}g \ge \mathbb{E}f - \eta \ge \delta - \eta.$$

Provided that ε is chosen to be small enough so that η is small enough (say, so that $\mathbb{E}g \ge \delta/2$), we deduce from Roth's theorem (the functional version, Theorem 9.6.1) $\Lambda_3(g) \gtrsim_\delta 1$. Therefore

$$p^{-3}N^{-2}\,|\{(x,d)\colon x, x+d, x+2d \in A\}| = \Lambda_3(f) \overset{(9.10)}{\ge} \Lambda_3(g) - \eta^{\Omega(1)} \gtrsim_\delta 1$$

provided that η is sufficiently small. We can now conclude that A must have a nontrivial 3-AP if N is large enough. Indeed, if A were 3-AP-free, then

$$|\{(x,d)\colon x, x+d, x+2d \in A\}| = |A| \le |S| = pN,$$

and so the above inequality would imply $p \lesssim_\delta N^{-1/2}$. However, this would be incompatible with the 3-linear forms condition on S, since the probability that random $x_0, x_1, y_0, y_1, z_0, z_1 \in \mathbb{Z}/N\mathbb{Z}$ satisfy

$$\left\{\begin{matrix} -y_0 - 2z_0, & x_0 - z_0, & 2x_0 + y_0, \\ -y_1 - 2z_0, & x_1 - z_0, & 2x_1 + y_0, \\ -y_0 - 2z_1, & x_0 - z_1, & 2x_0 + y_1, \\ -y_1 - 2z_1, & x_1 - z_1, & 2x_1 + y_1 \end{matrix}\right\} \subseteq S$$

lies in the interval $(1 \pm \varepsilon)p^{12}$, but this probability is at least $|S|/N^5 = p/N^4$ (the probability that all of the preceding 12 terms are equal to the same element of S). So, $(1+\varepsilon)p^{12} \ge pN^{-4}$, and hence $p \gtrsim N^{-4/11}$, which would contradict the earlier $p \lesssim_\delta N^{-1/2}$ if N is large enough. □

***Remark* 9.6.3.** The preceding proof generalizes to a proof of the relative Szemerédi theorem, assuming Szemerédi's theorem as a black box.

All the arguments in this chapter can be generalized to deduce the relative Szemerédi theorem (Theorem 9.2.7) from Szemerédi's theorem. The ideas are essentially the same, although the notation gets heavier.

Further Reading

The original paper by Green and Tao (2008) titled *The primes contain arbitrarily long arithmetic progressions* is worth reading. Their follow-up paper *Linear equations in primes* (2010a) substantially strengthens the result to asymptotically count the number of k-APs in the primes, though the proof was conditional on several claims that were subsequently proved, most notably the inverse theorem for Gowers uniformity norms (Green, Tao, and Ziegler 2012).

A number of expository articles were written on this topic shortly after the breakthroughs: Green (2007b, 2014), Tao (2007b), Kra (2006), Wolf (2013).

The graph-theoretic approach taken in this chapter is adapted from the article *The Green–Tao theorem: An exposition* by Conlon, Fox, and Zhao (2014). The article presents a full proof of the Green–Tao theorem that incorporates various simplifications found since the original work. The analytic number theoretic arguments, which were omitted from this chapter, can also be found in that article.

Chapter Summary

- **Green–Tao theorem**. The primes contain arbitrarily long arithmetic progressions. Proof strategy:
  - Embed the primes in a slightly larger set, the "almost primes," which enjoys certain pseudorandomness properties.
  - Show that every k-AP-free subset of such a pseudorandom set must have negligible size.
- **Relative Szemerédi theorem.** If $S \subseteq \mathbb{Z}/N\mathbb{Z}$ satisfies a ***k*-linear forms condition**, then every k-AP-free subset of S has size $o(|S|)$.
  - The 3-linear forms condition is a pseudorandomness hypothesis. It says that the associated tripartite graph has F-density close to random whenever $F \subseteq K_{2,2,2}$.
- Proof of the relative Szemerédi theorem uses the **transference principle** to transfer Szemerédi's theorem from the dense setting to the sparse pseudorandom setting.
  - First approximate $A \subseteq S$ by a dense set $B \subseteq \mathbb{Z}/N\mathbb{Z}$ (**dense model theorem**).
  - Then show that the normalized count of k-APs in A and B are similar (**sparse counting lemma**).
  - Finally conclude using Szemerédi's theorem that B has many k-APs, and therefore so must A.
- **Dense model theorem.** If a sparse set S is close to random in normalized cut norm, then every subset $A \subseteq S$ can be approximated by some dense $B \subseteq \mathbb{Z}/N\mathbb{Z}$ in normalized cut norm.
- **Sparse counting lemma.** If two graphs (one sparse and one dense) are close to normalized cut norm, then they have similar triangle counts, provided that the sparse graph lies inside a sparse pseudorandom graph satisfying the 3-linear forms condition (which says that the densities of $K_{2,2,2}$ and its subgraphs are close to random).

References

Ajtai, M, & **Szemerédi**, E. (1974)
Sets of lattice points that form no squares, Studia Sci. Math. Hungar. 9, 9–11 (1975). MR:369299

Ajtai, M., **Chvátal**, V., **Newborn**, M. M., & **Szemerédi**, E. (1982)
Crossing-free subgraphs, Theory and practice of combinatorics, North-Holland, 9–12. MR:806962

Alon, Noga (1986)
Eigenvalues and expanders, Combinatorica 6, 83–96. MR:875835

Alon, N. & **Milman**, V. D. (1985)
λ_1, *isoperimetric inequalities for graphs, and superconcentrators*, J. Combin. Theory Ser. B 38, 73–88. MR:782626

Alon, Noga & **Naor**, Assaf (2006)
Approximating the cut-norm via Grothendieck's inequality, SIAM J. Comput. 35, 787–803. MR:2203567

Alon, Noga & **Shapira**, Asaf (2008)
A characterization of the (natural) graph properties testable with one-sided error, SIAM J. Comput. 37, 1703–1727. MR:2386211

Alon, Noga & **Spencer**, Joel H. (2016)
The probabilistic method, fourth ed., Wiley. MR:3524748

Alon, Noga, **Rónyai**, Lajos, & **Szabó**, Tibor (1999)
Norm-graphs: variations and applications, J. Combin. Theory Ser. B 76, 280–290. MR:1699238

Alon, Noga, **Fischer**, Eldar, **Krivelevich**, Michael, & **Szegedy**, Mario (2000)
Efficient testing of large graphs, Combinatorica 20, 451–476. MR:1804820

Alon, Noga, **Krivelevich**, Michael, & **Sudakov**, Benny (2003a)
Turán numbers of bipartite graphs and related Ramsey-type questions, Combin. Probab. Comput. 12, 477–494. MR:2037065

Alon, Noga, Fernandez de la **Vega**, W., **Kannan**, Ravi, & **Karpinski**, Marek (2003b)
Random sampling and approximation of MAX-CSPs, J. Comput. System Sci. 67, 212–243. MR:2022830

Artin, Emil (1927)
Über die Zerlegung definiter Funktionen in Quadrate, Abh. Math. Sem. Univ. Hamburg 5, 100–115. MR:3069468

Atkinson, F. V., **Watterson**, G. A., & **Moran**, P. A. P. (1960)
A matrix inequality, Quart. J. Math. Oxford Ser. 11, 137–140. MR:118731

Babai, Lászlo & **Frankl**, Péter (2020)
Linear algebra methods in combinatorics, http://people.cs.uchicago.edu/~laci/CLASS/HANDOUTS-COMB/BaFrNew.pdf.

Baker, R. C., **Harman**, G., & **Pintz**, J. (2001)
The difference between consecutive primes. II, Proc. Lond. Math. Soc. 83, 532–562. MR:1851081

Balog, Antal & **Szemerédi**, Endre (1994)
A statistical theorem of set addition, Combinatorica 14, 263–268. MR:1305895

Balogh, József, **Morris**, Robert, & **Samotij**, Wojciech (2015)
Independent sets in hypergraphs, J. Amer. Math. Soc. 28, 669–709. MR:3327533

Balogh, József, **Hu**, Ping, **Lidický**, Bernard, & **Pfender**, Florian (2016)
Maximum density of induced 5-cycle is achieved by an iterated blow-up of 5-cycle, European J. Combin. 52, 47–58. MR:3425964

Balogh, József, **Morris**, Robert, & **Samotij**, Wojciech (2018)
The method of hypergraph containers, Proceedings of the International Congress of Mathematicians – Rio de Janeiro 2018. Vol. IV. Invited lectures, World Scientific Publishing, 3059–3092. MR:3966523

Bateman, Michael & **Katz**, Nets Hawk (2012)
New bounds on cap sets, J. Amer. Math. Soc. 25, 585–613. MR:2869028

Behrend, F. A. (1946)
On sets of integers which contain no three terms in arithmetical progression, Proc. Natl. Acad. Sci. USA 32, 331–332. MR:18694

Benson, Clark T. (1966)
Minimal regular graphs of girths eight and twelve, Canadian J. Math. 18, 1091–1094. MR:197342

Bergelson, V. & **Leibman**, A. (1996)
Polynomial extensions of van der Waerden's and Szemerédi's theorems, J. Amer. Math. Soc. 9, 725–753. MR:1325795

Bergelson, Vitaly, **Host**, Bernard, & **Kra**, Bryna (2005)
Multiple recurrence and nilsequences, Invent. Math. 160, 261–303. MR:2138068

Bilu, Yonatan & **Linial**, Nathan (2006)
Lifts, discrepancy and nearly optimal spectral gap, Combinatorica 26, 495–519. MR:2279667

Blakley, G. R. & **Roy**, Prabir (1965)
A Hölder type inequality for symmetric matrices with nonnegative entries, Proc. Amer. Math. Soc. 16, 1244–1245. MR:184950

Blasiak, Jonah, **Church**, Thomas, **Cohn**, Henry, **Grochow**, Joshua A., **Naslund**, Eric, **Sawin**, William F., & **Umans**, Chris (2017)
On cap sets and the group-theoretic approach to matrix multiplication, Discrete Anal., Paper No. 3. MR:3631613

Blichfeldt, H. F. (1914)
A new principle in the geometry of numbers, with some applications, Trans. Amer. Math. Soc. 15, 227–235. MR:1500976

Bloom, Thomas F. & **Sisask**, Olof (2020)
Breaking the logarithmic barrier in Roth's theorem on arithmetic progressions. arXiv:2007.03528

Bogolyubov, N. (1939)
Sur quelques propriétés arithmétiques des presque-périodes, Ann. Chaire Phys. Math. Kiev 4, 185–205. MR:20164

Bollobás, Béla (1976)
Relations between sets of complete subgraphs, Proceedings of the Fifth British Combinatorial Conference (Univ. Aberdeen, Aberdeen, 1975), 79–84. MR:396327

Bollobás, Béla (1998)
Modern graph theory, Springer. MR:1633290

Bollobás, Béla & **Thomason**, Andrew (1995)
Projections of bodies and hereditary properties of hypergraphs, Bull. London Math. Soc. 27, 417–424. MR:1338683

Bondy, J. A. & **Murty**, U. S. R. (2008)
Graph theory, Springer. MR:2368647

Bondy, J. A. & **Simonovits**, M. (1974)
Cycles of even length in graphs, J. Combin. Theory Ser. B 16, 97–105. MR:340095

Borgs, C., **Chayes**, J. T., **Lovász**, L., **Sós**, V. T., & **Vesztergombi**, K. (2008)
Convergent sequences of dense graphs. I. Subgraph frequencies, metric properties and testing, Adv. Math. 219, 1801–1851. MR:2455626

Bourgain, J. (1999)
On triples in arithmetic progression, Geom. Funct. Anal. 9, 968–984. MR:1726234

Bourgain, J., **Katz**, N., & **Tao**, T. (2004)
A sum-product estimate in finite fields, and applications, Geom. Funct. Anal. 14, 27–57. MR:2053599

Bourgain, J., **Glibichuk**, A. A., & **Konyagin**, S. V. (2006)
Estimates for the number of sums and products and for exponential sums in fields of prime order, J. Lond. Math. Soc. 73, 380–398. MR:2225493

Brown, W. G. (1966)
On graphs that do not contain a Thomsen graph, Canad. Math. Bull. 9, 281–285. MR:200182

Brown, W. G., **Erdős**, P., & **Sós**, V. T. (1973)
Some extremal problems on r-graphs, New directions in the theory of graphs (Proc. Third Ann Arbor Conf., Univ. Michigan, Ann Arbor, Mich., 1971), 53–63. MR:351888

Bukh, Boris (2015)
Random algebraic construction of extremal graphs, Bull. Lond. Math. Soc. 47, 939–945. MR:3431574

Bukh, Boris (2021)
Extremal graphs without exponentially-small bicliques. arXiv:2107.04167

Chang, Mei-Chu (2002)
A polynomial bound in Freiman's theorem, Duke Math. J. 113, 399–419. MR:1909605

Chatterjee, Sourav (2016)
An introduction to large deviations for random graphs, Bull. Amer. Math. Soc. 53, 617–642. MR:3544262

Chatterjee, Sourav (2017)
Large deviations for random graphs, Springer. MR:3700183

Chatterjee, Sourav & **Varadhan**, S. R. S. (2011)
The large deviation principle for the Erdős-Rényi random graph, European J. Combin. 32, 1000–1017. MR:2825532

Cheeger, Jeff (1970)
A lower bound for the smallest eigenvalue of the Laplacian, Problems in analysis (Papers dedicated to Salomon Bochner, 1969), 195–199. MR:402831

Chung, Fan R. K. (1997)
Spectral graph theory, American Mathematical Society. MR:1421568

Chung, F. R. K., **Graham**, R. L., **Frankl**, P., & **Shearer**, J. B. (1986)
Some intersection theorems for ordered sets and graphs, J. Combin. Theory Ser. A 43, 23–37. MR:859293

Chung, F. R. K., **Graham**, R. L., & **Wilson**, R. M. (1989)
Quasi-random graphs, Combinatorica 9, 345–362. MR:1054011

Conlon, David (2021)
Extremal numbers of cycles revisited, Amer. Math. Monthly 128, 464–466. MR:4249723

Conlon, David & **Fox**, Jacob (2013)
Graph removal lemmas, Surveys in combinatorics 2013, Cambridge University Press, 1–49. MR:3156927

Conlon, D. & **Gowers**, W. T. (2016)
Combinatorial theorems in sparse random sets, Ann. of Math. 184, 367–454. MR:3548529

Conlon, David & **Zhao**, Yufei (2017)
Quasirandom Cayley graphs, Discrete Anal., Paper No. 6. MR:3631610

Conlon, David, **Fox**, Jacob, & **Sudakov**, Benny (2010)
An approximate version of Sidorenko's conjecture, Geom. Funct. Anal. 20, 1354–1366. MR:2738996

Conlon, David, **Fox**, Jacob, & **Zhao**, Yufei (2014)
The Green-Tao theorem: an exposition, EMS Surv. Math. Sci. 1, 249–282. MR:3285854

Conlon, David, **Fox**, Jacob, & **Zhao**, Yufei (2015)
A relative Szemerédi theorem, Geom. Funct. Anal. 25, 733–762. MR:3361771

Conlon, David, **Kim**, Jeong Han, **Lee**, Choongbum, & **Lee**, Joonkyung (2018)
Some advances on Sidorenko's conjecture, J. Lond. Math. Soc. 98, 593–608. MR:3893193

Coppersmith, Don & **Winograd**, Shmuel (1990)
Matrix multiplication via arithmetic progressions, J. Symbolic Comput. 9, 251–280. MR:1056627

Croot, Ernie, **Lev**, Vsevolod F., & **Pach**, Péter Pál (2017)
Progression-free sets in $\mathbb{Z}_4^n$ are exponentially small, Ann. of Math. 185, 331–337. MR:3583357

Davidoff, Giuliana, **Sarnak**, Peter, & **Valette**, Alain (2003)
Elementary number theory, group theory, and Ramanujan graphs, Cambridge University Press. MR:1989434

Dickson, L. E. (1909)
On the congruence $x^n + y^n + z^n \equiv 0 \pmod p$, J. Reine Angew. Math. 135, 134–141. MR:1580764

Diestel, Reinhard (2017)
Graph theory, fifth ed., Springer. MR:3644391

Dodziuk, Jozef (1984)
Difference equations, isoperimetric inequality and transience of certain random walks, Trans. Amer. Math. Soc. 284, 787–794. MR:743744

Dvir, Zeev (2012)
Incidence theorems and their applications, Found. Trends Theor. Comput. Sci. 6, 257–393. MR:3004132

Edel, Yves (2004)
Extensions of generalized product caps, Des. Codes Cryptogr. 31, 5–14. MR:2031694

Elekes, György (1997)
On the number of sums and products, Acta Arith. 81, 365–367. MR:1472816

Elkin, Michael (2011)
An improved construction of progression-free sets, Israel J. Math. 184, 93–128. MR:2823971

Ellenberg, Jordan S. & **Gijswijt**, Dion (2017)
On large subsets of $\mathbb{F}_q^n$ with no three-term arithmetic progression, Ann. of Math. 185, 339–343. MR:3583358

Erdős, P. (1971)
On some extremal problems on r-graphs, Discrete Math. 1, 1–6. MR:297602

Erdős, P. & **Simonovits**, M. (1966)
A limit theorem in graph theory, Studia Sci. Math. Hungar. 1, 51–57. MR:205876

Erdős, P. & **Szemerédi**, E. (1983)
On sums and products of integers, Studies in pure mathematics, Birkhäuser, 213–218. MR:820223

Erdős, P., **Rényi**, A., & **Sós**, V. T. (1966)
On a problem of graph theory, Studia Sci. Math. Hungar. 1, 215–235. MR:223262

Erdős, Paul (1984)
On some problems in graph theory, combinatorial analysis and combinatorial number theory, Graph theory and combinatorics (Cambridge, 1983), Academic Press, 1–17. MR:777160

Erdős, P. (1946)
On sets of distances of n points, Amer. Math. Monthly 53, 248–250. MR:15796

Erdős, P. & **Stone**, A. H. (1946)
On the structure of linear graphs, Bull. Amer. Math. Soc. 52, 1087–1091. MR:18807

Erdős, Paul (1955)
Some remarks on number theory, Riveon Lematematika 9, 45–48. MR:73619

Erdős, Paul & **Turán**, Paul (1936)
On some sequences of integers, J. Lond. Math. Soc. 11, 261–264. MR:1574918

Even Zohar, Chaim (2012)
On sums of generating sets in $\mathbb{Z}_2^n$, Combin. Probab. Comput. 21, 916–941. MR:2981161

Finner, Helmut (1992)
A generalization of Hölder's inequality and some probability inequalities, Ann. Probab. 20, 1893–1901. MR:1188047

Ford, Kevin (2008)
The distribution of integers with a divisor in a given interval, Ann. of Math. 168, 367–433. MR:2434882

Fox, Jacob (2011)
A new proof of the graph removal lemma, Ann. of Math. 174, 561–579. MR:2811609

Fox, Jacob & **Pham**, Huy Tuan (2019)
Popular progression differences in vector spaces II, Discrete Anal., Paper No. 16. MR:4042159

Fox, Jacob & **Sudakov**, Benny (2011)
Dependent random choice, Random Structures Algorithms 38, 68–99. MR:2768884

Fox, Jacob & **Zhao**, Yufei (2015)
A short proof of the multidimensional Szemerédi theorem in the primes, Amer. J. Math. 137, 1139–1145. MR:3372317

Fox, Jacob, **Pham**, Huy Tuan, & **Zhao**, Yufei (2022)
Tower-type bounds for Roth's theorem with popular differences, J. Eur. Math. Soc. (JEMS).

Frankl, Peter & **Rödl**, Vojtěch (2002)
Extremal problems on set systems, Random Structures Algorithms 20, 131–164. MR:1884430

Freiman, G. A. (1973)
Foundations of a structural theory of set addition, American Mathematical Society, Providence, RI, Translated from the Russian. MR:360496

Friedgut, Ehud (2004)
Hypergraphs, entropy, and inequalities, Amer. Math. Monthly 111, 749–760. MR:2104047

Friedman, Joel (2008)
A proof of Alon's second eigenvalue conjecture and related problems, Mem. Amer. Math. Soc. 195, viii+100. MR:2437174

Frieze, Alan & **Kannan**, Ravi (1999)
Quick approximation to matrices and applications, Combinatorica 19, 175–220. MR:1723039

Fulton, William & **Harris**, Joe (1991)
Representation theory: A first course, Springer. MR:1153249

Füredi, Zoltán (1991)
On a Turán type problem of Erdős, Combinatorica 11, 75–79. MR:1112277

Füredi, Zoltan & **Gunderson**, David S. (2015)
Extremal numbers for odd cycles, Combin. Probab. Comput. 24, 641–645. MR:3350026

Füredi, Zoltán & **Simonovits**, Miklós (2013)
The history of degenerate (bipartite) extremal graph problems, Erdős centennial, János Bolyai Mathematical Society, 169–264. MR:3203598

Furstenberg, H. (1977)
Ergodic behavior of diagonal measures and a theorem of Szemerédi on arithmetic progressions, J. Analyse Math. 31, 204–256. MR:498471

Furstenberg, H. & **Katznelson**, Y. (1978)
An ergodic Szemerédi theorem for commuting transformations, J. Analyse Math. 34, 275–291. MR:531279

Galvin, David (2014)
Three tutorial lectures on entropy and counting. arXiv:1406.7872

Galvin, David & **Tetali**, Prasad (2004)
On weighted graph homomorphisms, Graphs, morphisms and statistical physics, American Mathematical Society, 97–104. MR:2056231

Goemans, Michel X. & **Williamson**, David P. (1995)
Improved approximation algorithms for maximum cut and satisfiability problems using semidefinite programming, J. Assoc. Comput. Mach. 42, 1115–1145. MR:1412228

Goodman, A. W. (1959)
On sets of acquaintances and strangers at any party, Amer. Math. Monthly 66, 778–783. MR:107610

Gowers, W. T. (1997)
Lower bounds of tower type for Szemerédi's uniformity lemma, Geom. Funct. Anal. 7, 322–337. MR:1445389

Gowers, W. T. (1998a)
A new proof of Szemerédi's theorem for arithmetic progressions of length four, Geom. Funct. Anal. 8, 529–551. MR:1631259

Gowers, W. T. (1998b)
Additive and combinatorial number theory, online lecture notes written by Jacques Verstraëte based on a course given by W. T. Gowers, www.dpmms.cam.ac.uk/~wtg10/.

Gowers, W. T. (2001)
A new proof of Szemerédi's theorem, Geom. Funct. Anal. 11, 465–588. MR:1844079

Gowers, W. T. (2006)
Quasirandomness, counting and regularity for 3-uniform hypergraphs, Combin. Probab. Comput. 15, 143–184. MR:2195580

Gowers, W. T. (2007)
Hypergraph regularity and the multidimensional Szemerédi theorem, Ann. of Math. 166, 897–946. MR:2373376

Gowers, W. T. (2008)
Quasirandom groups, Combin. Probab. Comput. 17, 363–387. MR:2410393

Gowers, W. T. (2010)
Decompositions, approximate structure, transference, and the Hahn-Banach theorem, Bull. Lond. Math. Soc. 42, 573–606. MR:2669681

Graham, Ronald L., **Rothschild**, Bruce L., & **Spencer**, Joel H. (1990)
Ramsey theory, second ed., Wiley. MR:1044995

Green, B. (2005a)
A Szemerédi-type regularity lemma in abelian groups, with applications, Geom. Funct. Anal. 15, 340–376. MR:2153903

Green, Ben (2005b)
Roth's theorem in the primes, Ann. of Math. 161, 1609–1636. MR:2180408

Green, Ben (2005c)
Finite field models in additive combinatorics, Surveys in combinatorics 2005, Cambridge University Press, 1–27. MR:2187732

Green, Ben (2007a)
Montréal notes on quadratic Fourier analysis, Additive combinatorics, American Mathematical Society, 69–102. MR:2359469

Green, Ben (2007b)
Long arithmetic progressions of primes, Analytic number theory: A tribute to Gauss and Dirichlet, American Mathematical Society, 149–167. MR:2362199

Green, Ben (2009a)
Additive combinatorics (book review), Bull. Amer. Math. Soc. 46, 489–497. MR:2507281

Green, Ben (2009b)
Additive combinatorics, lecture notes, `http://people.maths.ox.ac.uk/greenbj/notes.html`.

Green, Ben (2014)
Approximate algebraic structure, Proceedings of the International Congress of Mathematicians – Seoul 2014. Vol. 1, Kyung Moon Sa, 341–367. MR:3728475

Green, Ben & **Ruzsa**, Imre Z. (2007)
Freiman's theorem in an arbitrary abelian group, J. Lond. Math. Soc. 75, 163–175. MR:2302736

Green, Ben & **Tao**, Terence (2008)
The primes contain arbitrarily long arithmetic progressions, Ann. of Math. 167, 481–547. MR:2415379

Green, Ben & **Tao**, Terence (2010a)
Linear equations in primes, Ann. of Math. 171, 1753–1850. MR:2680398

Green, Ben & **Tao**, Terence (2010b)
An equivalence between inverse sumset theorems and inverse conjectures for the U^3 norm, Math. Proc. Cambridge Philos. Soc. 149, 1–19. MR:2651575

Green, Ben & **Tao**, Terence (2010c)
An arithmetic regularity lemma, an associated counting lemma, and applications, An irregular mind, János Bolyai Mathematical Society, 261–334. MR:2815606

Green, Ben & **Tao**, Terence (2017)
New bounds for Szemerédi's theorem, III: a polylogarithmic bound for $r_4(N)$, Mathematika 63, 944–1040. MR:3731312

Green, Ben & **Wolf**, Julia (2010)
A note on Elkin's improvement of Behrend's construction, Additive number theory, Springer, 141–144. MR:2744752

Green, Ben, **Tao**, Terence, & **Ziegler**, Tamar (2012)
An inverse theorem for the Gowers $U^{s+1}[N]$-norm, Ann. of Math. 176, 1231–1372. MR:2950773

Grothendieck, A. (1953)
Résumé de la théorie métrique des produits tensoriels topologiques, Bol. Soc. Mat. São Paulo 8, 1–79. MR:94682

Grzesik, Andrzej (2012)
On the maximum number of five-cycles in a triangle-free graph, J. Combin. Theory Ser. B 102, 1061–1066. MR:2959390

Guth, Larry (2016)
Polynomial methods in combinatorics, American Mathematical Society. MR:3495952

Guth, Larry & **Katz**, Nets Hawk (2015)
On the Erdős distinct distances problem in the plane, Ann. of Math. 181, 155–190. MR:3272924

Hardy, G. H. & **Ramanujan**, S. (1917)
The normal number of prime factors of a number n, Quart. J. Math. 48, 76–92. MR:2280878

Håstad, Johan (2001)
Some optimal inapproximability results, J. ACM 48, 798–859. MR:2144931

Hatami, Hamed & **Norine**, Serguei (2011)
Undecidability of linear inequalities in graph homomorphism densities, J. Amer. Math. Soc. 24, 547–565. MR:2748400

Hatami, Hamed, **Hladký**, Jan, **Kráľ**, Daniel, **Norine**, Serguei, & **Razborov**, Alexander (2013)
On the number of pentagons in triangle-free graphs, J. Combin. Theory Ser. A 120, 722–732. MR:3007147

Hilbert, David (1888)
Ueber die Darstellung definiter Formen als Summe von Formenquadraten, Math. Ann. 32, 342–350. MR:1510517

Hilbert, David (1893)
Über ternäre definite Formen, Acta Math. 17, 169–197. MR:1554835

Hoory, Shlomo, **Linial**, Nathan, & **Wigderson**, Avi (2006)
Expander graphs and their applications, Bull. Amer. Math. Soc. 43, 439–561. MR:2247919

Hosseini, Kaave, **Lovett**, Shachar, **Moshkovitz**, Guy, & **Shapira**, Asaf (2016)
An improved lower bound for arithmetic regularity, Math. Proc. Cambridge Philos. Soc. 161, 193–197. MR:3530502

Ireland, Kenneth & **Rosen**, Michael (1990)
A classical introduction to modern number theory, second ed., Springer-Verlag. MR:1070716

Jordan, Herbert E. (1907)
Group-characters of various types of linear groups, Amer. J. Math. 29, 387–405. MR:1506021

Kahn, Jeff (2001)
An entropy approach to the hard-core model on bipartite graphs, Combin. Probab. Comput. 10, 219–237. MR:1841642

Katona, G. (1968)
A theorem of finite sets, Theory of graphs (Proc. Colloq., Tihany, 1966), 187–207. MR:290982

Kedlaya, Kiran S. (1997)
Large product-free subsets of finite groups, J. Combin. Theory Ser. A 77, 339–343. MR:1429085

Kedlaya, Kiran S. (1998)
Product-free subsets of groups, Amer. Math. Monthly 105, 900–906. MR:1656927

Keevash, Peter (2011)
Hypergraph Turán problems, Surveys in combinatorics 2011, Cambridge University Press, 83–139. MR:2866732

Khot, Subhash, **Kindler**, Guy, **Mossel**, Elchanan, & **O'Donnell**, Ryan (2007)
Optimal inapproximability results for MAX-CUT and other 2-variable CSPs?, SIAM J. Comput. 37, 319–357. MR:2306295

Kleinberg, Robert, **Speyer**, David E., & **Sawin**, Will (2018)
The growth of tri-colored sum-free sets, Discrete Anal., Paper No. 12. MR:3827120

Kollár, János, **Rónyai**, Lajos, & **Szabó**, Tibor (1996)
Norm-graphs and bipartite Turán numbers, Combinatorica 16, 399–406. MR:1417348

Komlós, J. & **Simonovits**, M. (1996)
Szemerédi's regularity lemma and its applications in graph theory, Combinatorics, Paul Erdős is eighty, Vol. 2 (Keszthely, 1993), János Bolyai Mathematical Society, 295–352. MR:1395865

Komlós, János, **Shokoufandeh**, Ali, **Simonovits**, Miklós, & **Szemerédi**, Endre (2002)
The regularity lemma and its applications in graph theory, Theoretical aspects of computer science (Tehran, 2000), Springer, 84–112. MR:1966181

Konyagin, S. V. & **Shkredov**, I. D. (2015)
On sum sets of sets having small product set, Proc. Steklov Inst. Math. 290, 288–299. MR:3488800

Kővári, T., **Sós**, V. T., & **Turán**, P. (1954)
On a problem of K. Zarankiewicz, Colloq. Math. 3, 50–57. MR:65617

Kra, Bryna (2006)
The Green-Tao theorem on arithmetic progressions in the primes: An ergodic point of view, Bull. Amer. Math. Soc. 43, 3–23. MR:2188173

Krivelevich, M. & **Sudakov**, B. (2006)
Pseudo-random graphs, More sets, graphs and numbers, Bolyai Soc. Math. Stud., 15, Springer, 199–262. MR:2223394

Kruskal, Joseph B. (1963)
The number of simplices in a complex, Mathematical optimization techniques, University of California Press, 251–278. MR:154827

Lang, Serge & **Weil**, André (1954)
Number of points of varieties in finite fields, Amer. J. Math. 76, 819–827. MR:65218

Lee, Joonkyung (2019)
MathOverflow post, https://mathoverflow.net/q/189222/.

Leighton, Frank Thomson (1984)
New lower bound techniques for VLSI, Math. Systems Theory 17, 47–70. MR:738751

Li, J. L. Xiang & **Szegedy**, Balazs (2011)
On the logarithimic calculus and Sidorenko's conjecture. arXiv:1107.1153

Loomis, L. H. & **Whitney**, H. (1949)
An inequality related to the isoperimetric inequality, Bull. Amer. Math. Soc. 55, 961–962. MR:31538

Lovász, László (2009)
Very large graphs, Current developments in mathematics, 2008, International Press, 67–128. MR:2555927

Lovász, László (2012)
Large networks and graph limits, American Mathematical Society. MR:3012035

Lovász, László & **Szegedy**, Balázs (2006)
Limits of dense graph sequences, J. Combin. Theory Ser. B 96, 933–957. MR:2274085

Lovász, László & **Szegedy**, Balázs (2007)
Szemerédi's lemma for the analyst, Geom. Funct. Anal. 17, 252–270. MR:2306658

Lovett, Shachar (2012)
Equivalence of polynomial conjectures in additive combinatorics, Combinatorica 32, 607–618. MR:3004811

Lovett, Shachar (2015)
An exposition of Sanders' quasi-polynomial Freiman-Ruzsa theorem, Theory of Computing Library Graduate Surveys, vol. 6, 1–14.

Lovett, Shachar & **Regev**, Oded (2017)
A counterexample to a strong variant of the polynomial Freiman-Ruzsa conjecture in Euclidean space, Discrete Anal., Paper No. 8. MR:3651924

Lubetzky, Eyal & **Zhao**, Yufei (2017)
On the variational problem for upper tails in sparse random graphs, Random Structures Algorithms 50, 420–436. MR:3632418

Lubotzky, A., **Phillips**, R., & **Sarnak**, P. (1988)
Ramanujan graphs, Combinatorica 8, 261–277. MR:963118

Lubotzky, Alexander (2012)
Expander graphs in pure and applied mathematics, Bull. Amer. Math. Soc. 49, 113–162. MR:2869010

Mantel, W. (1907)
Problem 28, Wiskundige Opgaven 10, 60–61.

Marcus, Adam W., **Spielman**, Daniel A., & **Srivastava**, Nikhil (2015)
Interlacing families I: Bipartite Ramanujan graphs of all degrees, Ann. of Math. 182, 307–325. MR:3374962

Margulis, G. A. (1988)
Explicit group-theoretic constructions of combinatorial schemes and their applications in the construction of expanders and concentrators, Problemy Peredachi Informatsii 24, 51–60. MR:939574

Matiyasevich, Ju. V. (1970)
The Diophantineness of enumerable sets, Dokl. Akad. Nauk. SSSR. 191, 279–282. MR:258744

Matoušek, Jiří (2010)
Thirty-three miniatures: Mathematical and algorithmic applications of linear algebra, American Mathematical Society. MR:2656313

Meshulam, Roy (1995)
On subsets of finite abelian groups with no 3-term arithmetic progressions, J. Combin. Theory Ser. A 71, 168–172. MR:1335785

Minkowski, Hermann (1896)
Geometrie der Zahlen, Teubner. MR:249269

Morgenstern, Moshe (1994)
Existence and explicit constructions of $q + 1$ regular Ramanujan graphs for every prime power q, J. Combin. Theory Ser. B 62, 44–62. MR:1290630

Moshkovitz, Guy & **Shapira**, Asaf (2016)
A short proof of Gowers' lower bound for the regularity lemma, Combinatorica 36, 187–194. MR:3516883

Moshkovitz, Guy & **Shapira**, Asaf (2019)
A tight bound for hypergraph regularity, Geom. Funct. Anal. 29, 1531–1578. MR:4025519

Motzkin, T. S. (1967)
The arithmetic-geometric inequality, Inequalities (Proc. Sympos. Wright-Patterson Air Force Base, Ohio, 1965), Academic Press, 205–224. MR:223521

Motzkin, T. S. & **Straus**, E. G. (1965)
Maxima for graphs and a new proof of a theorem of Turán, Canadian J. Math. 17, 533–540. MR:175813

Mulholland, H. P. & **Smith**, C. A. B. (1959)
An inequality arising in genetical theory, Amer. Math. Monthly 66, 673–683. MR:110721

Nešetřil, Jaroslav & **Rosenfeld**, Moshe (2001)
I. Schur, C. E. Shannon and Ramsey numbers, a short story, Discrete Math. 229, 185–195. MR:1815606

Nikiforov, V. (2011)
The number of cliques in graphs of given order and size, Trans. Amer. Math. Soc. 363, 1599–1618. MR:2737279

Nikolov, N. & **Pyber**, L. (2011)
Product decompositions of quasirandom groups and a Jordan type theorem, J. Eur. Math. Soc. (JEMS) 13, 1063–1077. MR:2800484

Nilli, A. (1991)
On the second eigenvalue of a graph, Discrete Math. 91, 207–210. MR:1124768

Pellegrino, Giuseppe (1970)
Sul massimo ordine delle calotte in $S_{4,3}$, Matematiche (Catania) 25, 149–157 (1971). MR:363952

Peluse, Sarah (2020)
Bounds for sets with no polynomial progressions, Forum Math. Pi 8, e16. MR:4199235

Petridis, Giorgis (2012)
New proofs of Plünnecke-type estimates for product sets in groups, Combinatorica 32, 721–733. MR:3063158

Pippenger, Nicholas & **Golumbic**, Martin Charles (1975)
The inducibility of graphs, J. Combin. Theory Ser. B 19, 189–203. MR:401552

Plünnecke, Helmut (1970)
Eine zahlentheoretische Anwendung der Graphentheorie, J. Reine Angew. Math. 243, 171–183. MR:266892

Polymath, D. H. J. (2012)
A new proof of the density Hales-Jewett theorem, Ann. of Math. 175, 1283–1327. MR:2912706

Radhakrishnan, Jaikumar (2003)
Entropy and counting, Computational Mathematics, Modelling and Algorithms, Narosa.

Razborov, Alexander A. (2007)
Flag algebras, J. Symbolic Logic 72, 1239–1282. MR:2371204

Razborov, Alexander A. (2008)
On the minimal density of triangles in graphs, Combin. Probab. Comput. 17, 603–618. MR:2433944

Razborov, Alexander A. (2013)
Flag algebras: an interim report, The mathematics of Paul Erdős. II, Springer, 207–232. MR:3186665

Reiher, Christian (2016)
The clique density theorem, Ann. of Math. 184, 683–707. MR:3549620

Reingold, Omer, **Trevisan**, Luca, **Tulsiani**, Madhur, & **Vadhan**, Salil (2008)
New proofs of the Green-Tao-Ziegler dense model theorem: An exposition. arXiv:0806.0381

Rödl, V., **Nagle**, B., **Skokan**, J., **Schacht**, M., & **Kohayakawa**, Y. (2005)
The hypergraph regularity method and its applications, Proc. Natl. Acad. Sci. USA 102, 8109–8113. MR:2167756

Roth, K. F. (1953)
On certain sets of integers, J. Lond. Math. Soc. 28, 104–109. MR:51853

Ruzsa, I. Z. (1994)
Generalized arithmetical progressions and sumsets, Acta Math. Hungar. 65, 379–388. MR:1281447

Ruzsa, Imre Z. (1989)
An application of graph theory to additive number theory, Sci. Ser. A Math. Sci. 3, 97–109. MR:2314377

Ruzsa, Imre Z. (1999)
An analog of Freiman's theorem in groups, Astérisque 258, xv, 323–326. MR:1701207

Ruzsa, Imre Z. (2009)
Sumsets and structure, Combinatorial number theory and additive group theory, Birkhäuser Verlag, 87–210. MR:2522038

Ruzsa, Imre Z. & **Szemerédi**, Endre (1978)
Triple systems with no six points carrying three triangles, Combinatorics (Proc. Fifth Hungarian Colloq., Keszthely, 1976), Vol. II, 939–945. MR:519318

Sagan, Bruce E. (2001)
The symmetric group: Representations, combinatorial algorithms, and symmetric functions, second ed., Springer-Verlag. MR:1824028

Sah, Ashwin, **Sawhney**, Mehtaab, **Stoner**, David, & **Zhao**, Yufei (2019)
The number of independent sets in an irregular graph, J. Combin. Theory Ser. B 138, 172–195. MR:3979229

Sah, Ashwin, **Sawhney**, Mehtaab, **Stoner**, David, & **Zhao**, Yufei (2020)
A reverse Sidorenko inequality, Invent. Math. 221, 665–711. MR:4121160

Sah, Ashwin, **Sawhney**, Mehtaab, & **Zhao**, Yufei (2021)
Patterns without a popular difference, Discrete Anal., Paper No. 8. MR:4293329

Salem, R. & **Spencer**, D. C. (1942)
On sets of integers which contain no three terms in arithmetical progression, Proc. Natl. Acad. Sci. USA 28, 561–563. MR:7405

Sanders, Tom (2012)
On the Bogolyubov-Ruzsa lemma, Anal. PDE 5, 627–655. MR:2994508

Sanders, Tom (2013)
The structure theory of set addition revisited, Bull. Amer. Math. Soc. 50, 93–127. MR:2994996

Sárközy, A. (1978)
On difference sets of sequences of integers. I, Acta Math. Acad. Sci. Hungar. 31, 125–149. MR:466059

Saxton, David & **Thomason**, Andrew (2015)
Hypergraph containers, Invent. Math. 201, 925–992. MR:3385638

Schacht, Mathias (2016)
Extremal results for random discrete structures, Ann. of Math. 184, 333–365. MR:3548528

Schelp, Richard H. & **Thomason**, Andrew (1998)
A remark on the number of complete and empty subgraphs, Combin. Probab. Comput. 7, 217–219. MR:1617934

Schoen, Tomasz (2011)
Near optimal bounds in Freiman's theorem, Duke Math. J. 158, 1–12. MR:2794366

Schoen, Tomasz & **Shkredov**, Ilya D. (2014)
Roth's theorem in many variables, Israel J. Math. 199, 287–308. MR:3219538

Schoen, Tomasz & **Sisask**, Olof (2016)
Roth's theorem for four variables and additive structures in sums of sparse sets, Forum Math. Sigma 4, e5. MR:3482282

Schrijver, Alexander (2003)
Combinatorial optimization: Polyhedra and efficiency, Springer-Verlag. MR:1956924

Schur, I. (1916)
Uber die kongruenz $x^m + y^m \equiv z^m \pmod p$, Jber. Deutsch. Math.-Verein 25, 114–116.

Schur, J. (1907)
Untersuchungen über die Darstellung der endlichen Gruppen durch gebrochene lineare Substitutionen, J. Reine Angew. Math. 132, 85–137. MR:1580715

Serre, Jean-Pierre (1977)
Linear representations of finite groups, Springer-Verlag. MR:450380

Sheffer, Adam (2022)
Polynomial methods and incidence theory, Cambridge University Press. MR:4394303

Shkredov, I. D. (2006)
On a generalization of Szemerédi's theorem, Proc. Lond. Math. Soc. 93, 723–760. MR:2266965

Sidorenko, A. F. (1991)
Inequalities for functionals generated by bipartite graphs, Diskret. Mat. 3, 50–65. MR:1138091

Sidorenko, Alexander (1993)
A correlation inequality for bipartite graphs, Graphs Combin. 9, 201–204. MR:1225933

Simonovits, M. (1974)
Extermal graph problems with symmetrical extremal graphs. Additional chromatic conditions, Discrete Math. 7, 349–376. MR:337690

Singleton, Robert (1966)
On minimal graphs of maximum even girth, J. Combinatorial Theory 1, 306–332. MR:201347

Skokan, Jozef & **Thoma**, Lubos (2004)
Bipartite subgraphs and quasi-randomness, Graphs Combin. 20, 255–262. MR:2080111

Solymosi, József (2003)
Note on a generalization of Roth's theorem, Discrete and computational geometry, Algorithms and Combinatorics 25, Springer, 825–827. MR:2038505

Solymosi, József (2009)
Bounding multiplicative energy by the sumset, Adv. Math. 222, 402–408. MR:2538014

Soundararajan, K. (2007)
Additive combinatorics, online lecture notes, http://math.stanford.edu/~ksound/Notes.pdf.

Spielman, Daniel A. (2019)
Spectral and algebraic graph theory, textbook draft, cs-www.cs.yale.edu/homes/spielman/sagt/.

Stein, Elias M. & **Shakarchi**, Rami (2003)
Fourier analysis: An introduction, Princeton University Press. MR:1970295

Sudakov, B., **Szemerédi**, E., & **Vu**, V. H. (2005)
On a question of Erdős and Moser, Duke Math. J. 129, 129–155. MR:2155059

Szegedy, Balázs (2015)
An information theoretic approach to Sidorenko's conjecture. arXiv:1406.6738

Székely, László A. (1997)
Crossing numbers and hard Erdős problems in discrete geometry, Combin. Probab. Comput. 6, 353–358. MR:1464571

Szemerédi, E. (1975)
On sets of integers containing no k elements in arithmetic progression, Acta Arith. 27, 199–245. MR:369312

Szemerédi, Endre & **Trotter**, William T., Jr. (1983)
Extremal problems in discrete geometry, Combinatorica 3, 381–392. MR:729791

Tao, Terence (2006)
A variant of the hypergraph removal lemma, J. Combin. Theory Ser. A 113, 1257–1280. MR:2259060

Tao, Terence (2007a)
Structure and randomness in combinatorics, 48th Annual IEEE Symposium on Foundations of Computer Science (FOCS'07), 3–15.

Tao, Terence (2007b)
The dichotomy between structure and randomness, arithmetic progressions, and the primes, International Congress of Mathematicians. Vol. I, European Mathematical Society, 581–608. MR:2334204

Tao, Terence (2012)
The spectral proof of the Szemeredi regularity lemma, blog post, https://terrytao.wordpress.com/2012/12/03/.

Tao, Terence (2014)
A proof of Roth's theorem, blog post, https://terrytao.wordpress.com/2014/04/24/.

Tao, Terence (2016)
A symmetric formulation of the Croot-Lev-Pach-Ellenberg-Gijswijt capset bound, blog post, `https://terrytao.wordpress.com/2016/05/18/`.

Tao, Terence & **Vu**, Van (2006)
Additive combinatorics, Cambridge University Press. MR:2289012

Tao, Terence & **Ziegler**, Tamar (2008)
The primes contain arbitrarily long polynomial progressions, Acta Math. 201, 213–305. MR:2461509

Tao, Terence & **Ziegler**, Tamar (2015)
A multi-dimensional Szemerédi theorem for the primes via a correspondence principle, Israel J. Math. 207, 203–228. MR:3358045

Tarski, Alfred (1948)
A decision method for elementary algebra and geometry, RAND Corporation. MR:28796

Thomason, Andrew (1987)
Pseudorandom graphs, Random graphs '85 (Poznań, 1985), North-Holland, 307–331. MR:930498

Thomason, Andrew (1989)
A disproof of a conjecture of Erdős in Ramsey theory, J. Lond. Math. Soc. 39, 246–255. MR:991659

Turán, Paul (1934)
On a theorem of Hardy and Ramanujan, J. Lond. Math. Soc. 9, 274–276. MR:1574877

Turán, Paul (1941)
Eine Extremalaufgabe aus der Graphentheorie, Mat. Fiz. Lapok 48, 436–452 (Hungarian, with German summary).

Varnavides, P. (1959)
On certain sets of positive density, J. Lond. Math. Soc. 34, 358–360. MR:106865

Vinogradov, I. M. (1937)
The representation of an odd number as a sum of three primes., Dokl. Akad. Nauk. SSSR. 16, 139–142.

van der **Waerden**, B. L. (1927)
Beweis einer baudetschen vermutung, Nieuw Arch. Wisk. 15, 212–216.

Wenger, R. (1991)
Extremal graphs with no C^4's, C^6's, or C^{10}'s, J. Combin. Theory Ser. B 52, 113–116. MR:1109426

West, Douglas B. (1996)
Introduction to graph theory, Prentice Hall. MR:1367739

Wigderson, Avi (2012)
Representation theory of finite groups, and applications, Lecture notes for the 22nd McGill invitational workshop on computational complexity, `www.math.ias.edu/~avi/TALKS/Green_Wigderson_lecture.pdf`.

Williams, David (1991)
Probability with martingales, Cambridge University Press. MR:1155402

Wolf, J. (2015)
Finite field models in arithmetic combinatorics – ten years on, Finite Fields Appl. 32, 233–274. MR:3293412

Wolf, Julia (2013)
Arithmetic and polynomial progressions in the primes [after Gowers, Green, Tao and Ziegler], Astérisque 352, 389–427. MR:3087352

Zarankiewicz, K. (1951)
Problem 101, Colloq. Math. 2, 201.

Zhao, Yufei (2010)
The number of independent sets in a regular graph, Combin. Probab. Comput. 19, 315–320. MR:2593625

Zhao, Yufei (2014)
An arithmetic transference proof of a relative Szemerédi theorem, Math. Proc. Cambridge Philos. Soc. 156, 255–261. MR:3177868

Zhao, Yufei (2017)
Extremal regular graphs: Independent sets and graph homomorphisms, Amer. Math. Monthly 124, 827–843. MR:3722040

Index

additive combinatorics, 5
adjacency matrix, xv
Alon–Boppana bound, 119
AP, xv
  4-AP, 85, 207, 217, 221
$\asymp$, *see* asymptotics
asymptotics, xvii
Azuma's inequality, 139

Balog–Szemerédi–Gowers theorem, 258
  graph, 260
Behrend construction, 69
bipartite graph
  complete, xv
    constructions, 39
    forbidding, *see* Kővári–Sós–Turán theorem
  sparse
    forbidding, 33
Blichfeldt's theorem, 251
block model, *see* stochastic block model
blow-up, 28
Bogolyubov's lemma, 245
Bohr set, 216, 247, 252
Borel–Cantelli lemma, 140
bounded differences inequality, 139, 152
BSG, *see* Balog–Szemerédi–Gowers theorem 258

Cayley graph, 104
character, 107, 200
Cheeger's inequality, 103
Chernoff bound, 91
$\chi(\,)$, *see* chromatic number
chromatic number, xv, 27
Chung–Graham–Wilson theorem, *see* quasirandom graph
C_ℓ, *see* cycle
clique, xv
  forbidding, *see* Turán's theorem
coloring
  H-coloring, 179
  proper, *see* chromatic number
  Ramsey, 2
common difference, xv
  popular, 227
common graph, 168
concentration
  subgraph density, 139
constellation, 8
construction
  algebraic, 37, 39
  randomized, 37, 38
  randomized algebraic, 38, 46
  Turán, 37
convergence
  cut metric, 133, 140
  left, 136, 138, 140, 150
convolution, 201
corner, 67
counting lemma
  graph, 70
  graphon, 140
  inverse, 155
  sparse, 281, 287
  3-AP, *see* Roth's theorem, 205, 212
  triangle, 61, 141
Courant–Fischer min-max theorem, 96
covering lemma, 238
crossing number, 267
cut
  convergence, *see* convergence
  distance, 133
  metric, 133
  norm, 132, 140, 281
cycle, xv
  constructions, 45
  forbidding, 31

deg(x), *see* degree
degree, xv
dense model theorem, 281, 282
density
  clique, 183
  edge, 19, 53
  homomorphism, 135, 158, 277
  increment, 61, 206, 213
  integers, 5
  2-density, 39
dependent random choice, 33
diamond-free lemma, 65
discrepancy, 90, 131

distinct distances problem, 26
doubling
  constant, 232
  small, 232

$\mathbb{E}$, xv
$E(G)$, $e(G)$, $e(A,B)$, $e(A)$, *see* graph
edge expansion, 103
eigenvalue, 95
  abelian Cayley graph, 106
  bipartite, 98
  second, 119
    Alon–Boppana bound, 119
    Friedman's theorem, 123
    Ramanujan graph, 124
energy, 55, 222
  additive, 258
  increment, 223
  multiplicative, 271
entropy, 187
Erdős conjecture
  arithmetic progressions, 7
  distinct distances problem, 26
  unit distance problem, 24
Erdős–Stone–Simonovits theorem, 27, 74
Erdős–Turán conjecture, *see* Szemerédi's theorem
ESS theorem, *see* Erdős–Stone–Simonovits theorem
$\mathrm{ex}(n,H)$, *see* Turán problem
expander mixing lemma, 101
  bipartite, 102
  converse, 102
exponent
  abelian group, 240
extremal number, *see* Turán problem

finitary, 1
finite field model, 198, 203, 255
flag algebra, 170
forcing graph/conjecture, 97, 160
4-cycle
  forbidding, 22
    construction, 40
  minimum density, 93
Fourier, 106, 198, 210, 248
  inversion, 199, 210
  nonabelian, 115
  3-AP, 211
  uniform, 205, 207, 212, 217, 221
Freiman
  homomorphism, 241
  isomorphism, 241
  polynomial Freiman–Ruzsa conjecture, 255
  theorem, 233, 254
    abelian group, 234
    finite field model, 240
    groups with bounded exponent, 240
Furstenberg–Sárkőzy theorem, 8

$G(n,p)$, *see* random graph, Erdős–Rényi
$G[S]$, *see* induced subgraph
GAP, *see* progression, generalized arithmetic
Gauss sum, 108
Gowers uniformity norm, 209
graph, xv
graphon, 128
  associated, 129
  space, 133
    compact, 134, 148
  step, 129
Green–Tao theorem, 8, 274
Grothendieck's inequality, 117

H-free, xv, *see also* Turán problem
half-graph, 61, 129
hereditary graph property, 84
homomorphism
  Freiman, 241
  graph, xv, 135, 158
hypergraph, xv

induced subgraph, xv
infinitary, *see* finitary
invertible
  measure preserving map, 132

Kővári–Sós–Turán theorem, 22
  constructions, 39
  geometric applications, 24
K_r, *see* clique
$K_{s,t}$, *see* bipartite graph, complete
KST theorem, *see* Kővári–Sós–Turán theorem

lattice, 249
$\lesssim$, *see* asymptotics
linear forms condition, 278, 279

Mantel's theorem, 12
martingale, 146
  convergence theorem, 148
maximum cut, 145, 156
measure preserving map, 132
Minkowski
  first theorem, 251
  second theorem, 251
mixing
  quasirandom groups, 112
modeling lemma, 243
moment, 122
  graphon, 152

$\mathbb{N}$, xv
$N(x)$, *see* neighborhood
(n,d,λ)-graph, 101
  bipartite, 102
$[N]$, xv
neighborhood, xv

norm graph, 43
  projective, 44

$O(\ )$, *see* asymptotics
$\Omega(\ )$, *see* asymptotics
operator norm, 101

Paley graph, 92, 105
  eigenvalues, 107
Parseval, 199, 210
Perron–Frobenius theorem, 96
Plancherel, 199, 210
Plünnecke's Inequality, 235
point-line incidence, 268
polynomial method, 217
product set, 265
product-free, 109, 112
progression
  arithmetic, xv
  convex, 257
  generalized arithmetic, 233
progression, generalized arithmetic, 252
property testing, 83
pseudorandomness, 89
$\mathrm{PSL}(2,p)$, 92, 113

quasirandom
  Cayley graph, 116
  graph, 90, 129
    bipartite, 98
  group, 109, 114
    examples, 113

r-graph, *see* hypergraph
Ramanujan graph, 124
Ramsey's theorem, 2
random graph
  Erdős–Rényi, 90
  quasirandom, 92
rank
  slice, 218
regular
  pair, 54
  partition, 54
  set, 61
regularity lemma
  arithmetic, 222, 224
  bounded degree, 151
  equitable, 60
  graph, 55
  hypergraph, 86
  irregular pair, 59, 61
  lower bound, 59
  strong, 77
  weak, 142, 145
removal lemma
  graph, 74
  hypergraph, 85
  induced, 76
  infinite, 81
  triangle, 63, 173
representation theory, 109
Roth's theorem, 6, 9, 66, 211, 276
  Behrend construction, 69
  counting, 294
  finite field, 203, 217, 221
  relative, 279

Sárkőzy theorem, *see* Furstenberg–Sárkőzy theorem
Schur's theorem, 1
semidefinite
  program, 145, 171
  relaxation, 117
SET card game, 203
(7,4)-conjecture, 66
Shearer's inequality, 193
Sidorenko's conjecture, 159, 167, 170, 174, 189
$\sim$, *see* asymptotics
(6,3)-theorem, 66
spectral gap, 103
square, *see* Furstenberg–Sárkőzy theorem
Stanley sequence, 69
stochastic block model, 130, 137
subgraph, xv
  induced, xv
successive minimum, 250
sum-free, 1, 221
sum-product, 265
sumset, 231, 234, 235, 265
  iterated, 245
  restricted, 260
sunflower, 221
supersaturation, 21, 281
symmetrization
  Zykov, 17, 182
Szemerédi's regularity lemma, *see* regularity lemma, graph
Szemerédi's theorem, 6, 85, 275
  bounds, 7
  counting, 281
  multidimensional, 8
  polynomial, 8
  random, 280
  relative, 279

$\Theta(\ )$, *see* asymptotics
tower function, 59
trace method, 122
transference, 280
triangle
  density, 161
  forbidding, *see* Mantel's theorem
triangle inequality
  Ruzsa, 234
Turán
  density, 19
  graph, 14

hypergraph, 21, 186
number, 11
problem, 11
theorem, 14, 182

U^3 norm, 209, 256
undecidability, 158
unit distance problem, 24

$V(G)$, $v(G)$, *see* graph
van der Waerden's theorem, 5
vertex-transitive graph, 110

W-random graph, 137
W_G, *see* graphon, associated

Zarankiewicz problem, 22